AF851333

IntechOpen
intechopen.com

Built by scientists, for scientists

Explore all IntechOpen books

Meet the editor

Souraya Goumri-Said is a Full Professor of Physics at Alfaisal University, Riyadh, Saudi Arabia. She earned her Ph.D. in 2004 from the University of Bourgogne in France. Her research focuses on computational methodologies based on density functional theory, molecular dynamics, and analytical approaches for modeling functionalized materials, including perovskites, ceramics, and nanostructures, with applications in catalysis, photocatalysis, hydrogen storage, and photovoltaics. She has authored over 274 scientific papers, six book chapters, and holds two patents. Souraya is an editor for Open Physics (De Gruyter) and serves on the editorial board of Computational Condensed Matter (Elsevier). She received the 2022 LEWAS Award for Leadership Excellence and the 2024 Arab Environmental Excellence Award. She was listed among Stanford University's Top 2% Scientists in 2024.

Contents

Preface **IX**

Chapter 1 **1**
Introductory Chapter: Transition Metals – Unlocking the Path from Fundamentals to Frontiers
by Souraya Goumri-Said

Chapter 2 **5**
Laser-Assisted Synthesis of Metals, Metal Oxide Nanoparticles, and Metal–Organic Frameworks (MOFs): Applications in Photocatalysis, Batteries, and Beyond
by Bülend Ortaç, Saliha Mutlu, Ahmet Hakan Yilmaz, Nergis Arsu and Sevil Savaskan Yilmaz

Chapter 3 **49**
Graphitic Carbon Nitride as Photocatalysts for the Degradation of Organic Pollutants in Aqueous Suspension: A Review
by Mohammad Muneer, Ziyaur Rasool, Shreya Parashar, Abdulaziz M. Almohyawi, Ziad Moussa and Saleh A. Ahmed

Chapter 4 **71**
Research and Application of Sandwiched Dissimilar Metals
by Huisheng Cai, Siqi Yang and Huaqing Li

Chapter 5 **101**
Iron: From Basic Chemistry to Modern C−H Bond Functionalization Catalysts
by Hamad H. Al Mamari

Chapter 6 **129**
Exploring Cobalt, Nickel, and Palladium Complexes for the Hydrogenation Reactions: A Review
by Oluwasegun Emmanuel Olaoye and Fortunate Phenyo Sejie

Chapter 7 **151**
Individual Transition Single-Atom Catalysts for Selective Oxidation Reaction
by Shiyong Zhao, Jinfeng Xu, Jincheng Li and Chang Liu

Chapter 8 173
Magnetic Spinel Ferrite Nanomaterials: Fabrication and Application
by Hua Li, Wenqin Li, Jacques Robichaud and Yahia Djaoued

Chapter 9 225
Transition Metal Chalcogenides for Wastewater Treatment
by Sawaira Moeen, Iram Shahzadi, Ali Haider, Anum Shahzadi, Muhammad Ikram, Mohammed Benali Kanoun and Souraya Goumri-Said

Chapter 10 253
Palladium-Based Photonic Microresonator Hydrogen Sensors
by Kenan Cicek

Chapter 11 271
Metal-Metal Oxide, Metal-Organic Framework Nanocomposites for Energy Storage, Thermal Management, Radiation Shielding, and Magnetic Investigations with Quantum Computing
by Bülend Ortaç, Saliha Mutlu, Ahmet Hakan Yilmaz, Nergis Arsu and Sevil Savaskan Yilmaz

Preface

Scientists have been fascinated by transition metals for a long time because of their wide range of chemical behaviours, interesting electronic structures, and amazing adaptability. From classical catalysis and materials chemistry to quantum computing and environmental sustainability, these elements have always changed the way science and technology work. Transition Metals: From Fundamentals to Frontiers is a book that puts together a carefully chosen set of chapters written by top researchers. These chapters show the most recent developments and uses of transition metals.

Each chapter in this book has been carefully chosen to give readers a complete but focused overview of both basic ideas and cutting-edge discoveries in modern research. This book contains a lot of information, from basic chemical principles to cutting-edge technologies, that shows how important transition metals are to modern science and technology.

The first chapter of the book, ***Introductory Chapter: Transition Metals – Unlocking the Path from Fundamentals to Frontiers***, gives a conceptual framework and a roadmap for understanding the scientific importance and technological promise of transition metals in many fields.

The chapter ***Laser-Assisted Synthesis of Metals, Metal Oxide Nanoparticles, and Metal–Organic Frameworks (MOFs): Applications in Photocatalysis, Batteries, and Beyond*** examines how advanced laser-based methods make it possible to make nanomaterials with multiple functions. These materials can be used in a wide range of fields, from photocatalysis to energy storage and new quantum applications.

The third chapter, ***Graphitic Carbon Nitride as Photocatalysts for the Degradation of Organic Pollutants in Aqueous Suspension: A Review***, is about ways to clean up the environment that do not use metals. The research emphasises the photocatalytic properties of g-C_3N_4 for the enduring degradation of waterborne contaminants.

The authors of ***Research and Application of Sandwiched Dissimilar Metals*** give new information about layered metal systems. Their chapter talks about how to make these unique assemblies, how strong they are, and how they work in advanced structural, thermal, and electronic systems.

Next chapter, ***Iron: From Basic Chemistry to Modern C–H Bond Functionalization Catalysts*** is a work that looks at how iron chemistry has changed over time, from its basic principles to its current use as an efficient and sustainable catalyst in organic transformations, especially for selective C–H activation.

In ***Exploring Cobalt, Nickel, and Palladium Complexes for the Hydrogenation Reactions: A Review***, readers will find a full look at late transition metal complexes that are used in hydrogenation. The review covers the effects of ligands, the ways that catalytic hydrogenation works, and the latest developments in fine chemical and pharmaceutical synthesis.

Individual Transition Single-Atom Catalysts for Selective Oxidation Reaction looks into the new field of single-atom catalysis. This chapter talks about how isolated transition metal centres are the cutting edge of catalyst design because they have unique activity, selectivity, and atom-efficiency.

The next part of the volume is called ***Magnetic Spinel Ferrite Nanomaterials: Fabrication and Application***. It talks about how to make these materials, their magnetic properties, and how they can be used in electronics, biomedicine, and environmental sensors. It shows how these ferrite-based nanomaterials can be utilised in many different ways.

The chapter on ***Transition Metal Chalcogenides for Wastewater Treatment*** looks at how transition metal chalcogenides are structured, how they react, and how they can be used as catalysts. The main goal is to clean up the environment, especially to break down and get rid of pollutants in wastewater.

Palladium-Based Photonic Microresonator Hydrogen Sensors shows how palladium can be incorporated into cutting-edge photonic sensing platforms. This chapter shows how the unique way that palladium interacts with hydrogen can be used in microresonator devices for very sensitive detection systems. It connects materials chemistry with sensor technologies and energy-related applications.

The chapter ***Metal-Metal Oxide, Metal-Organic Framework Nanocomposites for Energy Storage, Thermal Management, Radiation Shielding, and Magnetic Investigations with Quantum Computing*** brings together classical nanomaterials and new uses. It illustrates hybrid material systems that have been designed to work in multiple areas, such as energy, defence, and quantum information.

These chapters show how wide and deep modern transition metal research is, from basic chemical knowledge to practical technologies that solve problems in the environment, industry, and technology around the world. This book is meant to be a reliable source for researchers, students, and professionals. It should encourage more research and new ideas in the ever-changing fields of transition metal chemistry and materials science.

I want to thank all of the authors who contributed to this collection from the bottom of my heart. Their hard work and knowledge made it possible. Their work together tells a story that is both unified and varied, covering both well-known facts and exciting new areas of transition metal research.

I really hope that this book will not only help readers understand transition metals better, but also encourage them to keep learning and coming up with new ideas in this interesting field.

Souraya Goumri-Said
College of Science and General Studies,
Alfaisal University,
Riyadh, Saudi Arabia

Chapter 1

Introductory Chapter: Transition Metals – Unlocking the Path from Fundamentals to Frontiers

Souraya Goumri-Said

1. Introduction

Transition metals occupy a pivotal place in chemistry, materials science, and technology due to their distinctively versatile electronic structures, rich chemical behaviors, and profound catalytic capabilities. These elements, located in groups 3–12 of the periodic table, are defined by partially filled *d* orbitals, endowing them with remarkable flexibility in oxidation states and bonding modes. This unique electronic configuration allows transition metals to play essential roles across a myriad of scientific disciplines, from fundamental chemistry and catalysis to advanced materials engineering and quantum technology [1, 2].

Historically, transition metals have fascinated chemists due to their vivid colors, magnetic properties, and centrality to life processes—iron in hemoglobin, cobalt in vitamin B_{12}, and copper in enzymes essential for respiration. However, contemporary research has significantly expanded their roles far beyond classical boundaries, delving into areas like single-atom catalysis, spintronics, and advanced environmental remediation technologies [3, 4]. Thus, a deeper understanding of these elements is not only of fundamental interest but also crucial for technological advancement.

This book, *Transition Metals: From Fundamentals to Frontiers*, aims to offer an insightful and comprehensive exploration of modern advances surrounding transition metals. The chapters within provide a detailed view of key aspects of their chemistry, unique properties, and emerging applications in cutting-edge scientific research and technology. As transition metals increasingly dominate diverse fields—including catalysis, energy storage, quantum computing, and environmental remediation—it is essential for readers to grasp both fundamental principles and the latest breakthroughs driving contemporary science and innovation.

2. Fundamental properties and significance of transition metals

The hallmark of transition metals lies in their partially filled *d* orbitals. This electronic characteristic underlies their notable chemical reactivity, allowing them to exhibit various oxidation states. For instance, iron can exist in oxidation states ranging from −2 to +6, each possessing distinct chemical behaviors and catalytic efficiencies [1]. Their electronic flexibility also leads to profound magnetic and optical properties, making them indispensable in magnetic storage devices, pigments, sensors, and electronic components.

Transition metals demonstrate extraordinary catalytic activity, attributed to their reversible oxidation-reduction capabilities and diverse ligand interactions. Recent advancements in single-atom catalysts have dramatically increased catalytic efficiency and selectivity, especially in oxidation and hydrogenation reactions, significantly advancing sustainable and economically viable chemical manufacturing [3, 4].

3. Emerging Frontiers and advanced applications

Today, the scientific frontier of transition metals encompasses sophisticated applications from quantum computing and nano-engineering to environmental sustainability. Magnetic transition metal–based spinel ferrites have become critical in high-density data storage, magnetic resonance imaging (MRI) contrast agents, and advanced electronics. Additionally, quantum computing technologies are increasingly exploiting metal-organic frameworks (MOFs) and metal–metal oxide composites to leverage their quantum coherence properties, enabling new computational paradigms [1].

Catalytic innovations are further driving developments, with systems based on cobalt, nickel, and palladium complexes significantly enhancing hydrogenation reactions. These catalysts demonstrate how fundamental chemistry can translate directly into practical and sustainable chemical production, particularly within pharmaceuticals and industrial synthesis [3].

Environmental remediation also represents a critical frontier. Materials such as graphitic carbon nitride (g-C_3N_4) and transition metal chalcogenides have emerged as highly efficient photocatalysts for water treatment and pollution mitigation, exemplifying sustainable and scalable approaches to global environmental challenges [5, 6].

The primary goal of this book is to articulate clearly the connection between fundamental transition-metal chemistry and frontier-level scientific applications. Understanding these links empowers researchers and students alike, fostering multidisciplinary innovation. Such integration of basic knowledge with cutting-edge applications exemplifies modern scientific inquiry, vividly illustrated by the diverse topics covered throughout this book.

Transition metals remain central to the evolution of science and technology. As researchers continue to explore their potential, new discoveries and innovative solutions to global challenges will undoubtedly emerge. This book collectively bridges fundamentals and frontiers, highlighting the versatility and extraordinary potential of transition metals. Readers are invited not only to appreciate existing knowledge but also to engage actively in the exciting and dynamic exploration of transition metals.

Author details

Souraya Goumri-Said
Department of Physics, College of Science and General Studies, Alfaisal University, Riyadh, Saudi Arabia

*Address all correspondence to: sosaid@alfaisal.edu

IntechOpen

© 2025 The Author(s). Licensee IntechOpen. This chapter is distributed under the terms of the Creative Commons Attribution License (http://creativecommons.org/licenses/by/4.0), which permits unrestricted use, distribution, and reproduction in any medium, provided the original work is properly cited.

References

[1] Manchado A, García-González Á, Nieto CT, Ledesma NG, Díez D, Garrido NM. Transition metal catalysis for the asymmetric synthesis of 2-arylethylamines: A review of the new millennium. Molecules. 2025;**30**(8):1721. DOI: 10.3390/molecules30081721

[2] Ganguly P, Kumar SR, Muscetta M, Padmanabhan NT, Clarizia L, Akande A, et al. New insights into the efficient charge transfer of ternary chalcogenides composites of TiO_2. Applied Catalysis B: Environmental. 2021;**282**:119612. DOI: 10.1016/j.apcatb.2020.119612

[3] Kment Š, Bakandritsos A, Tantis I, Kmentová H, Zuo Y, Henrotte O, et al. Single atom catalysts based on earth-abundant metals for energy-related applications. Chemical Reviews. 2024;**124**(21):11767-11847. DOI: 10.1021/acs.chemrev.4c00155

[4] Li S, Xin Z, Luo Y, Pan J, Liao G, Li Q, et al. Recent advances in the development of single atom catalysts for oxygen evolution reaction. International Journal of Hydrogen Energy. 2024;**82**:1081-1100. DOI: 10.1016/j.ijhydene.2024.08.026

[5] Okpara EC, Olatunde OC, Wojuola OB, Onwudiwe DC. Applications of transition metal oxides and chalcogenides and their composites in water treatment: A review. Environmental Advances. 2023;**11**:100341. DOI: 10.1016/j.envadv.2023.100341

[6] Bhanderi D, Lakhani P, Modi CK. Graphitic carbon nitride ($g\text{-}C_3N_4$) as an emerging photocatalyst for sustainable environmental applications: A comprehensive review. RSC Sustainability. 2024;**2**(2):265-287. DOI: 10.1039/D3SU00382E

Chapter 2

Laser-Assisted Synthesis of Metals, Metal Oxide Nanoparticles, and Metal–Organic Frameworks (MOFs): Applications in Photocatalysis, Batteries, and Beyond

Bülend Ortaç, Saliha Mutlu, Ahmet Hakan Yilmaz, Nergis Arsu and Sevil Savaskan Yilmaz

Abstract

This chapter reviews the synthesis, properties, and applications of metal and metal oxide nanoparticles, as well as MOFs, produced using high-power lasers. The integration of MOFs and nanometals is further explored in cutting-edge applications including photocatalysis, gas sensing, supercapacitors, lithium-ion batteries, and solar energy conversion. Special attention is given to defect engineering, 3D laser printing of MOF composites, and emerging photothermal and optoelectronic functionalities. These techniques enable precise control over nanoparticle size, morphology, and composition, and significantly reduce reaction times for MOF formation. Our experimental results on laser ablation and the nanoparticle size distributions are provided in detail. For example, with a pulsed laser ablation energy of 70 mJ, the zinc oxide nanoparticle's (ZnO NPs) size distribution produced an average nanoparticle size of 12.2 nm. Furthermore, the fast and scalable synthesis of MOFs employing laser irradiation techniques is extensively discussed for several uses in state-of-the-art technology. This work provides an explanation for high-power laser-induced rapid synthesis (LIRS) of Ni- and Zn-based MOFs achieved in a far shorter time and with higher yields than earlier reported techniques. Promising supercapacitor technologies, the chapter covers Cu-MOF/PANI- and Cu-MOF/PPy-based nanocomposites with high particular capacitance. The specific capacitance values are 160.5 F/g for Cu-MOF/PANI and 132.5 F/g for Cu-MOF/PPy nanocomposites, respectively. The activity of the Pt@UiO-66-NH_2 MOF nanocomposite was discovered among the investigated to be 257.38 mmol g^{-1} h^{-1}.

Keywords: metal-organic frameworks (MOFs), nanometal atoms, high-power laser techniques, photocatalysis, battery technology

1. Introduction

MOFs provide extensive specific surface area, significant porosity, customizable pore dimensions, and modifiable functionalities [1–3]. Researchers have extensively examined them in various conventional applications of porous materials, such as gas adsorption and storage [4], drug delivery [5], catalysis [6], ion exchange and separation [7], and sensing [8]. Furthermore, MOFs serve as an exceptional foundation for the synthesis of diverse nanostructured MOF derivatives through thermal transformation. People are increasingly using MOFs in catalysis, environmental cleanup, photocatalysis, and energy storage and conversion [9, 10]. These MOFs include porous carbon-based materials, metals, metal oxides, metal chalcogenides (sulfides and selenides), metal phosphides, and metal carbides, as well as different carbon-based composite materials. Nevertheless, the application of MOFs and their derivatives in photovoltaic systems has garnered significantly less focus than in their conventional uses; hence, the advancement of this field remains nascent [11]. The most extensively researched composite systems involve the integration of metal and metal oxide nanoparticles(Nps) with MOFs. **Figure 1a** presents different synthesis methods of metal-metal-oxide NP and MOFs, their some applications. The adaptability of synthetic methods for metal oxide (ceramic) and metallic NPs accounts for this.

1.1 Powerful lasers on MOFs and nanometals

A focused laser beam ablates material thermally or nonthermally. Laser energy at an ablation threshold breaks molecular in each substance. Higher laser wavelength absorption coefficients at specific wavelengths reduce light penetration and the ablation threshold. Standards range from tens of mJ/cm^2 to multiple J/cm^2 [12]. Laser-irradiated solvent-based colloidal solution targets shrink to the nanoscale. Metal particle characteristics depend on laser wavelength, pulse duration, energy, ablation time, fluency, and surfactants. Laser ablation produced Au, Ag, iron, Mg, Al, Ti, Pd, Pt, Cu, CuO, Cu_2O, ZrO_2, ZnO, Fe_3O_4, NiO, Al_2O_3, SnO_2, TiO_2, In_2O_3, Co_3O_4, SiO_2, Ni_2O_3, Cr_2O_3, Mn_2O_3, CoO, and recent studies have examined laser ablation's metal and oxide NP formation [13]. **Figure 1b** depicts laser ablation and laser-induced synthesis of metal-metal oxide NPs and MOFs for polymer nanocomposites. Its versatility and affordability make pulsed laser ablation of solids in solution appealing.

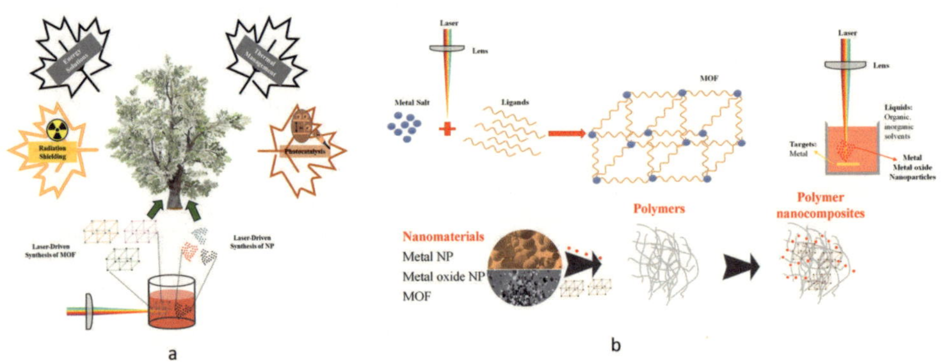

Figure 1.
Overview of laser-driven synthesis methods of metal-metal oxide NP and MOFs, their some important applications (a) and schematic illustrations of the synthesis of metal-metal oxide NP with the laser ablation method and MOFs with the laser-induced method, and the preparation of polymer nanocomposites (b).

Laser ablation nanoparticle synthesis is cheaper, faster, cleaner, and more flexible than chemical and thermal methods [14].

Nanomaterials with plasmonic resonance capabilities are being studied for optical modulation of NIR-spectrum plasmonic devices for nanophotonics and telecommunications. Plasmonic-enhanced NIR devices are necessary because most of the sun's spectrum is in this range. Metallic NPs and colloidal quantum dots (QDs) boost NIR plasmonic photogenerated currents. Nanosecond laser ablation of hexagonal crystalline 2H-MoS_2 powder in methanol creates 2D and 3D nanosheets and nanostructures [15]. To generate bimetallic NiAu and complex NiO:Au-NPs, Petrovic et al. [16] laser-ablated a Ni target in colloidal gold solution. Ablating the target with two Nd:YAG lasers at 1064 nm, at 150 ps and 20 ns pulse durations, we studied how pulse length, fluences, and ablation periods affect NP size and composition. Picosecond pulses stimulate colloidal solution processes that outperform nanosecond pulses at Ni target ablation and nanoparticle production. NP form, size, and content affect colloidal nickel species concentration.

Hydrothermally and solvothermally generated MOFs abound. These technologies enable high-purity product manufacturing through sustainable processes, resource management, and chemical removal. While making the cage, the basic components and solvent are heated in a sealed container at a certain temperature and duration. MOF production is still impossible industrially. Industrial manufacturing needs conditions. Low solvent consumption, cheap raw materials, low temperature and pressure, good yield, and impurity elimination [17]. Most MOFs are made via solvothermal techniques at high temperatures and long times. Laser- induced synthesis method boosts yields and speed production. Armon et al. used this method in 2019 to develop micro-patterning of a metal organic framework [18].

Mutlu et al. created a fast, controlled laser-driven MOF synthesis platform. The first Zn-based MOFs were created using laser irradiation, which took an hour less than earlier approaches with similar morphology [2]. In another study, Mutlu et al. synthesized NiMOF by the LIRS method using 1-(Anthracene-9-methyl)-1H-1,2,3-triazole-4,5-dicarboxylic acid (H2L) and 1,2-bis(4-pyridyl)ethane (bpe) compounds. Laser radiation may precisely alter material structures. LIRS with a 975 nm laser beam produced flowery Ni-MOF after 70 minutes [3].

Laser-assisted MOF synthesis is popular. Lasers accelerate MOF surface development in rapid laser-assisted synthesis, which is the sole substrate. Recent LIRS processing employs fast heating. Builds automated objects and materials rapidly and cleanly [19]. Laser ablation can reach 4000 K reaction temperature in nanoseconds (>1010 K s^{-1}). MOF porosity, open metal sites, and surface characteristics can be altered by defect engineering to improve separation, gas capture, catalysis, and mechanical performance [20]. LIRS of MOF derivatives at high laser temperatures produces more stable and long-lasting molecules than heating. Due to their dense crystalline structure, laser-induced MOF derivatives can resist severe conditions. Innovative laser pyrolysis of MOF compounds differs from conventional pyrolysis [21]. The wide variety of MOFs affords fascinating compounds. Laser processing has great potential for MOF derivative synthesis and development. Laser-driven MOF synthesis was compared to standard procedures. Organic ligands (H_2L and bpe) heat a water-based zinc salt solution with a 975 nm laser beam source because the salt rapidly reverts to its ionic form. Unlike laser ablation, this synthesis requires two ligands and aqueous salt. Laser-irradiated MOF crystal formations create derivatives in other laser manufacturing procedures. Similar to solvothermal synthesis, our laser beam-based study builds MOF structures faster. The optical, morphological, structural, photophysical, and photochemical properties of MOFs were examined.

Chemical properties, defect distribution, and existence affect MOF quality. Laser scribing and low-temperature selective solvothermal growth helped Wu et al. [22] build symmetric MSCs with higher electrochemical performance. Direct 3DLSG patterning uses polyimide (PI) film as a base and carbon source. Selectively growing 1D conductive Ni-catecholate (Ni-CAT) MOF nanorods creates a 3D conductive matrix. MOFs and graphitic materials like LIG are employed in supercapacitors, batteries, and other applications due to their unique characteristics. Lim et al. [23] developed a hybrid NO_2 detector using Cu_3HHTP_2, two-dimensional semiconducting MOFs, and laser-induced graphene. Hierarchical LIG@Cu_3HHTP_2 (HHTP = 2,3,6,7,10,11-hexahydroxytriphenylene) optimizes gas molecule mass transit and MOFs' enormous surface area and porosity for fast and sensitive NO_2 detection. LIG@Cu_3HHTP_2 has the fastest reactions and lowest detection limit of leading NO_2 sensors at ambient temperature.

Mid-IR laser sources in the 2.5–5 μm range are ideal for spectroscopy, surgery, countermeasures, and remote sensing. MIL-68(Al) and MIL-68(Fe) MOFs feature regular mesoporous architectures and ordered atom distributions, according to Le Zheng et al. [24] MOFs make wonderful mid-IR optical switches. MIL-68(Al) and MIL-68(Fe) reflection-type saturable absorber mirrors were hydrothermally manufactured. Ni-MOFs are effective broadband saturable absorbers producing mode-locked laser pulses in the NIR region (1.06–1.88 μm).

The zinc (Zn)-based MOF $[Zn_2(H_2L)_2(1,2\text{-bis(4-pyridyl)ethene})_4]_n$ mode-locked a thulium/holmium-doped fiber laser due to its increased optical absorption at 1925 nm. Saturable absorber was dropcast from arc-shaped fiber using Zn-MOF and isopropanol. The mode-locked laser has 3.251 mW peak average output power and a 1906.75 nm center wavelength. Essential repetition frequency and pulse duration were 12.89 MHz and 1.772 ps. Peak power was 142 W, pulse 252 pJ. It is thought that this is the first mode-locked pulse creation using a thulium/holmium-doped all-fiber laser and MOF. NIR fiber laser mode-locking uses MOF SA [25]. Ni-MOF, an SA, discovers an undiscovered O-band ultrafast mode-locked fiber laser. The Ni-MOF lasers are powerful. A stable mode-locked laser with 450 fs pulses was achieved. Laser repetition rate and SNR were 0.408 MHz and 52.5 dB at 1298.1 nm. At 88 mW pump power, the mode-locked laser had 0.69 mW average output power, 1.69 nJ pulse energy, and 3.53 kW pulse peak power. Harmonic mode-locked laser operation was achieved at 88 mW pump power. Harmonic mode-locked lasers with 125.7 mW pumps produced 2.45 MHz pulses. The author claims this is the first report of a new MOF SA material for praseodymium-doped fluoride fiber lasers [26].

Liu et al. [27] successfully crystallized a terbium-based luminous MOF (Tb(BTC)·G, where BTC = 1,3,5-benzenetricarboxylic acid and G = guest solvent) using femtosecond lasers. We believe this is the first direct MOF patterning method. Recent research has made MOF powders into beads, membranes, monoliths, gels, sponges, and nanofibers to address these challenges. 3D printing creates high-resolution macroscopic things with intricate patterns and geometries from digital models, garnering attention. 3D printing, including SLS and SLM, has been used to make robust MOF monoliths with changeable morphology for useful applications. SLS is an essential 3D printing process that sinters ceramics, metals, and thermoplastic polymers with a computer-controlled laser beam. Printer optimization prevents melting, keeping polymeric powders particle-like during sintering. This creates a porous, solid, powder-like monolith with easily accessible gaps between partially sintered polymer powders [28]. Temporary laser heating can melt or sinter powdered particles quickly, forming micro-voids. Small holes emerge when MOF NPs and polymer powders come into contact during SLS, making open channels easier to create. The polymer matrix allows MOF NPs to move freely and interact more with the outside world. Luz et al. [29] inkjet printed photoluminescent Ln-MOFs on paper and plastic. Lahtinen et al. SLS-made porous membranes [30]. Nylon-12 was a printed polymer matrix with HKUST-1 porous filler. Li et al. produced polymer-MOF mixed matrix films [31]. Using PA12 powder as the printed polymer matrix and five powdered MOFs (MOF-801, ZIF-67, ZIF-8, HKUST-1, and NH_2-MIL-101(Al)) as porous fill. Creating metal nanoparticles (M-NPs) with unique electrical, magnetic, catalytic, and optical properties is promising [32].

Jiang et al. [21] demonstrate that they can accomplish nanoscale metallurgy and the patterning of uniform M-NPs in a single step, a technique known as nanoscale laser metallurgy and patterning (nano-LaMP). The nano-LaMP method is very accurate, which makes it easier to make consistent gaps between M-NPs that are as small as 2 nm. This improves the plasmonic effect for finding molecules at very low concentrations (10–12 M). Using homogeneous particle sizes (3–200 nm) and interparticle gaps (2–50 nm), Fe, Co, Ni, Cu, Zn, Cd, In, Pb, and Bi M-NPs can be created. Air-synthesized M-NPs do not agglomerate or oxidize in humidity. Quickly created MNP patterns are repeatable and power-efficient. Nano-LaMP works on non-noble metals. Eight NPs of Bi, In, Zn, Cd, Pb, Fe, Co, and Ni were synthesized in air at zero oxidation using Cu and eight MOFs—CAU-7, CPM-5, Zn-MOF-74, Cd-MOF-74, Pb-TCPP, MIL-100, ZIF-12, and Ni-BTC-bipy. MOFs' chemistry permits exquisite MNP size and spacing control, and laser processing [21, 33] is efficient and scalable for device fabrication and large-scale manufacture. Nano-LaMP employs MOF. Organic linkers and metal ions generate porous crystalline solids [34].

When metal ions and organic linkers interact, secondary building units (SBUs) are made. These SBUs make it easier for light to be absorbed across a wavelength range, from ultraviolet to near-infrared. Armon et al. [18] describe a way to use the modulated laser-induced microbubble technique (M-LIMBT) to make tiny patterns in MOFs. This approach synthesizes and builds MOFs concurrently, improving resolution over inkjet printing and removing pre-preparation. Microbubbles form when laser light heats a precursor solution, creating nano-MOFs and vapor pressure. Previous laser-based studies have only shown how to post-process MOFs. This was done by using a crystal that had already been synthesized and micro-patterning it with metals [35]. The laser heats NPs near the substrate, raising the medium vapor pressure and generating microbubbles. In their model, a laser heats iron-based MOF for synthesis [36]. The metal precursor was iron (III) acetylacetonate, and the ligand was 1,3,5-benzenetricarboxylic acid (trimesic acid). The solution was made by mixing both precursors in room temperature N,N-dimethylformamide. Effective multiphase metal sulfide NP synthesis is essential for high-performance EES devices. Sulfur-impregnated nickel-based MOF-Ni laser irradiation can create ultra-small multiphase nickel sulfide NPs with conductive graphitic carbon in minutes. Altering the reactant ratio can change the phase of Ni_3S_2 and αNiS NPs. An MOF supplies thermodynamic carbon for nickel sulfide NP synthesis. A carbon composite with 61 wt% Ni_3S_2 showed exceptional cycle stability and a maximum specific capacity of 800°C/g due to its multiphase structure and high porosity. Over 10,000 cycles, composite hybrid supercapacitors have 42 Wh/kg energy density and 96% cycling stability. From sulfur-impregnated MOFs, an affordable laser-induced sulfurization approach may produce ultra-small and uniform metal sulfide NPs (NiSx, CoSx, and MnS/MnO) in graphitic carbon in minutes [37].

1.2 Metal oxide-polymer nanocomposites, energy solution

Metal/metal oxide nanoparticles produced using laser-based methods produce composites with outstanding chemical stability, electrical conductivity, and surface area when integrated with MOFs. These structures find their applications in countless areas, including energy storage, catalysis, sensor devices, and environmental aspects. This section will discuss the assessment of such structures and the fabrication of MOF-based metal/metal oxide nanocomposites, with emphasis on the energy field.

MOFs composites made from MOFs are innovative porous materials. NPs@ MOF composites combine metal or metal oxide nanoparticles' multifunctionality with MOFs' configurable porosity. Porous Coordination Polymers (PCPs), are of considerable interest to chemists because of their remarkably high surface areas and structural variety. The surface area-to-volume ratio of metal NPs is higher than that of bulk metals that are usually used in industrial catalysis [38], which makes them very appealing for catalysis. We use mostly well-defined porous materials. Molecular reactions with NPs are selective due to the split exterior and inside pores [39]. Their porousness changes their size, shape, and chemical properties. Multifunctional MOF composites can easily include graphene, metal NPs, and functional molecules. Because MOFs and their parts function better together, MOF composites perform better photo/electrochemically. Using chemicals or heat to create nanoarchitectures affects MOF characteristics [40].

Energy conversion efficiency and sustainability make fuel cells, water electrolysis, and metal-air batteries suitable replacements for fossil fuels. MIBs, supercapacitors, and metal-sulfur batteries are EES. These are vital to new energy technology that maximizes electricity utilization. Energy storage systems separate charges by metal cation intercalation and de-intercalation or reversible redox processes on electrode surfaces [41]. Low electrical conductivity makes MOFs poor energy storage electrodes. Electrodes contain conductive chemicals and binders to increase electrical conductivity. Add-ons produce "dead mass" inside electrodes.

Pyrolysis or solvothermal conversion can create MOF-based metal complexes. Due to their small particle size, high porosity, and high surface area, MOF-derived metal compounds provide ECS with active surface exposure and mass transfer. ZIF-67 pyrolysis produces yolk-shell Co_3O_4 in air, while MOF sulfidation produces bubble-like CoS_2 nanoprisms. Due to their intricate nanostructures, MOF metal complexes stored lithium well. Simple pyrolysis yielded porous RuO_2 [40], $CoSe_2$, and CoP with high ECS activity from MOFs [42]. MOF composites make complex metal complexes with different compositions and nanostructures. MIL-101/TiO_2 composite pyrolysis utilized platinum NPs on the exterior shell to cocatalyze Fe_2O_3@TiO_2/Pt. After 48 hours of visible-light exposure, Fe_2O_3@TiO_2/Pt showed considerable photochemical activity, releasing 30.0 μmol H_2 per mg due to component synergy. MOFs caused complex metal complexes, including Co/(NiCo)Se_2, ZnO/NiO, and NiCoS/Ti_3C_2Tx, with good photochemical and electrochemical properties [43]. Metallic/carbon hybrids are combined with MOF composites, employing their organic-inorganic hybrid characteristics. Metal NPs, complexes, and atomically scattered metal sites uniformly diffuse over porous carbon substrates in MOF-derived metal/carbon hybrids. Metal species (metal NPs, metal complexes, and atomically dispersed metal sites) uniformly distribute themselves on the porous carbon substrates in the MOF-derived metal/carbon hybrids. The carbon scaffolds offer high conductivity, large surface area, durability, and hierarchical porosity. Complex MOF-based multicomponent metal–carbon nanostructures with guests or substrates enhance photocatalytic and electrochemical performance. $CoSe_2$@B and N codoped carbon might enhance Na storage capacity to 580 mA h g^{-1} at 100 mA g^{-1} after 100 cycles by pyrolyzing and selenizing ZIF-67 in BCN nanotubes [44].

Polymers coat nano-MOF surfaces *via* covalent and coordination bonding. Zimpel et al. [45] exhibited covalent and coordination surfaces. Polymer-MOF surface collaboration is suggested. The Zr-fum NPs were made from $ZrCl_4$, fumaric acid, and formic acid. The authors proposed entropy increase may make polymers compete with formic acid. These polymers adhered to MOF surfaces after coordination. MOF composites or hybrids use metal NPs, QDs, POMs, molecular species, enzymes, silica, and polymers. MOF composites and hybrids combine porous design, chemical diversity, and structural adaptability with functional materials' catalytic, optical, electrical, magnetic, and mechanical strength to create new physical and chemical properties. Activity, selectivity, and stability make MOF composites catalytic. QDs with visible light response and MOFs with active metal cores and gas traps may boost photocatalysis. CsPbBr3@ZIF-8 and ZIF-67 zinc/cobalt ZIF-coated halide perovskite quantum dot composites were developed by Su and colleagues as CO_2 reduction photocatalysts [46]. Compared to pure $CsPbBr_3$, the composites dramatically reduced CO_2 with electron consumption rates of 15.498 and 29.630 mmol g^{-1} h^{-1}, respectively, compared to 11.14 mmol. Wheatley and colleagues [47] immobilized SnO_2-NPs in monoZIF-8, a monolithic MOF, to make hazardous waste photocatalysis SnO_2@monoZIF-8. In a 3-hour degradation of aqueous methylene blue (MB), 0.4 g of SnO_2@monoZIF-8 demonstrated 41.5% ± 3.0% degradation. Ruthenium(III)-substituted polyoxometalate $[SiW_{11}O_{39}Ru(H_2O)]$ [48], was added to copper(II)-bipyridine MOF pores by Duan and colleagues for C—C bond oxidative coupling [49].

N-phenyl-tetrahydroisoquinoline and nitromethane produced 90% CR-BPY1 after 24 hours of irradiation. MOF backbone and encapsulated molecular species work together. Jiang and colleagues created visible-light-driven H_2 by encapsulating [CoII(TPA)Cl][Cl] in MIL-125-NH_2's pores [50]. CoII complexes needed photosensitizers to make H_2. In virgin MIL-125-NH_2, hydrogen generation was restricted to 17 mmol/g·h. CoII@MIL-125-NH_2 composites significantly increased H_2 production to 553 mmol/g·h *via* improved photocatalytic activity.

EES devices use high-temperature inorganic materials with uncommon or hazardous components. Safe, affordable, sustainable energy storage systems can be made from organic materials. Unfortunately, inadequate cycling stability and charging rates limit EES utilization of these materials. Because they are unstable, conduct electricity poorly, and perform poorly commercially. MOFs outperform energy storage materials. MOFs conduct electricity due to precise block construction. To fully leverage MOFs for EES, multiple disciplines must collaborate. Electrochemists, synthetic and materials chemists, and others must collaborate on MOF-based EES devices.

Energy storage systems with high energy density, power, and durability are needed for the growing energy sector. Energy storage devices use toxic Pb or rare Li and Co inorganic materials at high temperatures. MOFs are unique EES. Ionic metal clusters form patterns with organic ligands. This creates crystal networks [51]. MOFs lose physical and chemical characteristics in alkaline aqueous solutions, restricting EES electrode use. Under highly alkaline circumstances, Zheng et al. [52] synthesize Co_3O_4 nanocube@Co-MOF, an alkaline-stable metal oxide@MOF composite, using a regulated and simple one-pot hydrothermal technique. The composite survives 15 days in 3.0 M KOH. Co_3O_4@Co-MOF has a unique structure, alkaline stability, and large surface area. With a specific capacitance of 1020 F g^{-1} at 0.5 A, it has good cycling stability, degrading only 3.3% after 5000 cycles at 5 A. A solid-state flexible device can achieve a maximum energy density of 21.6 mWh cm^{-3}.

Solar collectors convert thermal energy best [53]. Traditional solar collectors turn sunlight into water, glycols, and thermal oils using solid materials. Researchers generated nano/microparticle-containing heat transfer fluids to improve optical and thermal properties [54]. Water, oil, and ethylene glycol particles may contain metals, oxides, or carbon. Nanoparticles change fluid thermal and physical properties. Improved stability, convective heat transfer, and thermal conductivity [55]. Hussain et al. [56] study MOF dispersion-based fluids for solar-to-thermal energy conversion (STEC). The optical and thermal characteristics of MOFs in ethylene glycol were studied. The study examines ZIF8/EG, Cu-BTC/EG, and FeBTC/EG MOF dispersions with different particle ratios. The best STEC fluid was 0.3 wt% FeBTC/EG. Higher absorption and STEC efficiency than the other two fluids. The study shows that increasing MOF particle concentration affects SAR, highlighting the cost-efficiency trade-off. FeBTC/EG temporal dispersion stability is evaluated for STEC applications. MOF dispersion-based fluids are used in the novel research field STEC. It shows how MOF/EG could be utilized for STEC. It shows how MOF particle concentration affects performance and cost. MOF dispersion-based fluids boost STEC performance, proving FeBTC/EG's stability and efficiency. Sustainable energy MOF materials emphasize STEC dispersion stability and cost-efficiency.

MOFs have easy internal pore surface modification. Because M-NPs@MOFs catch substrates and do not multiply after catalysis, some have good organic catalytic activity. Recent research has synthesized M-NPs within MOFs to attain properties difficult with separate materials. M-NPs@MOF can store hydrogen and catalyze heterogeneously depending on the metal. At 77 K and ambient temperatures, Pd-NPs@MOF stores hydrogen better. Mg-NPs@MOF physisorbs hydrogen at low temperatures and chemisorbs at high temperatures. Combining adsorption approaches improves results. Submerging redox-active PCPs in $AgNO_3$, $NaAuCl_4$, and $Pd(NO_3)_2$ produced M-NPs [57].

1.3 Photocatalysis

MOFs' characteristics of high porosity, enormous surface area, and flexible structures render them attractive for energy storage, renewable energy generation, and environmental remediation. Applications of photocatalysis in particular represent an important area of research in which the MOFs' features of light absorption, carrier transport, and surface reaction can be optimally utilized. The photocatalyst activities of laser-grown MOF structures as well as derivatives of these systems along with their mechanisms will be presented in the current section.

Porous nanostructures, tunable semiconductor characteristics, and the capacity to integrate metals and metal oxides have made MOFs popular photocatalysts. MOFs photocatalyze *via* (LMCT), ligand-to-ligand charge transfer (LLCT), metal-to-ligand charge transfer (MLCT), metal-to-metal charge transfer (MMCT), and dual excitation. Photocatalysis involves light absorption, electron-hole separation, and surface redox reactions. It divides water into H_2 and O_2 and lowers CO_2 to carbonaceous fuels. Photocatalysis and electrocatalysis are renewable and environmentally friendly energy conversion methods. Traditional photocatalysts have limited light utilization due to poor light-harvesting components and restricted absorption spectrum. Long carrier transport distances usually generate electron-hole recombination, lowering photocatalytic efficiency. **Figure 2a** shows photocatalysis and dye spectral fluctuations. Photocatalyst production issues can reduce porosity, active site exposure, and stability. Three MOFs show promise for ECS system material development.

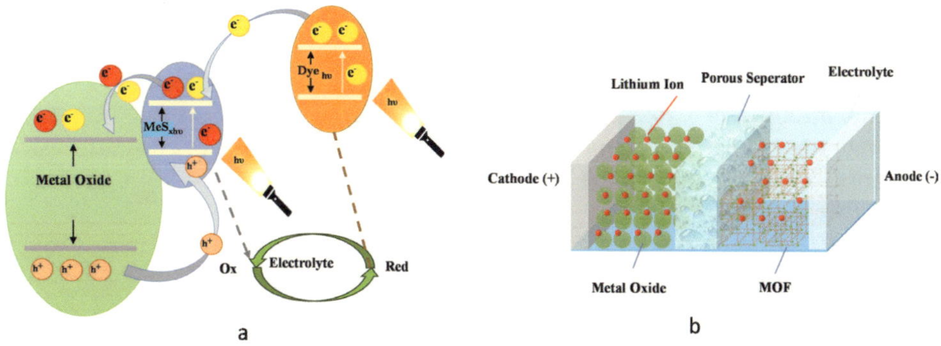

Figure 2.
Schematic illustration of the photocatalytic process and photocatalytically spectral changes to dye (a). Schematic representation of a lithium-ion battery design integrating a metal oxide with an MOF and the mechanism by which the porous separator in the electrolyte medium transports lithium ions from the metal oxide cathode to the anode (b).

Ordered and adjustable metal nodes and organic linkers give MOFs structural and compositional benefits over conventional materials. Industrial chemical synthesis using photocatalysis is efficient and sustainable. MOF catalysts are promising. These have complex crystals, great porosity, and flexibility. High light absorption, charge separation, and catalytic efficiency improve photocatalysis [58]. MOF substrates and single-atom catalytic centers increased photocatalytic gas reduction by Hao et al. [59]. Gas-permeable membranes were produced from structurally modified SA/MOF particles. These membranes help gas flow over porous gas–solid interfaces. This rapidly transforms humid CO_2 gas to HCOOH with good selectivity (about unity). The study team reached 15.76% apparent quantum efficiency at 420 nm. Also successful was photocatalytic O_2 to H_2O_2 conversion. Adaptable catalyst and reaction interface designs, effective and sustainable gas reduction reactions and demonstrates considerable potential for fostering a greener future [60]. Zhang et al. [61], applied porphyrinic Zr MOFs as photocatalysts for PET-RAFT polymerization, which adds and breaks electron chains. Solar energy is converted to heat by photoexciting semiconducting materials. Photothermal effect is an alternative approach for efficient solar energy conversion. Apart from e/h-coupled redox reactions on photocatalysts, thermal energy (heat) can be produced by the photothermal effect upon photoexcitation of semiconductor materials upon light irradiation, by locally heating the lattice to promote endothermic reactions [62]. Pore space engineering enhances MOFs' functionality. MOFs, NPs, and guest species like metal nanoparticles can increase catalytic performance. Optimizing NPs' surface plasmon resonance (SPR)-induced photothermal effect accelerates endothermic processes at ambient temperatures without heating. Chen et al. made Pt/PCN-224(M) composites from porphyrinic MOFs and Pt nanocrystals [63]. At room temperature, these composites photo-oxidize aromatic alcohols with 1 atm O_2. These composites work because photothermal activity and singlet oxygen production work together.

After 30 years of research, MOFs are one of the most studied catalytic chemicals. Pure MOFs offer enormous surface areas, high active site densities, fast catalytic rates per volume, post-synthetic addition of catalytic groups, and diverse functional groups. However, in large-scale applications such as fuel processing, gas emission reduction, and chemical synthesis, pure MOFs show low stability and regeneration at high temperatures. They anticipate a few MOF uses in these technologies.

The superior performance of MOFs in photocatalysis is due to their photoexcitable semiconductor properties and controlled pore architecture. Similarly, these properties make MOFs very suitable electrode materials for energy storage technologies—especially lithium-ion batteries and supercapacitors. The next section will discuss MOF structures in battery technologies, the advantages of laser-assisted synthesis, and sample applications from the literature.

1.4 Battery technology

Criteria such as high energy density, cycling stability, and electrochemical activity are quite critical for battery technologies. Metal-organic frameworks (MOFs) stand out with their modular structures and processability that can meet these needs. In recent years, MOF derivatives synthesized by laser provide efficient precursors for carbon, metal oxide, or metal sulfide compounds used in electrode structures. In this section, the use of MOFs and their derivatives in battery electrodes, synthesis details, and performance measurements will be discussed.

MOFs have been studied for fuel cells, supercapacitors, batteries, and water electrolysis for hydrogen and oxygen generation. Good carrier mobility, light conversion, and energy harvesting define many MOFs. Thus, MOFs can be used in microelectronic and optoelectronic devices [64]. Photocatalysis, catalysis, adsorption, supercapacitors, and batteries have improved rapidly due to global pollution and energy constraints. Perovskite solar cells (PSCs) set a record 25.2% efficiency in 2019 (KRICT/MIT) [65]. MOFs are applied to perovskite solar cell hole- or electron transport layers or nanocrystals.

To improve lithium-ion battery electrode materials, the MOF template technique synthesizes MOFs, their derivatives, carbon materials, and metal compounds with unique nanostructures [66]. Lithium cobalt oxide and lithium iron phosphate are lithium-ion cathodes. MOFs rarely produce these oxides. Low-redox-potential porous carbon or metal composites are the anode [67]. Techniques such as oxidation modification [68], coating modification [69], and etching modification [70] enhance the electrochemical performance and structure of graphene to address its deficiencies. Materials like MOF can make carbon composites with high nitrogen contents. The porous framework's high carbon content conducts electricity and decreases volume expansion, improving electrochemical performance.

The cathode material is a critical component of lithium-ion batteries, significantly influencing capacity, safety, power, thermal properties, and cycle life, while constituting 25–50% of the overall battery cost. Prevalent commercial cathode materials comprise lithium iron phosphate ($LiFePO_4$), ternary compounds ($LiNi_xCo_xM_{1-x-y}O_2$), lithium cobalt oxide ($LiCoO_2$), and lithium manganese oxide ($LiMn_2O_4$/$LiMnO_2$) [71]. Recently, G. Fang et al. described an easy liquid-phase deposition method for making porous metal oxide nanosheets on three-dimensional substrates using ZIF-L. These nanosheets have great electrochemical properties for lithium battery applications [72]. **Figure 2b** shows a schematic representation of a lithium-ion battery design integrating a metal oxide with a MOF. A permeable separator maintains the separation between the metal-organic framework and the lithium electrolyte in the lithium-ion battery. **Figure 2b** also shows the mechanism by which the porous separator in the electrolyte medium transports lithium ions from the metal oxide cathode to the anode. As the Li-ion battery charges and discharges, Li ions are transferred between the positive and negative electrodes through the separator. An equivalent number of electrons are returned through the external circuit. The inclusion of MOFs in the electrolyte improves the general electrochemical performance, structural stability, and ionic conductivity of the battery, thereby increasing its appeal for the development of next-generation high-capacity lithium-ion batteries. Huang et al. [73] explain a simple polyvinylpyrrolidone (PVP)-assisted cube-shaped ZIF-L synthesis and a programmed scaling-down method that generates uniform and monodispersed cuboids by altering PVP amounts. After post-processing, cuboidal ZIF-L forms porous metal oxides (ZnO, $ZnCo_2O_4$) and metal sulfides. This chemical affects semiconductors, energy, and environmental remediation. The porous cuboid $ZnCo_2O_4$ structure proves its lithium-ion battery electrochemical performance. Redox-active metal centers in MOFs, notably first-row transition metals like Fe, Co, Ni, and Mn, simplify electrochemistry.

Co_3O_4, SnO_2, Fe_3O_4, ZnO, and CuO are "conversion reaction" lithium-ion battery anodes with high theoretical capacity and chemical stability [74]. MOF structures can change chemically with metal oxide anodes. A two-step calcination method was used to manufacture spindle-shaped porous α-Fe_2O_3 utilizing MIL-88 as a template. Combining ZnO and carbon results in a 1200 $mAhg^{-3}$ reversible capacity at 75 mAg^{-3} current density. Precursors from Co-based metal–organic framework materials create porous carbon/CoS_2 composites [75]. Ultra-small CoS_2 and a nitrogen-rich porous carbon matrix boost C/CoS_2 lithium storage.

Renewable technology is needed owing to the energy issue and population expansion. Due to their cost, efficiency, and environmental friendliness, dye-sensitized solar cells (DSSCs) are desirable energy conversion technologies. ZIF-8 and Uio-66 MOFs and RGO co-modify the photoanode, while a graphene counter electrode produces full-carbon DSSCs. The photoanode's dye loading is 1.92×10^7 mol cm^2, a 120.1% increase from He et al.'s TiO_2 photoanode [76].

MOFs aid solar cell systems with porosity, stability, and guest compound attachment spaces. Solution processability is good. Nanoscale dispersed MOFs and other precursors form microporous films by spin-coating. Organics and metal salts accurately change optoelectronic characteristics. MOF size, surface area, and porosity may be easily changed to optimize solar cell performance. MOF-based strategic composition engineering in PSCs for device efficiency and stability is growing. MOFs' microporous framework controls perovskite layer development, improving crystallinity and quality. MOFs improve PSC band alignment and CTL conductivity [76]. For chemical and thermal stability, researchers synthesized MOF-doped PSCs [77]. Recent decades have seen MOF research emerge. Due to their unique physical and chemical properties, they are used in magnetism, fluorescence, nonlinear optics, adsorption, separation, catalysis, and hydrogen storage. Perovskite cells use gold, silver, and carbon electrodes, but absorption layer reactivity corrodes silver. ITO, FTO, and other conductive substrates are common. Use ITO-PET and ITO-PEN for flexible perovskite cells. An n-i-p perovskite solar cell uses TiO_2 (or SnO_2) as the electron transport layer atop FTO conductive glass. ETLs have perovskite light absorption layers or mesoporous TiO_2 and calcium titanium. Gold is evaporated from the PSC upper electrode after depositing the hole transport layer (HTL), usually Spiro-OMeTAD, on the light absorption layer [78]. The HTL is directly installed on ITO conductive glass in pin-structured PSCs. PSS or NiOx transfers holes, then perovskite light absorption, PCBM electron transport layers, and silver electrodes [79]. The design of solar cells separates and conducts electron-hole pairs after they receive light. Electrons approaching the negative electrode and holes approaching the positive electrode generate voltages and currents. Thus, electron-hole pair production, dissociation, transport, and recombination improve solar cell photoelectric conversion.

High theoretical capacity and gravimetric energy density make lithium-sulfur (LiS) batteries appealing for next-generation EES devices. New functionalized MOF membranes prevent lithium polysulfide transport. Bai and colleagues developed the Li^+ and polysulfide anion filter HKUST-1@GO ($Cu_3(BTC)_2$) in 2016. Metal ion electrochemically stable MOFs reduce deterioration and parasitic reactions at broad cutoff voltages. UiO-66 zirconium (Zr) MOFs are octahedral clusters of $Zr_6O_4(OH)_4$ with various 1,4-benzenedicarboxylic acid (H_2BDC) Zirconium-based MOFs' Lewis acidity allows polysulfide anions (Sn^{2-}, $4 < n \leq 8$) to stick and lithium ions (Li^+) to pass through lithium-sulfur batteries [80]. Zirconium-based MOFs have Lewis acidity, which makes it possible for polysulfide anions (Sn^{2-}, where $4 < n \leq 8$) to stick to them and for lithium ions (Li^+) to move through lithium-sulfur batteries [81].

Electrochemical storage is needed for renewable energy. Supercapacitors are becoming more popular because of their high specific capacitance, quick charge and discharge, extended lifespan, safe operation across temperatures, and little maintenance as demand for efficient and sustainable energy solutions rises. SCs outlast batteries and capacitors and have higher specific capacitance and power density. Their cycling power, stability, and reliability are excellent. The structure and content of electrode materials affect energy storage capacitor performance. For this, MOFs and other materials were explored. MOFs are being researched for energy storage to meet demand for sustainable, high-performance alternatives. These materials increase energy storage system capacity, efficiency, and sustainability. MOF-74 stands out for its structural stability and large surface area. MOFs carefully control SBUs and organic ligands with flexible frameworks. Researchers can customize pores' sizes, setup, and characteristics for application performance. Torabi and her team sacrificed MnNi-MOF-74 for great supercapacitor electrodes. Changing alkalization conditions gave MnNi-6 a new structure that increased electrochemical performance. The material's customized surface area and structure improve electrochemical interactions and energy storage. Its thermal stability made MnNi-6 a good high-temperature supercapacitor. With MnNi-6 increasing energy storage, sacrificial MOF-derived materials can produce reliable, high-performance supercapacitors [82].

Siddiki and colleagues are making a high-performance supercapacitor nanocomposite. Hydrothermally produced cerium-cobalt (Ce–Co) MOFs contain Ti_3C_2Tx, an MXene family component. The study explores merging Ce–Co MOFs with Ti_3C_2Tx to improve supercapacitor energy density, stability, and capacitive performance. Ce–Co MOFs/Ti_3C_2Tx nanocomposite is a promising energy storage alternative [83].

Supercapacitors store energy by EDLCs or rapid surface redox reactions. Hybrid autos, electronics, and renewable energy systems benefit from these devices' high-power density, rapid charge-discharge, and lifetime. Baig et al. found that porous crystalline materials with high surface areas provide good carbon and metal oxide electrode precursors [84].

Different topologies, porosity, and chemical compositions across domains make MOFs popular. Due to their molecular-level chemical composition management and highly porous structures that promote mass and charge transfer, electrochemical energy storage and conversion systems require MOF-based materials—pristine MOFs, MOF composites, and their derivatives. High-efficiency electrodes are needed for supercapacitors and fuel cells. Yang et al. began to study customized MOF-based materials for electrochemical energy storage and conversion [85]. Next, they discuss contemporary MOF-based supercapacitors and electrocatalytic oxygen reduction (ORR). The problems and perspectives of MOF-based materials utilized for supercapacitors and electrocatalytic ORR are examined. They concluded MOFs could change energy storage and conversion. Their unique designs, porosity, and composite and derivative production make them ideal for high-performance supercapacitors and electrocatalysts. This essay highlights MOF research to improve properties and overcome large-scale application issues [86].

Researchers are intrigued by MOFs' large surface areas, quick ion transport, high porosity, and plentiful redox metal centers. These electrode materials' design and production change MOF architecture and morphology (micro- or nanostructure), affecting electrochemical performance. High surface areas, porosity, and redox-active metal cores make MOFs ideal for electrochemical energy storage. Electrochemical performance is greatly affected by micro- to nanoscale design and shape. Awais et al. found that supercapacitor MOF customization boosted energy and power densities. The study suggests that hybrid materials and machine learning may speed MOF implementation in energy storage [87].

Pan et al. made Mn-based MOFs from manganese salt and 1,2,4,5-benzenetetracarboxylic acid. The MnO/C-600 nanocomposite electrode, made of Mn-based metal-organic frameworks, showed 322 mAh g^{-1} specific capacity at 0.1 A and retained 270 mAh g^{-1} after 360 cycles at 0.5 A g^{-1} [88].

2. Results and discussion

2.1 Nanometal atoms and MOFs with high-power laser techniques

2.1.1 Overview of laser-material interaction physics

Laser-induced engineering harnesses the interaction of irradiating photons with different materials; it has been used to manufacture materials and devices to harvest energy from the environment. Laser sources can be categorized as continuous-wave (CW) or pulse lasers based on their operation mode on a timescale [89]. CW lasers offer a continuous laser output with excellent coherence, a well-defined beam profile, and a stable energy output. However, their peak power is typically low, usually at the hundred-watt level. In contrast, pulsed lasers have varying durations for each pulse, ranging from micro-, nano-, pico-, femto-, and attoseconds, with the peak power reaching extremely high levels. Shorter-pulsed lasers cause localized superheating and ultrafast quenching within atomic lattices, leading to negligible thermal effects and better structures in the fabricated materials [90]. It is possible to produce metal nanoparticles of the desired size and properties with these laser-based methods [14].

2.1.1.1 Laser ablation

Laser ablation involves the removal of materials with thermal, chemical, and/or photochemical characteristics under intense laser irradiation. The dominant mechanism of ablation depends on the laser parameters, including photon energy, power, pulse duration, and repetition rate. This versatile technique allows the creation of controlled and precise patterned structures with micro/nano features on different materials, with significant applications in scientific and technical fields, especially in the preparation of micro/nanostructured functional materials [91].

Recent advancements in ultrafast laser ablation have demonstrated the feasibility of micro/nanostructures through advanced optics techniques such as temporal and spatial laser shaping. Ablation induced by picosecond/femtosecond laser pulses differs significantly from that induced by nanosecond or longer pulses. The energy of the incident laser is primarily absorbed by the electrons in the material skin layer, followed by nonlinear electronic heat conduction, which transfers the energy to the target interior [92]. This process leads to ultrafast melting and hydrodynamic expansion, the formation of metastable states, and subsequent material breaks, resulting in the removal and ablation of materials. Studies involving metal nanoparticle production by laser ablation are included in Section 2.1.

Laser-synthesized nano materials technique is a promising top-down approach for producing ultra-small nanoparticles, including metal, metal oxide, nitride, and semiconductor nanocrystals. This method offers advantages over others, enabling colloidal, pure, and contamination-free nanoparticles. The synthesis environment, laser energy, and ablation time directly affect achieving ultra-small size nanoparticles.

2.1.2 Applications of laser ablation

Recently, metal oxide NPs have garnered considerable attention from researchers owing to their extensive applicability across various domains, including sensing, catalysis, and optoelectronics. Copper oxide nanoparticles (CuONPs) are highly favored among metal oxide nanostructured materials due to their superior optical, electrical, physical, and magnetic properties. CuONPs are categorized as p-type semiconductors possessing a broad and straight band gap. A Q-Switched Nanosecond Nd:YAG laser was employed to accurately generate colloidal CuONPs *via* the laser ablation of a copper target submerged in distilled water. With a repetition rate of 10 Hz and a pulse duration of 10 ns at 532 nm, the Q-switched Nd: YAG system can deliver up to 1400 mJ of energy per pulse. Using a knife-edge technique, we measured the diameter of the laser spot on the target surface to be 2 mm. Scientists did some tests and found that when they raised the laser fluence from 28.6 to 51.5 × 10^3 J/cm^2 and kept the laser ablation time the same (30 minutes), the average nanoparticle size went down from 8.5 nm to 4.6 nm. Also, the average size of the CuO NPs shrank from 6.8 nm to 4.3 nm when the laser ablation time was increased from 10 minutes to 50 minutes at a constant incident laser fluence of 34.3 × 10^3 J/cm^2. The concentration of the CuONPs that were made went up as the laser fluence and ablation duration went up [93].

Poly(vinyl pyrrolidone) (PVP) nanofibers containing gold nanoparticles (Au-NPs) were synthesized using a combination of laser ablation and electrospinning methods. Ortac et al. [14] reported that they directly generated Au-NPs in a PVP solution *via* laser ablation, followed by the electrospinning of the PVP/Au-NPs solution to produce nanofibrous composites. In this investigation, PVP was dissolved using a 15% (w/v) polymer concentration in an ethanol:water (1:1, v/v) solvent system. A pulsed ND:YLF laser (Empower Q-Switched Laser, Spectra Physics, USA) was used to laser ablate a 15% (w/v) PVP solution to create gold nanoparticles (Au-NPs). A lens with a focal length of 50 mm focused laser radiation (wavelength: λ = 527 nm, average power: 16 W, pulse duration: t = 100 ns, pulse energy: E = 16 mJ for 1 kHz) onto a cylindrical gold block. The duration of the ablation was between 15 and 30 minutes. The PVP/Au-NP solution was then electrospun with the following settings: feed rate = 0.5 ml/h, applied voltage = 15 kV, and tip-to-collector distance = 12 cm. On a grounded stationary cylindrical metal collector that was wrapped in aluminum foil, electrospun nanofibers were applied. The electrospinning process was conducted at 23°C with 20% relative humidity using a Plexiglas cage to contain the electrospinning gear. In contrast, 15% (w/v) PVP was electrospun in ethanol:water (1:1, v/v) under the same circumstances without being exposed to laser ablation of NPs. The electrospun PVP/Au-NPs nanofibers were bead-free, with an average fiber diameter of 810 ± 480 nm, according to the SEM imaging. The Au-NPs were in spherical form, having an average diameter in the range of 5–20 nm, and they were mostly dispersed homogeneously in the PVP matrix, yet, some aggregations of the Au-NPs were also observed.

Blue luminous colloidal silicon nanocrystals (Si-NCs) were synthesized by a two-stage procedure. The first step was to make Si-NCs by ablation with a femtosecond pulsed laser on a silicon wafer that was submerged in deionized water, as done by Alkis et al. [94]. A commercial Ti:sapphire regenerative femtosecond laser amplifier system, operating at a wavelength of 800 nm, generated colloidal NPs from a silicon wafer, with a pulse duration of 200 fs, an average output power of 1.6 W, and a pulse repetition rate of 1 kHz, resulting in a pulse energy of 1.6 mJ. The purified silicon wafer was positioned in a glass container holding 20 mL of distilled deionized water. A plano-convex lens with a focal length of 50 mm was employed to concentrate the laser beam on the target. The liquid layer height above the silicon target is approximately 5 mm. The laser ablation was performed for approximately 5 minutes, with the laser beam swept across the target surface. TEM data reveal a particle size range between 5.5 and 100 nm. A good number of the nanoparticles have sizes between 5.5 and 40 nm. It should be mentioned that this work does not show any small particles with diameters smaller than 5.5 nm. Though these particles are not shown in the histogram, we also saw particles whose diameters ranged from 100 nm to 500 nm.

Koral et al. [95] describe how they used terahertz time-domain spectroscopy (THz-TDS) to look at laser-synthesized silver nanoparticles (Ag-NPs) in an organic liquid. Using a pulsed Nd:YLF laser with a wavelength of 527 nm, an average power of 16 W, a pulse duration of 110 ns, and a pulse energy of 16 mJ for 1 kHz, Ag-NPs were created. Using a 50 mm focal length lens, the laser-focused on the pure Ag (99.999%) target in 10 ml 2-propanol. For synthesis, the ablation was done for 5 minutes. Ag-NPs submerged in 2-propanol undergo ablation, giving them a yellow tint. The researchers produced colloidal Ag-NPs with an average diameter of 10 nm in a twopropanol solution using nanosecond pulsed laser ablation. THz-TDS measurements were conducted on various volumetric concentrations of Ag-NP suspensions contained in a 2-mm path length quartz cuvette. A technique has been devised to extract the optical properties of nanofluids, which exhibit dispersive and highly absorptive characteristics, by analyzing the changes in amplitude and phase exclusively around the principal peak of the THz waveform. Amorphous metal nanoparticles (A-NPs) have gained attention because their disordered structure and shrinking methods (like nanoscaling) enhance properties over crystalline counterparts. Traditional synthesis often causes contamination and size limits, restricting use. Recently, laser ablation in liquid (LAL) and laser fragmentation in liquid (LFL) have attracted attention as environmentally friendly, scalable methods for producing ultrapure crystalline NPs. They have also shown a lot of promise for making A-NPs [85]. There are some chemical reactions that happen during NP synthesis, like oxidation and carbonization, but pulsed laser synthesis of NPs is mostly a physical process. This makes it easier to make high-purity NPs that are not tainted by surfactants. This means they can be used for specific tasks that require extreme purity. Flemban synthesized ZnO NPs of varying sizes using the pulsed laser ablation in liquid (PLAL) technique at energy levels of 70, 90, and 130 mJ. Laser ablation pulses with energies of 70, 90, and 130 mJ were utilized to investigate the influence of pulse energy on the NPs. Focusing the laser beam through a convex lens on a small area about 70 mm across, they were able to achieve laser fluences of 18.2 J/mm^2, 23.3 J/mm^2, and 33.7 J/mm^2 at pulsed laser energies of 70 mJ, 90 mJ, and 130 mJ for 30 minutes on the target metal surface. A combination of spherical and hexagonal nanoparticles was acquired. Moreover, the dimensions of the ZnO NPs diminished with elevated pulsed laser ablation energy, resulting in a high concentration of ZnO NPs at greater energies. The size distribution of the ZnO Nps produced with a pulsed laser ablation energy of 70 mJ yielded an average nanoparticle size of 12.2 nm. The average particle sizes of ZnO NPs for each sample are 905, 876, and 519 nm at 70, 90, and 130 mJ, respectively [96]. In_2O_3 NPs can be utilized on polystyrene substrates. Laser ablation in liquid employs varying quantities of laser energy to fabricate photodetectors. This research involved the synthesis of In_2O_3 NPs, which were subsequently deposited onto polystyrene substrates with a pulsed Nd:YAG laser with a wavelength of 1064 nm and a pulse duration of 10 ns. The PS plates were fabricated using a technique known as photo-assisted electrolytic etching. n-type (111) silicon substrates were subjected to photoelectrochemical etching (PECE) to fabricate the porous silicon base. A 20% hydrofluoric acid (HF: C_2H_5OH) solution served as the medium for etching silicon plates, which were subsequently cut into 2x^2 cm square segments. An electric current density of approximately 15 mA/cm^2 was employed for 12 minutes during the etching procedure. A halogen light source was employed to illuminate the silicon chips, facilitating the etching process. Subsequently, the drop-casting technique was employed to incorporate In_2O_3-NPs onto the sculpted PS substrate. The object was irradiated with a pulsed Q-switched Nd-YAG laser operating at a wavelength of 1064 nm. The pulse duration is 10 nanoseconds, and the repetition rate is 1 hertz. The formation of In_2O_3-NPs was validated by the solution's color transition to yellow following a brief ablation period. This research highlights that a reduction in the size of In_2O_3-NPs correspondingly increases the surface-to-volume ratio and photoresponse of these nanostructures. Furthermore, their electrical characteristics and sensitivity as detectors in the visible spectrum exhibited considerable improvement. The In_2O_3-NPs/PS photodetector exhibited superior responsivity values of 0.20386982, 0.243676027, and 0.276599668 A/W at 650 nm and 0.0913793 A/W at 900 nm when subjected to an 800 mJ laser [97].

Salim et al. [98] demonstrated the manipulation and fine-tuning of the optical and electrical properties of Ag–WO_3 core-shell NPs by varying the Ag core concentration using the laser ablation in liquid technique. The study highlighted the influence of core concentration on the structural, morphological, optical, and electrical characteristics of Ag–WO_3 core-shell NPs. An Nd:YAG pulsed laser beam with a wavelength of 1064 nm was used to vaporize a 99.9% pure square silver target plate that was placed in 3 ml of distilled water in a glass container with no other chemicals present. A repetition rate of 1 Hz, a focal length of 10, and a pulse width of 15 ns were employed. Prior to immersion and irradiation, they purified the silver target by submerging it in acetone and rinsing it with distilled water. The laser ablation operation was performed on the silver target utilizing a laser fluence of 6.12 J/cm^2 and varying numbers of laser pulses (200, 250, 300, 350, and 400). The experiment was performed at various concentrations of silver. Leading to the generation of NPs as a suspension (0.36, 0.76, 1.2, 1.6, and 1.96 μg/mL). In the second step, the silver target was replaced with a tungsten target in the solution to generate the Ag–WO_3 colloidal core-shell. Utilizing a 1064 nm wavelength, 1 Hz frequency, and a 10 focal length lens with a consistent laser fluence of 76.34 J/cm^2 for 300 laser pulses, respectively. Ag–WO_3 core-shell NPs had an average size of 30.5–89 nm. At an Ag concentration of 1.6 μg/mL, the electrical characteristics demonstrated an increase in electrical conductivity from (5.89×10^{-4})

$(\Omega\ cm)^{-1}$ to (9.91×10^{-4}) $(\Omega\ cm)^{-1}$, as well as a decrease in average activation energy values of (0.155 eV) and (0.084 eV).

By altering the laser ablation duration, we have manufactured magnesium metallic NPs ranging from 20 to 30 nm using the pulsed laser ablation in liquid medium (PLAL) technology. In the laser ablation method, laser-induced breakdown spectroscopy (LIBS) is used to look at the physicochemical properties of laser-induced Mg plasma, specifically the spectral line intensities and plasma parameters (ne and Te). The energy of the laser pulse directed at the target was 90 mJ per pulse. The laser beam was concentrated on the target surface utilizing a lens with a focal length of 70 mm. The liquid depth was calibrated to 0.7 using the knife-edge approach to enhance ablation efficiency. Ablation was performed with varying irradiation durations of 10, 20, and 30 minutes [99].

Amendola et al. [100] concentrated on the laser ablation of an iron target in tetrahydrofuran (THF, inhibitor-free), acetonitrile (AN), dimethylformamide (DMF), dimethylsulfoxide (DMSO), toluene (TOL), and ethanol (EtOH) organic solvents. Laser ablation was conducted using 9 ns Nd:YAG (Quantel YG981E) laser pulses at 1064 nm, which were focused with a 10 cm lens on a 99.99% iron plate that was positioned at the base of a cell containing the solvent. They employed pulses with a repetition rate of 10 Hz and a power of approximately 5 J cm^{-2}. They demonstrate that the content and structure of NPs produced by laser ablation of metallic iron are significantly affected by the solvent used. Specifically, NPs can be synthesized with compositions that range from iron carbide to iron oxides or metallic iron, while the nanostructures may vary from isolated NPs to those embedded in an organic matrix or a core@shell configuration consisting of iron@iron-oxide or iron@graphite. The ablation process produces a plasma plume whose varying reactivity dictates the structural disparities of the resultant products. All of the NPs made by LASiS are magnetic, which shows that laser ablation in liquids is a possible way to make magnetic nanostructures in organic solutions. Laser ablation was conducted using 9 ns Nd:YAG (Quantel YG981E) laser pulses at 1064 nm, focused through a 10 cm lens onto a 99.99% iron plate situated at the base of a cell containing the solvent. Pulses of approximately 5 J cm^{-2} were employed at a repetition rate of 10 Hz. Approximately 300 nanoparticles were examined, and the core-shell structure was present in 79% of them. The results of the high-resolution TEM (HRTEM) analysis indicate that the particles' core and shell are both crystalline. However, the cores exhibit an interplanar distance of 0.20 nm, while the shells have interplanar distances of 0.25 or 0.48 nm. Only interplanar distances of 0.25 or 0.48 nm are observed in the fraction of non-core/shell particles.

2.2 MOFs, metal and metal oxide-polymer nanocomposites, energy solution

2.2.1 Preparation of nanocomposites

There exist several techniques for manufacturing metal NP/MOF composites. The MOF was mixed with the metal precursor by solid grinding [101] or solution infiltration [102]. The metal NP/MOFs were then made by reducing the metal precursor. This was the first effort at loading MOFs with NPs. More recently, a chemical vapor deposition (CVD) approach has been devised to better regulate the final NP size whereby porous MOFs are loaded with a metal NP precursor in the vapor phase and then subsequently decomposed to produce metal NPs within the MOF pores [103]. The disadvantage of the CVD method, which is considered a complex process not suitable for mass production, is that although the protective agents and/or surfactants used to synthesize the metal NPs remain as contaminants, metal NP/MOF composites were produced by growing and encapsulating MOF around previously synthesized metal NPs [104].

One study, for instance, produced Au seed/MOF-199.50 mg of MOF-199 was mixed in a flask with 30 mL of 0.1% $HAuCl_4$ solution at 45°C for 2.5 h. To get Au seed/ MOF-199, 220 μL of 10% (W/W) trisodium citrate was added and agitated for an additional 40 min and then cooled to room temperature [105].

2.2.2 Applications of nanocomposites

Gold nanoparticles, referred to as Au/UiO-66, were synthesized within a metal-organic framework using a dual solvent approach (**Table 1**). Gold nanoparticles (Au-NPs) were highly disseminated within the pores of the MOF UiO-66, without any deposition on the exterior surface or agglomeration. Using benzyl alcohol as a substrate, Au/UiO-66 demonstrated exceptional catalytic efficacy, yielding benzaldehyde as the sole product. The reactions conducted under solvent-free conditions provided an eco-friendly approach utilizing atmospheric pressure O_2 as the oxidant. Water was demonstrated to be advantageous for the reaction, and the hydrophilic nature of the MOF's inner surface significantly enhanced its promoting effect. The catalyst demonstrated significant stability and could be reused a minimum of eight times without noticeable aggregation or leaching. UiO-66 was produced utilizing zirconium tetrachloride ($ZrCl_4$, 5 mmol) and benzene-1,4-dicarboxylic acid (BDC, 5 mmol) in 150 mL of DMF at 120°C with stirring for 24 hours. The white precipitate was filtered, then washed with DMF to eliminate the surplus unreacted organic linker. The product was immersed in a dichloromethane solution for solvent exchange to eliminate DMF, followed by filtering and subsequent drying under vacuum overnight at ambient temperature. Encapsulation of Au^{3+}/UiO-66 by the dual solvent approach involved suspending 0.2 g of UiO-66 in 40 mL of dry n-hexane, a hydrophobic solvent, and sonicating for 20 minutes until homogeneity was achieved.

MOF composite	Metal and metal oxide nanoparticles	Ref.
MIL-101(Cr)	Cu/Pd	[106]
MOF-5	Pd	[107]
HKUST-1	CuO/CeO_2	[108]
MOF-5	Ru	[109]
MOF-5	Au/ZnO–Au/TiO_2–Au	[110]
MOF-5/MIL-53	Al/Au	[110]
DUT-5/UiO-66/MOF-253	Pt	[111]
UiO-67	Pd	[112]
MOF-5	Pd	[113]
MOF-177	Pt	[114]
MOF-5	Pd	[115]
UiO-66	Au	[116]

Table 1.
List of MOF-supported metal nanoparticle catalysts.

Following a time of stirring, 0.16 mL of an aqueous solution containing 0.068 mmol of $HAuCl_4 \cdot 4H_2O$ was incrementally added over 20 minutes while maintaining vigorous stirring. The resultant solution was agitated incessantly for 3 hours. The product, which precipitated at the bottom of the sample vial after mixing, was separated from the supernatant by decantation and air-dried at room temperature. The synthesized sample was subjected to vacuum drying at ambient temperature overnight. To prepare Au/UiO-66, after drying Au^{3+}/UiO-66, 40 mL of dry n-hexane was added, and the mixture was sonicated for 20 minutes until it achieved a pseudo-homogeneous state. A $NaBH_4$ solution (5 mL, 0.6 mol/L) was subsequently introduced under vigorous stirring to produce a dark purple suspension of the catalyst. Au/UiO-66 was extracted *via* centrifugation, washed with water, and subsequently dried under vacuum overnight at ambient temperature [116].

It has been shown that MIL-88 (Fe) can be used to efficiently make Fe_2O_0 with spindle-shaped mesoporous structures. A look at the product's microstructure showed that it had a lot of holes and was made up of uniform NPs about 20 nm in size [116]. Moreover, an additional benefit of utilizing MOFs as templates is the straightforward production of carbon–metal/metal oxide hybrids. A standard iron-based metal-organic framework, MIL-88-Fe, characterized by the formula $Fe_3O(H_2O)_2Cl(BDC)_3 \cdot nH_2O$, was selected as the template for the synthesis of porous α-Fe_2O_3. The synthesis of MIL-88-Fe was achieved through a modified solvothermal method, utilizing $FeCl_3 \cdot 6H_2O$ and 1,4-benzene dicarboxylate (BDC) as the initial materials. Spindle-like porous α-Fe_2O_3 was synthesized using an iron-based MOF template. In evaluations as an anode material for lithium batteries, this spindle-like porous α-Fe_2O_3 demonstrates significantly improved Li storage capabilities. The particle, measuring approximately 0.8 μm in length and 0.4 μm in width, consisted of aggregated Fe_2O_3 nanoparticles, each with dimensions of less than 20 nm. The porous α-Fe_2O_3 demonstrated a capacity of 911 mAh g^{-1} after 50 cycles at a rate of 0.2°C. A comparable capacity of 424 mAh g^{-1} could be achieved even when cycled at 10°C [117]. In 2011, Lu and his colleagues wrote about how Cr-MIL-101 MOFs could be used as models to make crystal-like ordered hybrid inorganic solid nanoparticles. Using Cr-MIL-101 as a model, they talk about a surface coating-reinforced framework method for making ordered microporous materials using the nanocasting method. Cr-MIL-101 features two types of mesoporous cavities with free widths of ca. 2.9 and 3.4 nm accessible through two microporous windows of ca. 1.2 and 1.6 nm, and a significant amount of unsaturated chromium sites (ca. 3.0 mmol g^{-1}) [118].

Dou et al. [119] demonstrated that the laser conversion method is applicable for synthesizing Tb_2O_3 with various phase types by utilizing different Tb-MOFs as precursors, including $Tb_6O_8(C_4H_2O_4)_6$ (Tb-MOF-Fum), $Tb(C_{27}H_{12}O_6)(H_2O)_2(C_6H_{12}O)$ (Tb-MOF-BTB), and $Tb_{16}(C_{24}H_{12}O_6N_3)_{16} \cdot (C_4H_9N_{20})_{24} \cdot$-$(C_4H_9N_{20})_{91}(H_2O)_{108}$ (Tb-MOF-TATB). The diversity of topologies in these Tb-MOFs enables accurate manipulation of the volumetric density of Tb metals. When laser light hits Tb-MOF-Fum, which has a Tb metal volumetric density of 61.3 mmol cm^{-3}, it directly makes C-type Tb_2O_3. On the other hand, laser light hits Tb-MOF-BTB and TATB, which have lower Tb metal volumetric densities of 0.095 and 17.3 mmol cm^{-3}, and they both make B-type Tb_2O_3. The terbium sesquioxide NPs made from laser-converting Tb-MOFs are all the same size and work better in the oxygen evolution reaction (OER). This B-type Tb_2O_2 comes from Tb-MOF-TATB. It has particles about 3 nm in size and the lowest overpotential of 331 mV in the oxygen evolution reaction, making it the best of all the terbium sesquioxides.

Mutlu et al. [2, 3] synthesized ZnMOF and NiMOF in 70 minutes using the LIRS. The yield of NiMOF, whose crystal structure is quite different from the ZnMOF produced by laser, is 85–90%. Thus, it was observed that the primitive triclinic structure without center symmetry was reached in a much shorter time with the LGS method. Additional research employing this methodology has demonstrated significant efficiency within a short timeframe. **Table 2** compares the synthesized MOFs according to their processing times and yields.

Batibay et al. [129] synthesized different kinds of polymeric nanocomposite thin films and hydrogel nanocomposites by using thiol-ene photo-click reactions and 2-mercapto-thioxanthone (TX-SH) as a single Type II photoinitiator. The photocurable nanocomposite thin film formulations, with/without Ag salt and NBPOSS derivatives in the presence of photoinitiator which was dissolved in a few drops of DMF, and thin films containing only photoinitiator were prepared. 40 μm-thick nanocomposite films were coated on 2 × 5 × 0.5 cm aluminum metal plates by roller coater. These coated aluminum plates were cured by a mini-UV curing device in 250 passes (1 pass = 2 s). Thus, the photochemical *in situ* formation of Ag-NPs and its effect on the polymeric matrix for the thin films having different amounts of NBPOSS derivatives were investigated. Consequently, they produced *in situ* metallic Ag-NPs using photo irradiation. Photo-initiated mechanism of the TX-SH mediated in the presence of 1NBPOSS *via* thiol-ene click reactions (**Figure 3a**) and SEM images of nanocomposite thin films are located in **Figure 3b**. The researchers have examined the influence of NPs on the characteristics of nanocomposites.

In addition, Batibay et al. [129] prepared POSS-containing nanocomposites using hydrophilic and hydrophobic monomer formulations with and without Ag-NPs and tuned the hydrophobic surface to superhydrophilic. Firstly, the hydrophilic monomer system was taken into consideration to examine the benefit of POSS-NH_2 while TX-O-CH_2COOH was used both as a photoinitiator and photo-reducing agent; hence, the experiments were conducted with PEGMEA/PEGDA multifunctional acrylate systems for the preparation of polymeric thin films. Formulations were coated on the aluminum substrate after sonication and were cured using the Mini-Cure unit, and passes were applied starting from 1 to 250. As expected, $AgNO_3$ salt containing polymeric films changed color from clear to brown and silver for prolonged irradiation of the nano hybrid film, as shown in **Figure 4**.

MOF name	Synthesis method	Synthesis time	Yield	Reference
UiO-66	Solvothermal	24 hours	70%	[120]
MOF-808	Solvothermal	7 days	70%	[121]
DUT-67	Microwave	48 hours	70%	[122]
PCN-222	continuous process	48 hours	46%	[123]
MOF-14 ($Cu_3(BTB)_2$, BTB = 4,40,400-benzenetribenzoate)	Mechanochemical (by ball milling)	20 hours	%49	[124]
Zn-MOF	Laser-driven	70 minutes	70%	[2]
Ni-MOF	Laser-driven	70 minutes	90%	[3]
Al-AZB	solvothermal	16 hours	87.0%	[125]
Zn(II)–MOF	solvothermal	3 days	57.2%	[126]
Cu(AZPY)-MOF	solvothermal	24 hours	45.2%	[127]
DUT-8(Co)	solvothermal	48 hours	96%	[128]

Table 2.
Synthesis times and yields of various MOFs according to their synthesis methods.

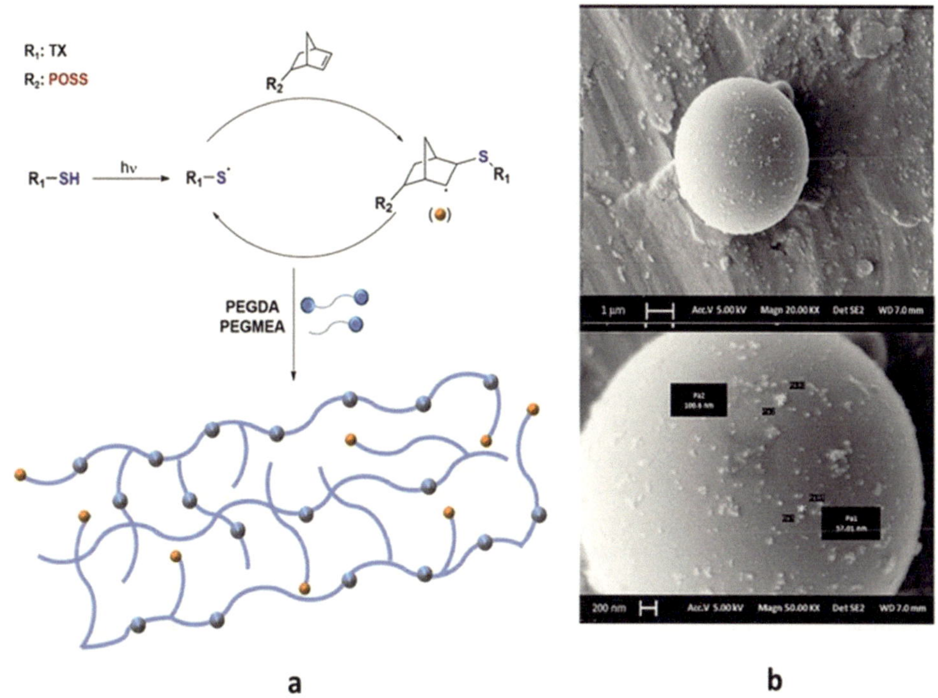

Figure 3.
Schematic representation of photo-initiated mechanism of the TX-SH mediated in the presence of 1NBPOSS via thiol-ene click reactions (a) and SEM images of nanocomposite thin films (b) Reprinted with permission from ref [129], Copyright 2023 Elsevier.

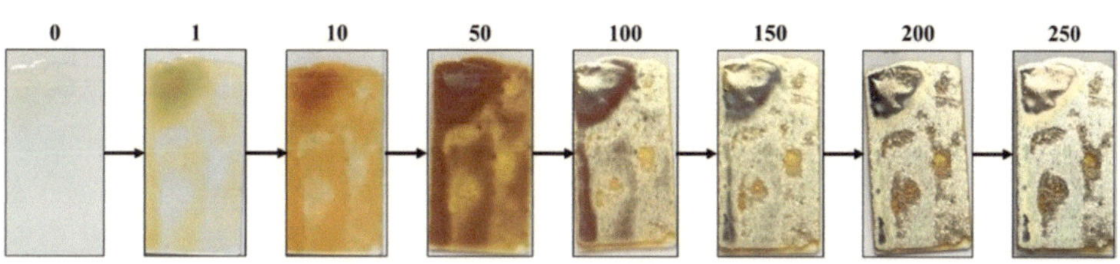

Figure 4.
The change of the colour of AgNps and POSS-NH2 containing PEGMEA/PEGDA hybrid nanocomposite films (E2PP) Reprinted with permission from ref [130], Copyright 2021 Elsevier.

A wide variety of methods are available for nanoparticle synthesis, either in solution or in a polymer matrix. The *in situ* photochemical method [131] is the most convenient among these methods because it offers several advantages, including speed, ease of use, environmental friendliness, controllability with light intensity, and affordability. In this method, photoinitiators play a great role in the synthesis of shape- and size-controlled NPs. When light shines on the formulation, produced radicals are mainly responsible for the photoreduction of metallic salts to metallic NPs, and simultaneously, the same or other initiating radicals attack the monomer to initiate the polymerization. The type of photoinitiator plays a crucial role in determining the size, shape, and distribution properties of the NPs in the polymer matrix, as the produced radicals vary. Nanocomposites that depend on metal NPs as the inorganic part and polymer as the organic part present unique properties [132].

Bimetallic nanoparticles (BNp) are also seen to have increasing importance in addition to metallic Nps both scientifically and technologically [133]. BNPs are made up of a combination of two different metals, and they can be synthesized in different shapes, sizes, and structures. Adding silver-gold bimetallic NPs with a core-shell structure gives systems special physical and optical properties that are not possible with systems made of only one metal [134]. They can serve as catalysts due to their decrease in size and increase in surface area.

Kırtay et al. [135] synthesized ZnO NPs photochemically in epoxydiacrylate/TPGDA monomer formulation, which displayed both shape and size changes compared to the previously used formulation containing PEG acrylate. The shape of NPs changed from hollow doughnuts to rod shapes. 5 mg (0.5% wt) of photoinitiator and 40 mg (4% wt) of $Zn(CH_3COO)_2 \cdot 2H_2O$ were included in a mixture of 191 mg (19% wt) of tri(propylene glycol) diacrylate (TPGDA) and 764 mg (76.4% wt) of epoxy diacrylate (EA) monomers. The completed mixture was subjected to ultrasonic mixing for 2 minutes to ensure uniform distribution of the metal salt. Formulations were applied to aluminum plates with a 40 μm kbar roller, and the films were then cured in a small UV-Cure Unit. **Figure 5** shows the mechanism of photoreduction of Zn^{2+} salt to ZnO NPs in the presence of TX-SH and SEM images.

Mutlu et al. [136] synthesized Au NPs, Ag NPs, and MnO NPs using the one-component Type II photoinitiator thioxanthone-anthracene (TX-A) through *in situ* photoreduction in both solution and polymer matrix under atmospheric conditions (**Figure 6**). They examined the dielectric characteristics of PEGMEA/PEGDA nanocomposite films incorporating Au NPs (**Figure 6a**), Ag NPs (**Figure 6b**), and MnO NPs (**Figure 6c**) synthesized using photochemical methods. The formulation of the nanoparticle-free film (F1) was achieved by dissolving TX-A (0.5 mg, 0.05 wt%) in a minimal amount of DMF, which was then included in a monomer mixture of

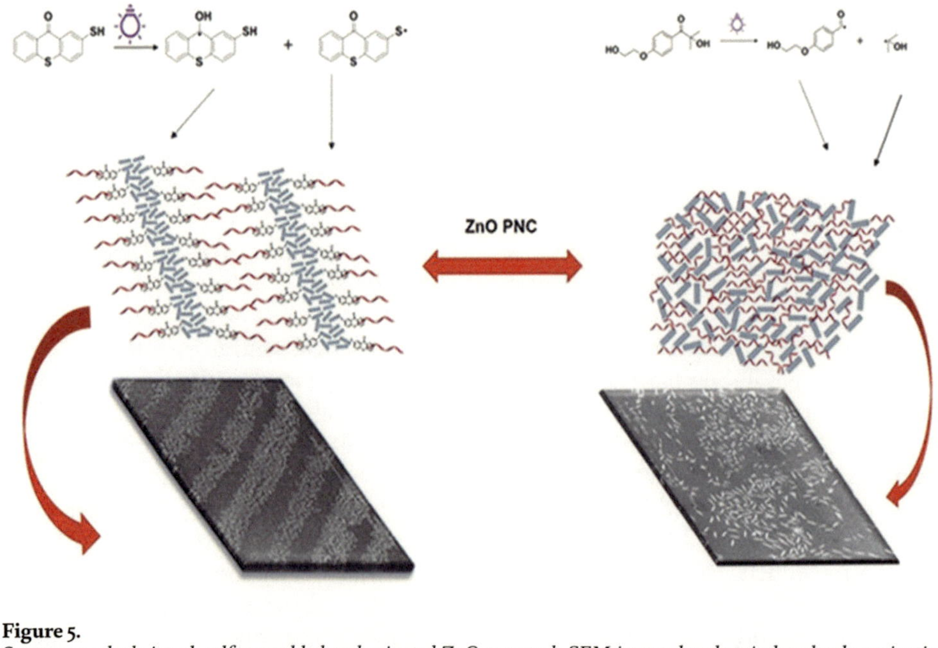

Figure 5.
Spontaneously designed, self-assembled and oriented ZnO nanorods SEM images by photoinduced polymerization of epoxydiacrylate formulation in the presence of TX-SH. Reprinted with permission from ref [135], Copyright 2022 Elsevier.

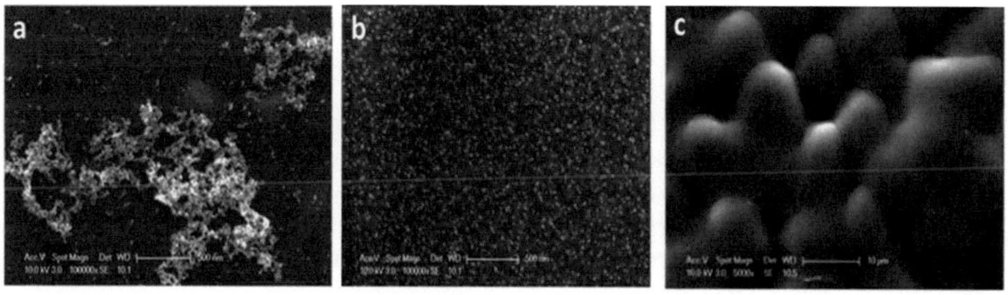

Figure 6.
SEM images of Au NPs (a), Ag NPs (b) and MnO NPs (c) prepared with TX-A (0.05 wt%) and HAuCI4 (4 wt%) in the DMF solution [136], Copyright 2021 Elsevier.

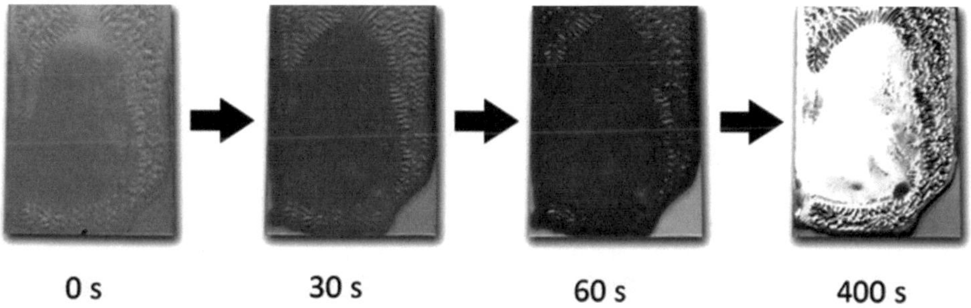

Figure 7.
The color changes of TX-SH/PEGMEA+PEGDA/HAuCl4 containing film upon irradiation time: Au NPs containing gold mirror film formation. Reprinted with permission from ref [137], Copyright 2017 Wiley.

PEGMEA (19.2 wt%) and PEGDA (76.75 wt%). This formula served as the basis for the subsequent three formulae, each including distinct metal salts: $HAuCl_4$ (40 mg, 4% wt%), $AgNO_3$ (40 mg, 4% wt%), and $MnCl_2$ (40 mg, 4% wt%). Each formulation was sonicated for 5 minutes to achieve optimum salt dispersion. Formulated mixtures were coated onto a 20 × 50 mm, 1.0 mm-thick aluminum plate using a 40 μm roller coater.

Çeper et al. [137] prepared Au-NPs that were at or near the surface of the polymer. They controlled the size distribution of the Au-NPs to prevent adhesion by using a thiol-functionalized initiator for UV curing polyethyleneglycol-based acrylate formulations (**Figure** 7). Together with $HAuCl_4$ (40 mg, 4 wt%), 2-mercaptothioxanthone, thioxanthone, and thioxanthone-thiophenol binary complexes were mixed with PEGDA (191 mg, 19.1 wt%). poly(ethylene glycol) methyl ether (PEGMEA; 764 mg, 76.4 wt%). monomers. The last formulations were ultrasonically mixed for 5 minutes in an ice bath and then degassed using an ultrasonic instrument in order to get a homogeneous dispersion of the metal salt. Eventually, they were applied to the glass surface using a roller coater, suitable for a coating thickness of 40 μm, and then cured using Mini-UV-cure equipment.

2.3 Photocatalysis

2.3.1 Preparation of photocatalysis samples

One can prepare a photocatalyst in several ways. We separate these techniques in this review into the electrospinning, solid phase, liquid phase, and vapor phase technologies. Simple and fit for mass production on an industrial level are solid-phase techniques. TiO_2 nanoparticles are prepared using such techniques even if they have drawbacks like high energy consumption, the creation of big particles, and the inclusion of contaminants [138]. After which solid nanoparticles are produced by condensation and deposition by either physical or chemical means, gas-phase techniques employ gaseous source materials or transform raw materials into their gaseous states in different ways. Physical and chemical forms define two categories of gas-phase techniques. Highly pure, tiny, particle-containing, well-dispersed, gas-phase-prepared nano-TiO_2 is produced [139]. Because they employ basic equipment, are straightforward to run, entail low reaction temperatures, and consume less energy than other techniques, liquid-phase procedures are favorable. Sol-gel, liquid-phase precipitation, hydrothermal synthesis, microemulsion, and other operational technologies are among the further divisions of these techniques.

2.3.2 Applications of photocatalysis

Ultrathin MOFs are great candidates for photocatalysis because they have more accessible active sites and lower mass transfer resistance. Zhou and his team [140] made PCN-134-2D by changing the surface of Zr-BTB nanosheets (BTB = 1,3,5-tris(4-carboxyphenyl)benzene) with TCPP modulators. The PCN-134-2D that was already formed kept its nanosheet structure and showed photoactive porphyrin species on its surfaces. On the other hand, a bottom-up approach synthesized PCN-134-3D using Zr-BTB layers intercalated with TCPP pillars. The photocatalytic efficiencies of PCN-134-2D and PCN-134-3D were examined in the oxidation of dihydroartemisinic acid to artemisinin by photoinduced singlet oxygen ($^1O_2$). Using PCN-134-2D as the catalyst, they achieved a nearly complete conversion after 1 hour, while PCN-134-3D achieved a significantly lower conversion (72%) after 3 hours. The high performance of PCN-134-2D was mostly because it had many catalytically active porphyrin sites on its surfaces that were easily accessible to substrate molecules and did not create any mass transfer resistance. The altered electronic characteristics of ultrathin nanostructures may enhance their photocatalytic efficacy. Alshareef and his colleagues [141] created a metal-organic framework (MOF) in the shape of a nanowire by using organic phosphonate linkers (TiPNW) and titanium-oxo nodes. The photocatalytic hydrogen production ability of TiPNW was examined following the incorporation of Pt NPs as the co-catalyst. Upon exposure to visible light ($\lambda > 400$ nm), a hydrogen generation rate of 1260 mmol h^{-1} g^{-1} was attained. TiPNP, on the other hand, had an H_2 production rate of 501 mmol h^{-1} g^{-1} because its structural units were all the same. The variation in photocatalytic activity stemmed from their differing charge mobility. The one-dimensional nanostructure imparted to TiPNW a decreased charge diffusion distance, facilitating the passage of photogenerated electrons to the superficially catalytically active sites. Some MOFs that work well for CO_2 cycloaddition are Zn-based IRMOFs (or MOF-5), ZIF-8, Cu-based HKUST-1 and MOF-505, Hf-based NU-1000, Co- or Mg-based MOF-74, Cr- or Fe-based MIL-101, In-based MIL-68, Zr-based UiO-66, and Al-based USTC-253. A series of cation-exchanged MOFs utilizing alkali metals were employed for CO_2 cycloaddition. A group of scientists led by Wang [142] created a Zr-TCPP MOF that looks like a hollow nanotube. They called it HNTM, and it has MOF pores and extra cavities inside it that help make H_2 when Pt and Ir single atoms are attached to it. The as-formed HNTM-Pt/Ir catalyst exhibited a substantial H_2 generation rate of up to 201.9 mmol g^{-1} h^{-1}. The bulk Zr-TCPP, which was also changed with Pt and Ir single atoms, only produced 31.7 mmol g^{-3} h^{-3} of hydrogen per hour. Zhu and his colleagues [141] did a different study where they compared the photocatalytic hydrogen production efficiency of a metal-organic framework based on titanium and phosphonate (HM-TiPPh) to that of its bulk equivalent (M-TiPPh). Adding mesopores (30 nm) on purpose made HM-TiPPh produce 945 mmol g^{-1} h^{-1} of hydrogen per hour, which was significantly higher than M-TiPPh's rate of 698 mmol g^{-1} h^{-1}. Recently, Eder and colleagues [143] produced a variety of MIL-125-Ti MOFs with varying pore topologies. Pristine X% NH_2-MILs had micropores that were less than 1 nm in size. Here, X is the percentage of BDC linkers in MIL-125-Ti that were replaced with NH_2-BDC. Heat treatment at 300°C formed cavities of 5–10 nm when the proportion of NH2-BDC was minimal, such as 2–10% NH_2-MIL. At elevated NH_2-BDC ratios (e.g., 50–80% NH_2-MIL), interconnecting fracture-type mesopores (4–5 nm) were prominent. Following the incorporation of Pt NPs as the co-catalyst onto their surfaces, the NH_2-MILs were then utilized in photocatalytic hydrogen production. In comparison to unaltered NH_2-MIL devoid of mesopores, the annealed 2–10%NH_2-MIL featuring cavity-type mesopores exhibited a 40–50% increase in hydrogen production rate. The annealed 50–80% NH_2-MIL, characterized by fracture-type mesopores, demonstrated a four- to fivefold increase in rate. The interconnectedness of fracture-type mesopores significantly enhanced the accessibility of internal active sites, thereby markedly improving catalytic performance.

Defects can influence photocatalytic performance by altering the energy band structure of semiconductive MOFs. Jiang and colleagues examined the photochemical behavior of UiO-66-NH_2 materials in relation to their defect quantity. A series of UiO-66-NH_2-Xs (where X denotes the modulator equivalent to the linker during synthesis) were synthesized using an acetic acid-modulated approach, resulting in an increase in defect quantity proportional to the modulator's added amount. Following the integration of Pt NPs as the co-catalyst, the resultant Pt@UiO-66-NH_2-Xs were then utilized in photocatalytic hydrogen generation. The H_2 production rate exhibited a positive correlation with the initial fault quantity. Nonetheless, an additional increase in defect quantity led to a reduction in the H_2 production rate. Jiang and colleagues [144] produced two variants of platinum nanoparticle-decorated UiO-66-NH_2 materials, namely Pt/UiO-66-NH_2 and Pt@UiO-66-NH_2, for photocatalytic water splitting. The sole distinction between them was the positioning of Pt NPs, specifically being supported on surfaces vs. enclosed within the interiors of UiO-66-NH_2. Notably, the activity of Pt@UiO-66-NH_2 (257.38 mmol g^{-1} h^{-1}) surpassed that of Pt/UiO-66-NH_2 (50.26 mmol g^{-1} h^{-1}), as the charge transfer from UiO-66-NH_2 to uniformly distributed Pt NPs was more efficient than to the superficially supported counterparts. The photocatalytic capabilities of porous ZnO cuboids were evaluated by degrading Methyleneblue (MB), rhodamine B (RhB), and Methylorange (MO) under ultraviolet light. For photocatalytic degradation, a specified quantity of porous ZnO cuboids was dispersed in 15 mL of dye solutions (15 mg/L, MB, RhB, and MO). Before irradiation, they stirred the dispersion in the dark for approximately 30 minutes to achieve absorption-desorption equilibrium between the dyes and the photocatalyst. They then irradiated under a xenon/mercury-xenon lamp (BL-GHX-V, Shanghai Bilon Co. Ltd., 500 W) equipped with a UV filter. Using a UV-Vis spectrophotometer (Evolution 220, Thermo Fisher Scientific), they determined the dye concentrations in accordance with Beer's law, utilizing the absorption peaks at 664 nm, 553 nm, and 464 nm for MB, RhB, and MO. Porous ZnO cuboids have a lot of photocatalytic activity for breaking down dyes (MB, RhB, and MO).

A study was conducted to assess the photocatalytic efficacy of Zn-MOF, utilizing spectrophotometric analysis to quantify the breakdown of MB dye by exposure to UV radiation. Using spectrophotometric monitoring, they analyzed the absorption spectra of pure MB without Zn-MOF and MB in its presence to quantify the dye degradation [2].

The absorption spectra of pure MB without Zn-MOF and MB with Zn-MOF demonstrate a swift decline in the strength of the characteristic peak at 664 nm, as measured during the time of irradiation, as illustrated in **Figure 8**. The observation indicates the progression of the period. The visible range spectral bands show a quick drop in the intensity of the characteristic peak at 664 nm (**Figure 8a**). This suggests that chromophore groups are present during the irradiation periods of 664 and 612 nm. The visible range spectral bands seen at 664 nm and 612 nm wavelengths are linked to the sulfur-nitrogen-conjugated system and show that chromophore groups are inside the MB molecules [145]. So, about 83% of the MB's degradation happened in just 1 minute of irradiation, during which time the color of the MB solution gradually changed. Approximately 83% of the degradation solution comprises ZnMOF (**Figure 8b**). Upon extending the irradiation to 60 minutes, a negligible amount of MB degradation occurred within the first minute of irradiation in the solution containing Zn-MOF. Upon extending the irradiation to 60 minutes, a negligible change was seen [2].

Quantum dot NPs within MOFs have been thoroughly investigated as prospective photocatalysts, including the CdS/UiO-66-NH_2 composite for the selective visible light oxidation of benzyl alcohol to benzaldehyde utilizing molecular oxygen as an oxidant [146], CdS@MIL-101(Fe) nanocomposites for the selective oxidation of benzyl alcohol to benzaldehyde under mild conditions with visible light [147], and CdS/Zn-MOF composites for photocatalytic water splitting under visible-light irradiation [148]. MOFs and their mixtures, especially the eco-friendly Fe-MOFs, help advanced oxidation processes (AOPs) get rid of organic molecules in water and wastewater by reacting with hydroxyl radicals *via* oxidation [149]. Fe-MOFs containing 1,4-piperazinediylbis(methylene) phosphonic acid, specifically STA-12(Fe), break down organic dyes very well in water when activated by H_2O_2 and exposed to natural sunlight. They can also be used over and over again, which suggests they could be useful for cleaning water [147]. Researchers have investigated the efficacy of MOFs as heterogeneous photocatalysts for the degradation of chemical warfare agents. Notably, Zr-based MOFs, specifically PCN-57 analogs containing benzothiadiazole and benzoselenadiazole, have selective photocatalytic activity for changing 2-chloroethyl ethyl sulfide (CEES) into 2-chloroethyl sulfoxide (CEESO), which is

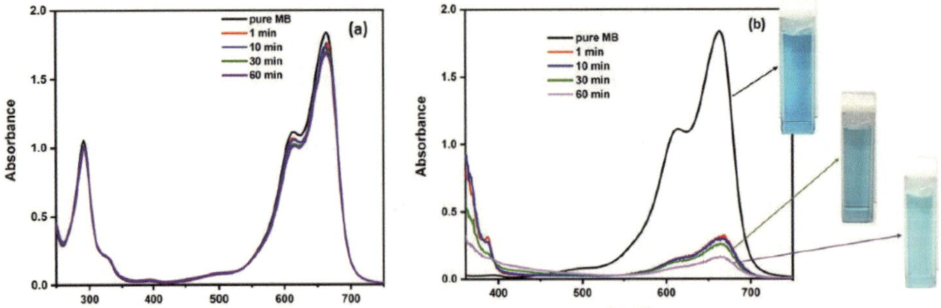

Figure 8.
UV-Vis absorption spectrum of the photodegradation process of the pure MB solution (10 ppm) depending on the duration of irradiation (a). UV-Vis absorption spectrum of the photodegradation of the MB solution (10 ppm) with ZnMOF (2.5 mg/10 mL) depending on the irradiation time and the change of solution color (b). Reprinted with permission from ref [2], Copyright 2024 MDPI.

not harmful. Additionally, a Zr_6-based MOF called NU-1000 that has been post-synthetically modified and functionalized with the photosensitizer BODIPY (boron-dipyrromethene) ligand shows improved singlet oxygen generation, which makes it easier to detoxify the sulfur mustard simulant CEES to CEESO, with a half-life of about 2 minutes. Jiang et al. experimentally demonstrated that structural flaws in UiO-66-NH_2 enhanced the photocatalytic hydrogen generation by Pt@UiO-66-NH_2 in a CH_3CN-H_2O solution [150]. Ryu et al. documented the initial application of MOFs as the electron transporting layer (ETL) in 2018, demonstrating remarkable efficiencies of 18.94% and 17.43% for rigid and flexible devices, respectively, by modifying a PCBM ([6,6]-phenyl-C61-butyric acid) layer with a nTi-MOF. Nguyen et al. recently synthesized a Co(MOF)-doped mesoporous TiO_2, achieving a power conversion efficiency (PCE) of 15.75% [151]. In their study, Zhang et al. showed an ETL that used a ZnO layer derived from MOF. This layer had an efficiency of 18.1%, little hysteresis, and less performance degradation over time in normal conditions [152]. Ahmadian-Yazdi et al. [153] recently reported that adding a Zn-MOF interlayer between the dense TiO_2 blocking layer and the active perovskite layer improved the crystallinity of the perovskite grains, which led to bigger grains and better photovoltaic performance. Researchers have also reported the utilization of MOF compounds in hole-carrying layers. In a new paper, Zhou et al. use a MOF material as a dopant in the Spiro-MeOTAD HTL to get the best hole mobilities, which results in a 19.47% power conversion efficiency [154]. Similarly, a Ni-MOF with a central layer of core-shell CuO@NiO achieved favorable efficiencies and remarkable temporal stability [155]. According to Zhou et al. [156], the interconnected MOF micropores make it easier for perovskite to crystallize in the even cavities. This results in better morphology and fewer grain boundaries, which leads to a 20.87% efficiency and very stable performance over time.

Because MOFs and perovskite materials are similar metal-organic composites, MOFs have recently shown promise as useful additives for improving perovskite crystallization and fixing flaws. At the moment, PSCs primarily use MOFs as interfacial modification layers, additives, or direct electron transport layers. When organic linkers or metal clusters are excited by light, some MOFs can work as semiconductors and charge carrier transport systems. The porous architecture in MOFs offers supplementary channels for the transfer of photogenerated electrons, thereby enhancing charge carrier movement and improving light-to-electricity conversion efficiency. The pore architecture in Zr-MOFs is deemed appropriate for perovskite infiltration. Zr-MOF UiO-66 (NH_2) serves as a cross-layer additive linking the ETL and the perovskite layer in a conventional $MAPbI_3$ PSC construction. The champion power conversion efficiency (PCE) of the Zr-MOF-modified $MAPbI_3$ PSCs is 18.25%, and the PCE of unpackaged devices retains 99% of the initial value after 800 hours of storage in an atmosphere with humidity exceeding 60%. The following factors enhance the results. The SnO_2 particles are efficiently integrated into the porous structure of Zr-MOF, while the oxygen-containing groups in Zr-MOF can also mitigate oxygen vacancies in SnO_2, enhancing the crystal quality of SnO_2 and the charge transport of the ETL. Moreover, the porous scaffold in Zr-MOF enhances the longitudinal crystallinity of perovskite grains. The organic functional groups in Zr-MOF effectively passivate the Pb and I defects. This stops the ions from moving inside the perovskite structure, which makes the PSC device more stable. The research conducted by Jing Zhang et al. [157] presents a straightforward multi-layered and multifunctional passivation method for PSCs utilizing large-particle MOFs (MOFs), which is also relevant for other PSCs employing SnO_2 as the ETL. On top of that, MOFs' many functional groups and porous structure make it easier to modulate different perovskite optoelectronic devices, like LEDs and photodetectors. Zhang et al. [158] put ionic liquids inside polyoxometalate-based MOFs to improve the conductivities of both ions and electrons. This makes it easier to store lithium. In a different common case study, Salcedo-Abraira et al. [159] used MIL-100(Fe) to surround the conducting polymer poly(3,4-ethylenedioxythiophene). The polymer-MOF composite exhibited a significantly enhanced electrical conductivity of 7.51×10^3 Scm^{-1}. The charge-discharge capacity of $LiFeO_4$@CAN at 20°C is 93.6 $mAhg^{-1}$. A straightforward mono-dentate ligand substitution method was investigated to develop three novel tunable heteroleptic copper redox mediators with the general formula $[Cu(bnbip)(X)]^{1+/2+}$, where bnbip denotes the tridentate ligand 2,6-bis(1-(naphthalen-2-ylmethyl)-1H-benzo[d]imidazol-2-yl)pyridine, and X represents 4-tert-butylpyridine (tbp), 4-chloropyridine (cp), and 5-methoxy-1-(pyridin-4-yl)-1H-indole (mpi). The copper mediator with a bulky high donor ligand (mpi) achieved an optimal redox potential of 0.72 V against NHE, which enhances the Voc and Jsc of the DSSC devices due to its reduced reorganization energy and increased driving force. Six aqueous gel electrolytes were synthesized, designated as G-1, G-2, G-3 (without additive) and G-4, G-5, G-6 (with Co-MOF additive), to investigate the influence of three distinct heteroleptic copper redox couples and Co-MOF in aqueous gelrite polymer electrolytes [160].

Tang et al. discovered that the incorporation of Cu-BTC markedly enhances the dye adsorption capacity of the RGO/TiO_2 composite [161]. Ismail's group recently looked into how charges move between MOFs and TiO_2 in the photoanode and were able to get 86% of the charge to move from TiO_2 to $Ru_3(BTC)_2$ [162]. Because of this, the large BET surface area and close interaction between MOFs and TiO_2 make the photoanode domain a promising area for MOFs. Nonetheless, MOFs are predominantly insulators or semiconductors, which impedes the superior electronic properties of MOF-based photoanodes. Victoria A. Gómez Andrade et al. [163] successfully synthesized the powder nanocrystalline NH_2-MIL-101(Fe) (1) and NH_2-MIL-125(Ti)(2) MOFs. These MOF films, when used as the ETL in solar cells, demonstrate their effectiveness. The primary function of the MOF films is to capture electrons from the perovskite film and transfer them to the FTO bottom electrode. The perovskite is necessary for the crystallization of the perovskite layer above, which changes the photoactive layer's optoelectronic quality and the photovoltaic parameters in a big way. To their knowledge, this is the first time they have successfully used a Fe MOF as an ETL in a perovskite solar cell, yielding favorable results. Although the efficiencies were not exceptionally high, this is a promising initial effort to illustrate the application of these materials as the ETL in a three-layer architecture.

Lee et al. [164] documented highly effective inverted p-i-n PSCs utilizing perovskite/Zr-MOF heterojunctions. The perovskite precursors incorporated two zirconium-based MOFs, UiO-66 and MOF-808. Thereafter, the hybrid films were directly deposited onto the NiOx HTL to construct the p-i-n-structured PSCs. As a result, PCE increased from 15.79% (pristine) to 17.01%. The MOFs and perovskites enhance the durability of the perovskite layer. The structured porosity of MOFs diminishes the nucleation barrier of perovskite; hence, it modulates the crystallization process. Yuan et al. [165] showed a terpyridyl chromium Cr-MOF with an π-conjugated aromatic structure. It is called $[Cr(III)TPy2]TPy_2]_2Pb_5I_{16}$,, and it acts as an A-site cation in the Pb–I framework. The $[Cr(III)TPy_2]_2Pb_5I_{16}$ perovskite facilitated the development of a coherent heterojunction, enhancing carrier transport. Furthermore, Cr-MOF effectively inhibited the moisture-induced degradation of the perovskite layer. The $CsPbI_2Br$ PSC attained a power conversion efficiency of 17.02% and demonstrated improved stability in ambient conditions over 1000 hours. In ZIF-8@MAI or ZIF-8@FAI-based perovskite solar cells, the formamidinium iodide or methylammonium iodide is released to mitigate vacancy defects in the perovskite layer. Additionally, ZIF-8 prevents the volatilization of organic constituents in perovskite and protects it from moisture exposure. An efficiency of 19.13% and a consistent PCE exceeding 93% of the initial efficiency at maximum power input for 150 hours can be attained.

3. Future directions

3.1 Photocatalysis

Functional groups or linkers improve MOF catalytic activity, charge transfer, and light absorption. Adding well-dispersed metal or non-metal active nanostructures to MOF cages improves co-catalytic behavior. MOF band gaps can be electronically modified to maximize visible-light photocatalysis. Mixed-ligand frameworks or heterometallic centers serve multiple catalytic functions. Under light irradiation, MOF charge separation and transfer mechanisms are studied and optimized. MOFs and their composites, including metal oxides and nanometals, must be studied for photophysical properties, stability, and photodegradability. Mechanistic studies were needed to understand MOFs and their blends in photocatalytic processes. Combining traditional cocatalysts like metal-metal oxide nanoparticles, semiconductors, graphene, TiO_2, conductive polymers, or QDs with MOF cages can improve photocatalyst efficiency. To enhance stability and reusability, cocatalysts should be immobilized inside or on MOF cages. Tuning reaction parameters including pH, temperature, and solvent duration can boost MOF-based photocatalyst efficiency. Density functional theory (DFT) and machine learning are used to predict MOF structure and its use in certain applications.

3.2 Technology of batteries

MOF-based battery research should focus on material performance, scalability, and commercial battery integration. Mass MOF production and use should be studied for environmental impacts. Doping MOFs with conjugated organic binders, conductive metal-metal oxide nanoparticles, polymers, or conductive additives improves electrical conductivity. For enhanced ionic conductivity and ion transport, MOFs require specific pore sizes and functional groups. High specific surface area and active sites form energy-dense and energy-storing MOFs. Hybridizing MOFs with other battery materials, such as carbon nanotubes, graphene, or metals and metal oxides, improves stability and conductivity. Improved structural and chemical stability of MOFs ensures durability throughout several charge/discharge cycles. Maximizing charge transfer channels inside MOF structures supports fast charge and discharge rates. To compete with Li-ion and Na-ion batteries, MOF-based electrodes will be carefully increased in capacity and voltage range. Bimetal and metal oxide nanoparticles should boost MOF storage and redox activity. MOFs should be developed as solid electrolytes with high ionic conductivity and low interfacial resistance for solid-state batteries with high capacity and low volume expansion materials, such as lithium-sulfur or lithium-metal batteries. Good-quality MOFs must be synthesized using scalable, inexpensive methods. Designing MOF electrodes to replace binders or additives reduces battery weight and complexity. MOF materials should be compatible with existing battery manufacturing. This needs confirmation. Understanding

MOFs' electrochemical and ion storage mechanisms can assist in customizing them for specific uses. Strategies must be developed to maximize MOF pore structure and reduce diffusion resistance. MOF structures can improve redox-active site reversibility and kinetics. To improve MOF-based battery sustainability, recyclable MOFs should be developed first. Greener synthesis should reduce manufacturing's environmental effects. Take into account the environmental effects and energy efficiency of MOF-based battery systems.

4. Conclusion

Due to their tunable chemical environment, enhanced porosity, and modular architecture, MOFs are potential for many new applications. This includes several applications that are regularly inspected. Porous materials have been studied for medication delivery, gas adsorption and storage, catalysis, ion exchange and separation, and sensing. MOFs are also ideal for thermally transforming a variety of nanostructured MOF derivatives. MOFs can be used to strategically assemble catalytic metal sites in exact shapes. Adjustable pore size and structured design enable selective conversion of reactants into desired products. MOFs can use light or electricity to perform chemical reactions like water splitting or CO_2 reduction by incorporating photoactive linkers or redox-active metal centers. Metal-organic MOFs have been extensively studied as photocatalysts. MOFs are porous nanostructures, can contain metals and metal oxides as co-catalysts and have adjustable semiconductor characteristics. Multiple charge transfer processes make MOFs photocatalytic. MMCT, LMCT, and LLCT. Both ligand and metal can be stimulated in two ways. MOFs improve ion transport and stabilize electrodes, which may boost supercapacitor and battery cycle life and capacity.

Acknowledgements

The authors thank Ibrahim ERDEMOGLU for their support. In addition, the authors would like to express gratitude to Karadeniz Technical University and Bilkent University UNAM for their kind hospitality.

Author details

Bülend Ortaç[1], Saliha Mutlu[1,2], Ahmet Hakan Yilmaz[3], Nergis Arsu[4]
and Sevil Savaskan Yilmaz[1,2]*

1 UNAM-National Nanotechnology Research Center and Institute of Materials Science and Nanotechnology, Bilkent University, Ankara, Turkey

2 Faculty of Sciences, Department of Chemistry, Karadeniz Technical University, Trabzon, Turkey

3 Faculty of Sciences, Department of Physics, Karadeniz Technical University, Trabzon, Turkey

4 Department of Chemistry, Davutpasa Campus, Yildiz Technical University, Istanbul, Turkey

*Address all correspondence to: sevily@ktu.edu.tr

IntechOpen

© 2025 The Author(s). Licensee IntechOpen. This chapter is distributed under the terms of the Creative Commons Attribution License (http://creativecommons.org/licenses/by/4.0), which permits unrestricted use, distribution, and reproduction in any medium, provided the original work is properly cited.

References

[1] Yaghi OM, O'Keeffe M, Ockwig NW, Chae HK, Eddaoudi M, Kim J. Reticular synthesis and the design of new materials. Nature. 2003;**423**:705-714. DOI: 10.1038/nature01650

[2] Mutlu S, Ortaç B, Ozbey DH, Durgun E, Savaskan Yılmaz S, Arsu N. Laser-driven rapid synthesis of metal-organic frameworks and investigation of UV-NIR optical absorption, luminescence, photocatalytic degradation, and gas and ion adsorption properties. Polymers. 2024;**16**:217. DOI: 10.3390/polym16020217

[3] Mutlu S, Ortaç B, Karatutlu A, Gorkan T, Durgun E, Söyler D, et al. Engineering of the high-power laser-induced synthesis of Ni-based metal-organic framework: Investigation of its optical properties, computational methodology, electrocatalytic performances, and glucose-sensing ability. Advanced Materials Technologies. 2024;**10**:2401245. DOI: 10.1002/admt.202401245

[4] Suh MP, Park HJ, Prasad TK, Lim D-W. Hydrogen storage in metal–organic frameworks. Chemical Reviews. 2012;**112**:782-835. DOI: 10.1021/cr200274s

[5] Wang L, Zheng M, Xie Z. Nanoscale metal–organic frameworks for drug delivery: A conventional platform with new promise. Journal of Materials Chemistry B. 2018;**6**:707-717. DOI: 10.1039/C7TB02970E

[6] Wang Q, Astruc D. State of the art and prospects in metal–organic framework (MOF)-based and MOF-derived nanocatalysis. Chemical Reviews. 2020;**120**:1438-1511. DOI: 10.1021/acs.chemrev.9b00223

[7] Zhao S-N, Song X-Z, Song S-Y, Zhang H. Highly efficient heterogeneous catalytic materials derived from metal-organic framework supports/precursors. Coordination Chemistry Reviews. 2017;**337**:80-96. DOI: 10.1016/j.ccr.2017.02.010

[8] Li H-Y, Zhao S-N, Zang S-Q, Li J. Functional metal–organic frameworks as effective sensors of gases and volatile compounds. Chemical Society Reviews. 2020;**49**:6364-6401. DOI: 10.1039/C9CS00778D

[9] Hussain MZ, van der Linden B, Yang Z, Jia Q, Chang H, Fischer RA, et al. Bimetal–organic framework derived multi-heterostructured TiO_2/Cux O/C nanocomposites with superior photocatalytic H_2 generation performance. Journal of Materials Chemistry A. 2021;**9**:4103-4116. DOI: 10.1039/D0TA10853G

[10] Huang Z, Yang Z, Jia Q, Wang N, Zhu Y, Xia Y. Bimetallic Co–Mo sulfide/carbon composites derived from polyoxometalate encapsulated polydopamine-decorated ZIF nanocubes for efficient hydrogen and oxygen evolution. Nanoscale. 2022;**14**:4726-4739. DOI: 10.1039/D1NR07913A

[11] Bon V. Metal-organic frameworks for energy-related applications. Current Opinion in Green and Sustainable Chemistry. 2017;**4**:44-49. DOI: 10.1016/j.cogsc.2017.02.005

[12] Vahur S, Treshchalov A, Lohmus R, Teearu A, Niman K, Hiiop H, et al. Laser-based analytical techniques in cultural heritage science – Tutorial review. Analytica Chimica Acta. 2024;**1292**:342107. DOI: 10.1016/j.aca.2023.342107

[13] Stankic S, Suman S, Haque F, Vidic J. Pure and multi metal oxide nanoparticles: Synthesis, antibacterial and cytotoxic properties. Journal of Nanobiotechnology. 2016;**14**:73. DOI: 10.1186/s12951-016-0225-6

[14] Deniz AE, Vural HA, Ortaç B, Uyar T. Gold nanoparticle/polymer nanofibrous composites by laser ablation and electrospinning. Materials Letters. 2011;**65**:2941-2943. DOI: 10.1016/j.matlet.2011.06.045

[15] Oztas T, Sen HS, Durgun E, Ortaç B. Synthesis of colloidal 2D/3D MoS_2 nanostructures by pulsed laser ablation in an organic liquid environment. The Journal of Physical Chemistry C. 2014;**118**:30120-30126. DOI: 10.1021/jp505858h

[16] Petrović S, Milovanović D, Salatić B, Peruško D, Kovač J, Dražić G, et al. Composition and structure of NiAu nanoparticles formed by laser ablation of Ni target in Au colloidal solution. Materials Chemistry and Physics. 2015;**166**:223-232. DOI: 10.1016/j.matchemphys.2015.10.007

[17] Ryu UJ, Jee S, Rao PC, Shin J, Ko C, Yoon M, et al. Recent advances in process engineering and upcoming applications of metal–organic frameworks. Coordination Chemistry Reviews. 2021;**426**:213544

[18] Armon N, Greenberg E, Edri E, Kenigsberg A, Piperno S, Kapon O, et al. Simultaneous laser-induced synthesis and micro-patterning of a metal organic framework. Chemical Communications. 2019;**55**:12773-12776. DOI: 10.1039/C9CC05990C

[19] Guo S, Gao M, Zhang W, Liu F, Guo X, Zhou K. Recent advances in laser-induced synthesis of MOF derivatives. Advanced Materials. 2023;**35**:2303065. DOI: 10.1002/adma.202303065

[20] Jangizehi A, Schmid F, Besenius P, Kremer K, Seiffert S. Defects and defect engineering in soft matter. Soft Matter. 2020;**16**:10809-10859. DOI: 10.1039/D0SM01371D

[21] Jiang H, Jin S, Wang C, Ma R, Song Y, Gao M, et al. Nanoscale laser metallurgy and patterning in air using MOFs. Journal of the American Chemical Society. 2019;**141**:5481-5489. DOI: 10.1021/jacs.9b00355

[22] Wu H, Zhang W, Kandambeth S, Shekhah O, Eddaoudi M, Alshareef HN. Conductive metal–organic frameworks selectively grown on laser-scribed graphene for electrochemical microsupercapacitors. Advanced Energy Materials. 2019;**9**:1900482. DOI: 10.1002/aenm.201900482

[23] Lim H, Kwon H, Kang H, Jang JE, Kwon H-J. Semiconducting MOFs on ultraviolet laser-induced graphene with a hierarchical pore architecture for NO_2 monitoring. Nature Communications. 2023;**14**:3114. DOI: 10.1038/s41467-023-38918-3

[24] Zheng L, Wei C, Zhou H, Sun F, Gao A, Wang D, et al. Mid-infrared optical switches enabled by metal-organic frameworks for compact high-power nanosecond laser sources at 3 μm. Optics Express. 2022;**30**:12409. DOI: 10.1364/OE.455854

[25] Ahmad H, Chiam JW, Samion MZ, Thambiratnam K, Mutlu S, Yılmaz SS, et al. Zn-MOF as a saturable absorber for thulium/holmium-doped fiber laser. Physica Scripta. 2024;**99**:105552. DOI: 10.1088/1402-4896/ad7896

[26] Ahmad H, Nizamani B, Samion MZ, Mutlu S, Yılmaz SS, Arsu N, et al.

O-band femtosecond mode-locked laser by using Ni-MOF in praseodymium-doped fluoride fiber. Optical Materials. 2024;**157**:116175. DOI: 10.1016/j.optmat.2024.116175

[27] Liu Y, Chai N, Wang X, Luo Z, Zhao J, Xie C, et al. Femtosecond laser induced *in-situ* crystallization of Tb-based luminescent metal organic framework. Optics Express. 2021;**29**:39304. DOI: 10.1364/OE.411315

[28] Stichel T, Frick T, Laumer T, Tenner F, Hausotte T, Merklein M, et al. A round Robin study for selective laser sintering of polymers: Back tracing of the pore morphology to the process parameters. Journal of Materials Processing Technology. 2018;**252**:537-545. DOI: 10.1016/j.jmatprotec.2017.10.013

[29] da Luz LL, Milani R, Felix JF, Ribeiro IRB, Talhavini M, Neto BAD, et al. Inkjet printing of lanthanide–organic frameworks for anti-counterfeiting applications. ACS Applied Materials & Interfaces. 2015;**7**:27115-27123. DOI: 10.1021/acsami.5b06301

[30] Lahtinen E, Hänninen MM, Kinnunen K, Tuononen HM, Väisänen A, Rissanen K, et al. Porous 3D printed scavenger filters for selective recovery of precious metals from electronic waste. Advanced Sustainable Systems. 2018;**2**:1800048. DOI: 10.1002/adsu.201800048

[31] Li R, Yuan S, Zhang W, Zheng H, Zhu W, Li B, et al. 3D printing of mixed matrix films based on metal–organic frameworks and thermoplastic polyamide 12 by selective laser sintering for water applications. ACS Applied Materials & Interfaces. 2019;**11**:40564-40574. DOI: 10.1021/acsami.9b11840

[32] Zou N, Zhou X, Chen G, Andoy NM, Jung W, Liu G, et al. Cooperative communication within and between single nanocatalysts. Nature Chemistry. 2018;**10**:607-614. DOI: 10.1038/s41557-018-0022-y

[33] Zhang D, Gökce B, Barcikowski S. Laser synthesis and processing of colloids: Fundamentals and applications. Chemical Reviews. 2017;**117**:3990-4103. DOI: 10.1021/acs.chemrev.6b00468

[34] Zhou HCJ, Kitagawa S. Metal-organic frameworks (MOFs). Chemical Society Reviews. 2014;**43**:5415-5418. DOI: 10.1039/c4cs90059f

[35] Yu J, Cui Y, Wu C-D, Yang Y, Chen B, Qian G. Two-photon responsive metal–organic framework. Journal of the American Chemical Society. 2015;**137**:4026-4029. DOI: 10.1021/ja512552g

[36] Torres N, Galicia J, Plasencia Y, Cano A, Echevarría F, Desdin-Garcia LF, et al. Implications of structural differences between Cu-BTC and Fe-BTC on their hydrogen storage capacity. Colloids and Surfaces A: Physicochemical and Engineering Aspects. 2018;**549**:138-146. DOI: 10.1016/j.colsurfa.2018.04.016

[37] Van Lam D, Dung DT, Kim J-H, Kim H, Lee S-M. Laser-induced sulfurization of nickel-based metal-organic frameworks for highly stable phase-engineered energy materials. Chemical Engineering Journal. 2022;**437**:135237. DOI: 10.1016/j.cej.2022.135237

[38] Lloyd L. No title. In: Handbook of Industrial Catalysts. Boston: Springer, US; 2011

[39] Astruc D, Lu F, Aranzaes JR. Nanoparticles as recyclable catalysts: The frontier between homogeneous and heterogeneous catalysis. Angewandte Chemie International Edition. 2005;**44**:7852-7872. DOI: 10.1002/anie.200500766

[40] Qiu T, Liang Z, Guo W, Gao S, Qu C, Tabassum H, et al. Highly exposed ruthenium-based electrocatalysts from bimetallic metal-organic frameworks for overall water splitting. Nano Energy. 2019;**58**:1-10. DOI: 10.1016/j.nanoen.2018.12.085

[41] Winter M, Brodd RJ. What are batteries, fuel cells, and supercapacitors? (Chem. Rev. 2003, 104, 4245–4269. Published on the web 09/28/2004.). Chemical Reviews. 2005;**105**:1021-1021. DOI: 10.1021/cr040110e

[42] Guo W, Liang Z, Zhao J, Zhu B, Cai K, Zou R, et al. Hierarchical cobalt phosphide hollow nanocages toward electrocatalytic ammonia synthesis under ambient pressure and room temperature. Small Methods. 2018;**2**:1800204. DOI: 10.1002/smtd.201800204

[43] Park S-K, Kim JK, Chan Kang Y. Metal–organic framework-derived $CoSe_2$/(NiCo)Se_2 box-in-box hollow nanocubes with enhanced electrochemical properties for sodium-ion storage and hydrogen evolution. Journal of Materials Chemistry A. 2017;**5**:18823-18830. DOI: 10.1039/C7TA05571D

[44] Yao J, Chen J, Shen K, Li Y. Phase-controllable synthesis of MOF-templated maghemite–carbonaceous composites for efficient photocatalytic hydrogen production. Journal of Materials Chemistry A. 2018;**6**:3571-3582. DOI: 10.1039/C7TA10284D

[45] Zimpel A, Al Danaf N, Steinborn B, Kuhn J, Höhn M, Bauer T, et al. Coordinative binding of polymers to metal–organic framework nanoparticles for control of interactions at the biointerface. ACS Nano. 2019;**13**:3884-3895. DOI: 10.1021/acsnano.8b06287

[46] Kong Z-C, Liao J-F, Dong Y-J, Xu Y-F, Chen H-Y, Kuang D-B, et al. Core@Shell CsPbBr 3 @zeolitic imidazolate framework nanocomposite for efficient photocatalytic CO_2 reduction. ACS Energy Letters. 2018;**3**:2656-2662. DOI: 10.1021/acsenergylett.8b01658

[47] Mehta JP, Tian T, Zeng Z, Divitini G, Connolly BM, Midgley PA, et al. Sol–gel synthesis of robust metal–organic frameworks for nanoparticle encapsulation. Advanced Functional Materials. 2018;**28**:1705588. DOI: 10.1002/adfm.201705588

[48] Shi D, He C, Qi B, Chen C, Niu J, Duan C. Merging of the photocatalysis and copper catalysis in metal–organic frameworks for oxidative C–C bond formation. Chemical Science. 2015;**6**:1035-1042. DOI: 10.1039/C4SC02362E

[49] Lu K, Aung T, Guo N, Weichselbaum R, Lin W. Nanoscale metal–organic frameworks for therapeutic, imaging, and sensing applications. Advanced Materials. 2018;**30**:1707634. DOI: 10.1002/adma.201707634

[50] Li Z, Xiao J-D, Jiang H-L. Encapsulating a Co(II) molecular photocatalyst in metal–organic framework for visible-light-driven H_2 production: Boosting catalytic efficiency via spatial charge separation. ACS Catalysis. 2016;**6**:5359-5365. DOI: 10.1021/acscatal.6b01293

[51] Hong CN, Crom AB, Feldblyum JI, LukatskayaMR.Metal-organicframeworks for fast electrochemical energy storage: Mechanisms and opportunities. Chem. 2023;**9**:798-822. DOI: 10.1016/j.chempr.2023.02.016

[52] Zheng S, Li Q, Xue H, Pang H, Xu Q. A highly alkaline-stable metal oxide@metal–organic framework composite for high-performance electrochemical energy storage. National Science Review. 2020;**7**:305-314. DOI: 10.1093/nsr/nwz137

[53] Dalvi VH, Panse SV, Joshi JB. Solar thermal technologies as a bridge from fossil fuels to renewables. Nature Climate Change. 2015;**5**:1007-1013. DOI: 10.1038/nclimate2717

[54] Amjad M, Jin H, Du X, Wen D. Experimental photothermal performance of nanofluids under concentrated solar flux. Solar Energy Materials and Solar Cells. 2018;**182**:255-262. DOI: 10.1016/j.solmat.2018.03.044

[55] Sreekumar S, Ganguly A, Khalil S, Chakrabarti S, Hewitt N, Mondol JD, et al. Thermo-optical characterization of novel MXene/carbon-dot hybrid nanofluid for heat transfer applications. Journal of Cleaner Production. 2024;**434**:140395. DOI: 10.1016/j.jclepro.2023.140395

[56] Irfan Hussain M, Ashraf GA, Farooq MU, Ahmad MS, Jazib Ali MUK. MOFs metal oxide-based nanocomposites. In: Metal-Organic Frameworks-Based Hybrid Materials for Environmental Sensing and Monitoring. London, United Kingdom: Routledge Taylor & Francis; 2022. p. 10

[57] Moon HR, Lim D-W, Suh MP. Fabrication of metal nanoparticles in metal–organic frameworks. Chemical Society Reviews. 2013;**42**:1807-1824. DOI: 10.1039/C2CS35320B

[58] Hu L, Chen J, Wei Y, Wang M, Xu Y, Wang C, et al. Photocatalytic degradation effect and mechanism of Karenia mikimotoi by non-noble metal modified TiO_2 loading onto copper metal organic framework (SNP-TiO_2@Cu-MOF) under visible light. Journal of Hazardous Materials. 2023;**442**:130059. DOI: 10.1016/j.jhazmat.2022.130059

[59] Hao Y-C, Chen L-W, Li J, Guo Y, Su X, Shu M, et al. Metal-organic framework membranes with single-atomic centers for photocatalytic CO_2 and O_2 reduction. Nature Communications. 2021;**12**:2682. DOI: 10.1038/s41467-021-22991-7

[60] Liu K-G, Bigdeli F, Panjehpour A, Larimi A, Morsali A, Dhakshinamoorthy A, et al. Metal organic framework composites for reduction of CO_2. Coordination Chemistry Reviews. 2023;**493**:215257. DOI: 10.1016/j.ccr.2023.215257

[61] Zhang L, Shi X, Zhang Z, Kuchel RP, Namivandi-Zangeneh R, Corrigan N, et al. Porphyrinic zirconium metal–organic frameworks (MOFs) as heterogeneous photocatalysts for PET-RAFT polymerization and stereolithography. Angewandte Chemie International Edition. 2021;**60**:5489-5496. DOI: 10.1002/anie.202014208

[62] Xiao J-D, Jiang H-L. Metal–organic frameworks for photocatalysis and photothermal catalysis. Accounts of Chemical Research. 2019;**52**:356-366. DOI: 10.1021/acs.accounts.8b00521

[63] Chen Y-Z, Wang ZU, Wang H, Lu J, Yu S-H, Jiang H-L. Singlet oxygen-engaged selective photo-oxidation over Pt nanocrystals/porphyrinic MOF: The roles of photothermal effect and Pt electronic state. Journal of the American Chemical Society. 2017;**139**:2035-2044. DOI: 10.1021/jacs.6b12074

[64] Du L, Xing L, Zhang G, Sun S. Metal-organic framework derived carbon materials for electrocatalytic oxygen reactions: Recent progress and future perspectives. Carbon. 2020;**156**:77-92. DOI: 10.1016/j.carbon.2019.09.029

[65] Wang H, Shao Y, Mei S, Lu Y, Zhang M, Sun J, et al. Polymer-derived heteroatom-doped porous carbon materials. Chemical Reviews. 2020;**120**:9363-9419. DOI: 10.1021/acs.chemrev.0c00080

[66] Gao C, Jiang Z, Wang P, Jensen LR, Zhang Y, Yue Y. Optimized assembling of MOF/SnO_2/graphene leads to superior anode for lithium ion batteries. Nano Energy. 2020;**74**:104868. DOI: 10.1016/j.nanoen.2020.104868

[67] Liu K, Liu Y, Lin D, Pei A, Cui Y. Materials for lithium-ion battery safety. Science Advances. 2018;**4**:eaas9820. DOI: 10.1126/sciadv.aas9820

[68] Cheng Q, Zhang Y. Multi-channel graphite for high-rate lithium ion battery. Journal of the Electrochemical Society. 2018;**165**:A1104-A1109. DOI: 10.1149/2.1171805jes

[69] Moradi B, Wang D, Botte GG. Carbon-coated Fe_3O_4 nanospindles as solid electrolyte interface for improving graphite anodes in lithium ion batteries. Journal of Applied Electrochemistry. 2020;**50**:321-331. DOI: 10.1007/s10800-019-01393-0

[70] Kim J, Park K, Woo H, Gil B, Park Y-S, Kim IS, et al. Selective removal of nanopores by triphenylphosphine treatment on the natural graphite anode. Electrochimica Acta. 2019;**326**:134993. DOI: 10.1016/j.electacta.2019.134993

[71] Li W, Dolocan A, Oh P, Celio H, Park S, Cho J, et al. Dynamic behaviour of interphases and its implication on high-energy-density cathode materials in lithium-ion batteries. Nature Communications. 2017;**8**:14589. DOI: 10.1038/ncomms14589

[72] Fang G, Zhou J, Liang C, Pan A, Zhang C, Tang Y, et al. MOFs nanosheets derived porous metal oxide-coated three-dimensional substrates for lithium-ion battery applications. Nano Energy. 2016;**26**:57-65. DOI: 10.1016/j.nanoen.2016.05.009

[73] Huang J, Fang G, Liu K, Zhou J, Tang X, Cai K, et al. Controllable synthesis of highly uniform cuboid-shape MOFs and their derivatives for lithium-ion battery and photocatalysis applications. Chemical Engineering Journal. 2017;**322**:281-292. DOI: 10.1016/j.cej.2017.03.136

[74] Wang Q, Zou R, Xia W, Ma J, Qiu B, Mahmood A, et al. Facile synthesis of ultrasmall CoS_2 nanoparticles within thin N-doped porous carbon Shell for high performance lithium-ion batteries. Small. 2015;**11**:2511-2517. DOI: 10.1002/smll.201403579

[75] Zhao Z, Li H, Cheng X, Ren R, Meng Z, Wang X. Multifunctional FeP/spongy carbon modified separator with enhanced polysulfide immobilization and conversion for flame-retardant lithium-sulfur batteries. Chemistry Select. 2021;**6**:7098-7102. DOI: 10.1002/slct.202102399

[76] He Y, Zhang Z, Wang W, Fu L. Metal organic frameworks derived high-performance photoanodes for DSSCs. Journal of Alloys and Compounds. 2020;**825**:154089. DOI: 10.1016/j.jallcom.2020.154089

[77] Makhanya N, Oboirien B, Ren J, Musyoka N, Sciacovelli A. Recent advances on thermal energy storage using metal-organic frameworks (MOFs). Journal of Energy Storage. 2021;**34**:102179. DOI: 10.1016/j.est.2020.102179

[78] Jiang Q, Zhang L, Wang H, Yang X, Meng J, Liu H, et al. Enhanced electron extraction using SnO_2 for high-efficiency planar-structure HC(NH2)2PbI3-based perovskite solar cells. Nature Energy. 2016;**2**:16177. DOI: 10.1038/nenergy.2016.177

[79] Chen Y, Yang Z, Wang S, Zheng X, Wu Y, Yuan N, et al. (frank): Design of an inorganic mesoporous hole-transporting

layer for highly efficient and stable inverted perovskite solar cells. Advanced Materials. 2018;**30**:1805660. DOI: 10.1002/adma.201805660

[80] DeCoste JB, Peterson GW, Jasuja H, Glover TG, Huang Y, Walton KS. Stability and degradation mechanisms of metal–organic frameworks containing the $Zr6O_4(OH)_4$ secondary building unit. Journal of Materials Chemistry A. 2013;**1**:5642. DOI: 10.1039/c3ta10662d

[81] Li W, Zheng G, Yang Y, Seh ZW, Liu N, Cui Y. High-performance hollow sulfur nanostructured battery cathode through a scalable, room temperature, one-step, bottom-up approach. National Academy of Sciences of the United States of America. 2013;**110**:7148-7153. DOI: 10.1073/pnas.1220992110

[82] Torabi E, Kazemi A, Tamtaji M, Manteghi F, Rohani S, Goddard WA. Sacrificial MOF-derived MnNi hydroxide for high energy storage supercapacitor electrodes via DFT-based quantum capacitance study. Heliyon. 2025;**11**:e41261. DOI: 10.1016/j.heliyon.2024.e41261

[83] Siddiqui R, Rani M, Shah AA, Siddique S, Ibrahim A. Enhanced electrochemical performance with exceptional capacitive retention in Ce–Co MOFs/Ti_3C_2Tx nanocomposite for advanced supercapacitor applications. Heliyon. 2024;**10**:e36540. DOI: 10.1016/j.heliyon.2024.e36540

[84] Baig MM, Khan MA, Gul IH, Rehman SU, Shahid M, Javaid S, et al. A review of advanced electrode materials for supercapacitors: Challenges and opportunities. Journal of Electronic Materials. 2023;**52**:5775-5794. DOI: 10.1007/s11664-023-10532-5

[85] Liang S-X, Zhang L-C, Reichenberger S, Barcikowski S. Design and perspective of amorphous metal nanoparticles from laser synthesis and processing. Physical Chemistry Chemical Physics. 2021;**23**:11121-11154. DOI: 10.1039/D1CP00701G

[86] Yang B, Li B, Xiang Z. Advanced MOF-based electrode materials for supercapacitors and electrocatalytic oxygen reduction. Nano Research. 2023;**16**:1338-1361. DOI: 10.1007/s12274-022-4682-y

[87] Ali A, Ahmed S, Jiang W, Park G, Oh SJ. Harnessing MOF intrinsic properties for enhanced supercapacitor performance. Current Opinion in Electrochemistry. 2025;**50**:101640. DOI: 10.1016/j.coelec.2024.101640

[88] Pan G, Hu Y, Wang Z, Li H, Wu D, Zhang L, et al. Metal–organic framework-derived MnO/C nanocomposite with lamellar porous structure for high-performance aqueous zinc-ion battery. Energy Technology. 2024;**12**:2301109. DOI: 10.1002/ente.202301109

[89] Chong TC, Hong MH, Shi LP. Laser precision engineering: From microfabrication to nanoprocessing. Laser & Photonics Reviews. 2010;**4**:123-143. DOI: 10.1002/lpor.200810057

[90] Malinauskas M, Žukauskas A, Hasegawa S, Hayasaki Y, Mizeikis V, Buividas R, et al. Ultrafast laser processing of materials: From science to industry. Light: Science & Applications. 2016;**5**:e16133-e16133. DOI: 10.1038/lsa.2016.133

[91] Semaltianos NG. Nanoparticles by laser ablation. Critical Reviews in Solid State and Materials Sciences. 2010;**35**:105-124. DOI: 10.1080/10408431003788233

[92] Xie J, Yan J, Zhu D, He G. Atomic-level insight into the formation of

subsurface dislocation layer and its effect on mechanical properties during ultrafast laser micro/nano fabrication. Advanced Functional Materials. 2022;**32**:2108802. DOI: 10.1002/adfm.202108802

[93] Fouad N, Mahmoud A, Abdel Samad F, El-Salam YA, Ashour M, Apsari R, et al. Utilizing the 2nd harmonic Nd: YAG laser ablation in liquid for the production of copper oxide quantum dots: Influence of laser fluence and ablation duration. Physica B: Condensed Matter. 2024;**694**:416453. DOI: 10.1016/j.physb.2024.416453

[94] Alkis S, Okyay AK, Ortaç B. Post-treatment of silicon nanocrystals produced by ultra-short pulsed laser ablation in liquid: Toward blue luminescent nanocrystal generation. The Journal of Physical Chemistry C. 2012;**116**:3432-3436. DOI: 10.1021/jp211521k

[95] Koral C, Ortac B, Altan H. Terahertz time-domain study of silver nanoparticles synthesized by laser ablation in organic liquid. IEEE Transactions on Terahertz Science and Technology. 2016;**6**:525-531. DOI: 10.1109/TTHZ.2016.2572360

[96] Flemban TH. Synthesis, characterization, and analysis of zinc oxide nanoparticles using varying pulsed laser ablation energies in liquid. Journal of Experimental Nanoscience. 2023;**18**:2175817. DOI: 10.1080/17458080.2023.2175817

[97] Abdulhasan W, Rasheed M, Nayef UM. Diverse energy laser ablation in liquid for indium oxide nanoparticles on porous silicon: Enhancing photodetector performance. Journal of Physics: Conference Series. 2024;**2857**:012058. DOI: 10.1088/1742-6596/2857/1/012058

[98] Salim ET, Saimon JA, Muhsin MS, Fakhri MA, Amin MH, Azzahrani AS, et al. Mesoporous Ag@WO3 core–shell, an investigation at different concentrated environment employing laser ablation in liquid. Scientific Reports. 2024;**14**:5473. DOI: 10.1038/s41598-024-55146-x

[99] ElFaham MM, Okil M, Mostafa AM. Fabrication of magnesium metallic nanoparticles by liquid-assisted laser ablation. Journal of the Optical Society of America B. 2020;**37**:2620. DOI: 10.1364/JOSAB.398543

[100] Amendola V, Riello P, Meneghetti M. Magnetic nanoparticles of iron carbide, iron oxide, iron@iron oxide, and metal iron synthesized by laser ablation in organic solvents. The Journal of Physical Chemistry C. 2011;**115**:5140-5146. DOI: 10.1021/jp109371m

[101] Jiang H-L, Liu B, Akita T, Haruta M, Sakurai H, Xu Q. Au@ZIF-8: CO oxidation over gold nanoparticles deposited to metal-organic framework. Journal of the American Chemical Society. 2009;**131**:11302-11303. DOI: 10.1021/ja9047653

[102] Houk RJT, Jacobs BW, El Gabaly F, Chang NN, Talin AA, Graham DD, et al. Silver cluster formation, dynamics, and chemistry in metal-organic frameworks. Nano Letters. 2009;**9**:3413-3418. DOI: 10.1021/nl901397k

[103] Hermannsdörfer J, Friedrich M, Miyajima N, Albuquerque RQ, Kümmel S, Kempe R. Ni/Pd@MIL-101: Synergistic catalysis with cavity-conform Ni/Pd nanoparticles. Angewandte Chemie International Edition. 2012;**51**:11473-11477. DOI: 10.1002/anie.201205078

[104] Khaletskaya K, Reboul J, Meilikhov M, Nakahama M, Diring S, Tsujimoto M, et al. Integration of porous coordination polymers and gold nanorods into core–shell mesoscopic

composites toward light-induced molecular release. Journal of the American Chemical Society. 2013;**135**:10998-11005. DOI: 10.1021/ja403108x

[105] Zhou H, Zhu A, Wang C, Guo X, Ying Y, Wu Y, et al. Preparation of gold nanoparticles loaded MOF-199 for SERS detection of 5-hydroxyindole-3-acetic acid in serum. Spectrochimica Acta Part A: Molecular and Biomolecular Spectroscopy. 2024;**304**:123280. DOI: 10.1016/j.saa.2023.123280

[106] El-Shall MS, Abdelsayed V, Khder AERS, Hassan HMA, El-Kaderi HM, Reich TE. Metallic and bimetallic nanocatalysts incorporated into highly porous coordination polymer MIL-101. Journal of Materials Chemistry. 2009;**19**:7625. DOI: 10.1039/b912012b

[107] Kleist W, Maciejewski M, Baiker A. MOF-5 based mixed-linker metal–organic frameworks: Synthesis, thermal stability and catalytic application. Thermochimica Acta. 2010;**499**:71-78. DOI: 10.1016/j.tca.2009.11.004

[108] Zamaro JM, Pérez NC, Miró EE, Casado C, Seoane B, Téllez C, et al. HKUST-1 MOF: A matrix to synthesize CuO and CuO–CeO_2 nanoparticle catalysts for CO oxidation. Chemical Engineering Journal. 2012;**195-196**:180-187. DOI: 10.1016/j.cej.2012.04.091

[109] Schröder F, Esken D, Cokoja M, van den Berg MWE, Lebedev OI, Van Tendeloo G, et al. Ruthenium nanoparticles inside porous [Zn 4 O$(bdc)_3$] by hydrogenolysis of adsorbed [Ru(cod)(cot)]: A solid-state reference system for surfactant-stabilized ruthenium colloids. Journal of the American Chemical Society. 2008;**130**:6119-6130. DOI: 10.1021/ja078231u

[110] Müller M, Turner S, Lebedev OI, Wang Y, van Tendeloo G, Fischer RA. Au@MOF-5 and Au/MO x @MOF-5 (M = Zn, Ti; x = 1, 2): Preparation and microstructural characterisation. European Journal of Inorganic Chemistry. 2011;**2011**:1876-1887. DOI: 10.1002/ejic.201001297

[111] Liu H, Chang L, Chen L, Li Y. *In situ* one-step synthesis of metal–organic framework encapsulated naked Pt nanoparticles without additional reductants. Journal of Materials Chemistry A. 2015;**3**:8028-8033. DOI: 10.1039/C5TA00030K

[112] Chen L, Chen H, Luque R, Li Y. Metal-organic framework encapsulated Pd nanoparticles: Towards advanced heterogeneous catalysts. Chemical Science. 2014;**5**:3708-3714. DOI: 10.1039/C4SC01847H

[113] Park T-H, Hickman AJ, Koh K, Martin S, Wong-Foy AG, Sanford MS, et al. Highly dispersed palladium(II) in a defective metal–organic framework: Application to C–H activation and functionalization. Journal of the American Chemical Society. 2011;**133**:20138-20141. DOI: 10.1021/ja2094316

[114] Proch S, Herrmannsdörfer J, Kempe R, Kern C, Jess A, Seyfarth L, et al. Pt@MOF-177: Synthesis, room-temperature hydrogen storage and oxidation catalysis. Chemistry – A European Journal. 2008;**14**:8204-8212. DOI: 10.1002/chem.200801043

[115] Gao S, Zhao N, Shu M, Che S. Palladium nanoparticles supported on MOF-5: A highly active catalyst for a ligand- and copper-free Sonogashira coupling reaction. Applied Catalysis A: General. 2010;**388**:196-201. DOI: 10.1016/j.apcata.2010.08.045

[116] Zhu J, Wang PC, Lu M. Selective oxidation of benzyl alcohol under

solvent-free condition with gold nanoparticles encapsulated in metal-organic framework. Applied Catalysis A: General. 2014;**477**:125-131. DOI: 10.1016/j.apcata.2014.03.013

[117] Xu X, Cao R, Jeong S, Cho J. Spindle-like mesoporous α-Fe_2O_3 anode material prepared from MOF template for high-rate lithium batteries. Nano Letters. 2012;**12**:4988-4991. DOI: 10.1021/nl302618s

[118] Meng Y, Wang G-H, Bernt S, Stock N, Lu A-H. Crystal-like microporous hybrid solid nanocast from Cr-MIL-101. Chemical Communications. 2011;**47**:10479. DOI: 10.1039/c1cc13699b

[119] Dou X, Liu J, Gong X, Jiang H, Deng H. Laser driven conversion of MOFs to rare earth metal oxide nanoparticles. APL Materials. 2022;**10**:041110. DOI: 10.1063/5.0085497

[120] Li Y, Liu Y, Gao W, Zhang L, Liu W, Lu J, et al. Microwave-assisted synthesis of UIO-66 and its adsorption performance towards dyes. CrystEngComm. 2014;**16**:7037-7042. DOI: 10.1039/C4CE00526K

[121] Furukawa H, Gándara F, Zhang Y-B, Jiang J, Queen WL, Hudson MR, et al. Water adsorption in porous metal–organic frameworks and related materials. Journal of the American Chemical Society. 2014;**136**:4369-4381. DOI: 10.1021/ja500330a

[122] Reinsch H, Waitschat S, Chavan SM, Lillerud KP, Stock N. A facile "green" route for scalable batch production and continuous synthesis of zirconium MOFs. European Journal of Inorganic Chemistry. 2016;**2016**:4490-4498. DOI: 10.1002/ejic.201600295

[123] Feng D, Gu Z, Li J, Jiang H, Wei Z, Zhou H. Zirconium-metalloporphyrin PCN-222: Mesoporous metal–organic frameworks with ultrahigh stability as biomimetic catalysts. Angewandte Chemie International Edition. 2012;**51**:10307-10310. DOI: 10.1002/anie.201204475

[124] Klimakow M, Klobes P, Thünemann AF, Rademann K, Emmerling F. Mechanochemical synthesis of metal–organic frameworks: A fast and facile approach toward quantitative yields and high specific surface areas. Chemistry of Materials. 2010;**22**:5216-5221. DOI: 10.1021/cm1012119

[125] Mogale R, Akpomie KG, Conradie J, Langner EHG. Dye adsorption of aluminium- and zirconium-based metal organic frameworks with azobenzene dicarboxylate linkers. Journal of Environmental Management. 2022;**304**:114166. DOI: 10.1016/j.jenvman.2021.114166

[126] Ma M-L, Qin J-H, Ji C, Xu H, Wang R, Li B-J, et al. Anionic porous metal–organic framework with novel 5-connected vbk topology for rapid adsorption of dyes and tunable white light emission. Journal of Materials Chemistry C. 2014;**2**:1085-1093. DOI: 10.1039/C3TC32001D

[127] Yang S, Veerana M, Yu N, Ketya W, Park G, Kim S, et al. Copper(II)-MOF containing glutarate and 4,4′-azopyridine and its antifungal activity. Applied Sciences. 2021;**12**:260. DOI: 10.3390/app12010260

[128] Ehrling S, Senkovska I, Bon V, Evans JD, Petkov P, Krupskaya Y, et al. Crystal size versus paddle wheel deformability: Selective gated adsorption transitions of the switchable metal–organic frameworks DUT-8(Co) and DUT-8(Ni). Journal of Materials Chemistry A. 2019;**7**:21459-21475. DOI: 10.1039/C9TA06781G

[129] Batibay GS, Arsu N. Synthesis of photoactive nanocomposite hydrogels via in-situ photoinitiated thiol-ene click polymerization of acrylates containing norbornene-POSS and norbornene-POSS/Ag NPs. Progress in Organic Coatings. 2023;**175**:107364. DOI: 10.1016/j.porgcoat.2022.107364

[130] Batibay GS, Ozcelik Kazancioglu E, Kose E, Arsu N. Reversible tuning of UV-cured nano hybrid coating from hydrophobic to superhydrophilic. Progress in Organic Coatings. 2021;**152**:106113. DOI: 10.1016/j.porgcoat.2020.106113

[131] Scaiano JC. Photochemical and free-radical processes in benzil-amine systems. Electron-donor properties of. alpha.-aminoalkyl radicals. The Journal of Physical Chemistry. 1981;**85**:2851-2855. DOI: 10.1021/j150619a031

[132] Dizman HM, Eroglu GO, Kuruca SE, Arsu N. Photochemically prepared monodisperse gold nanoparticles as doxorubicin carrier and its cytotoxicity on leukemia cancer cells. Applied Nanoscience. 2021;**11**:309-320. DOI: 10.1007/s13204-020-01589-3

[133] Sharma G, Gupta VK, Agarwal S, Kumar A, Thakur S, Pathania D. Fabrication and characterization of Fe@MoPO nanoparticles: Ion exchange behavior and photocatalytic activity against malachite green. Journal of Molecular Liquids. 2016;**219**:1137-1143. DOI: 10.1016/j.molliq.2016.04.046

[134] Boote BW, Byun H, Kim J-H. Silver–gold bimetallic nanoparticles and their applications as optical materials. Journal of Nanoscience and Nanotechnology. 2014;**14**:1563-1577. DOI: 10.1166/jnn.2014.9077

[135] Kirtay H, Akbulut Ö, Arsu N. Spontaneously designed, self-assembled and oriented ZnO nanorods by photoinduced polymerization of epoxydiacrylate formulation in the presence of 2-mercapto-thioxanthone. Progress in Organic Coatings. 2022;**172**:107074. DOI: 10.1016/j.porgcoat.2022.107074

[136] Mutlu S, Metin E, Aydin Yuksel S, Bayrak U, Nuhoglu C, Arsu N. In-situ photochemical synthesis and dielectric properties of nanocomposite thin films containing Au, Ag and MnO nanoparticles. European Polymer Journal. 2021;**144**:110238. DOI: 10.1016/j.eurpolymj.2020.110238

[137] Çeper T, Arsu N. Photochemically prepared gold/polymer nanocoatings: Formation of gold Mirror. Macromolecular Chemistry and Physics. 2017;**218**:1700030. DOI: 10.1002/macp.201700030

[138] Liu X, He Y, Yang B, Yan Q, Yang J. Highly efficient photo-degradation of gaseous organic pollutants catalyzed by diatomite-supported titanium dioxide. Catalysts. 2020;**10**:380. DOI: 10.3390/catal10040380

[139] Mendoza JA, Lee DH, Kang J-H. Photocatalytic removal of gaseous nitrogen oxides using WO_3/TiO_2 particles under visible light irradiation: Effect of surface modification. Chemosphere. 2017;**182**:539-546. DOI: 10.1016/j.chemosphere.2017.05.069

[140] Wang Y, Feng L, Pang J, Li J, Huang N, Day GS, et al. Photosensitizer-anchored 2D MOF nanosheets as highly stable and accessible catalysts toward artemisinin production. Advanced Science. 2019;**6**:2-9. DOI: 10.1002/advs.201802059

[141] Zhu Y, Yin J, Abou-Hamad E, Liu X, Chen W, Yao T, et al. Highly stable phosphonate-based MOFs with

engineered bandgaps for efficient photocatalytic hydrogen production. Advanced Materials. 2020;**32**:1906368. DOI: 10.1002/adma.201906368

[142] Kim Y-J, Park D-W. Functionalized IRMOF-3: An efficient heterogeneous catalyst for the cycloaddition of allyl glycidyl ether and CO_2. Journal of Nanoscience and Nanotechnology. 2013;**13**:2307-2312. DOI: 10.1166/jnn.2013.6878

[143] Naghdi S, Cherevan A, Giesriegl A, Guillet-Nicolas R, Biswas S, Gupta T, et al. Selective ligand removal to improve accessibility of active sites in hierarchical MOFs for heterogeneous photocatalysis. Nature Communications. 2022;**13**:282. DOI: 10.1038/s41467-021-27775-7

[144] Ma X, Wang L, Zhang Q, Jiang H. Switching on the photocatalysis of metal–organic frameworks by engineering structural defects. Angewandte Chemie International Edition. 2019;**58**:12175-12179. DOI: 10.1002/anie.201907074

[145] Yang Q, Xu Q, Yu S, Jiang H. Pd Nanocubes@ZIF-8: Integration of plasmon-driven photothermal conversion with a metal–organic framework for efficient and selective catalysis. Angewandte Chemie International Edition. 2016;**55**:3685-3689. DOI: 10.1002/anie.201510655

[146] Sharma VK, Feng M. Water depollution using metal-organic frameworks-catalyzed advanced oxidation processes: A review. Journal of Hazardous Materials. 2019;**372**:3-16. DOI: 10.1016/j.jhazmat.2017.09.043

[147] Zhu G, Wang S, Yu Z, Zhang L, Wang D, Pang B, et al. Application of Fe-MOFs in advanced oxidation processes. Research on Chemical Intermediates. 2019;**45**:3777-3793. DOI: 10.1007/s11164-019-03820-5

[148] Farrokhi A, Jafarpour M, Alipour M. Solar-driven advanced oxidation process catalyzed by metal–organic frameworks for water depollution. Polyhedron. 2019;**170**:325-333. DOI: 10.1016/j.poly.2019.06.005

[149] Goswami S, Miller CE, Logsdon JL, Buru CT, Wu Y-L, Bowman DN, et al. Atomistic approach toward selective photocatalytic oxidation of a mustard-gas simulant: A case study with heavy-chalcogen-containing PCN-57 analogues. ACS Applied Materials & Interfaces. 2017;**9**:19535-19540. DOI: 10.1021/acsami.7b07055

[150] Atilgan A, Islamoglu T, Howarth AJ, Hupp JT, Farha OK. Detoxification of a sulfur mustard simulant using a BODIPY-functionalized zirconium-based metal–organic framework. ACS Applied Materials & Interfaces. 2017;**9**:24555-24560. DOI: 10.1021/acsami.7b05494

[151] Ryu U, Jee S, Park J-S, Han IK, Lee JH, Park M, et al. Nanocrystalline titanium metal–organic frameworks for highly efficient and flexible perovskite solar cells. ACS Nano. 2018;**12**:4968-4975. DOI: 10.1021/acsnano.8b02079

[152] Zhang Y-N, Li B, Fu L, Li Q, Yin L-W. MOF-derived ZnO as electron transport layer for improving light harvesting and electron extraction efficiency in perovskite solar cells. Electrochimica Acta. 2020;**330**:135280. DOI: 10.1016/j.electacta.2019.135280

[153] Ahmadian-Yazdi M-R, Gholampour N, Eslamian M. Interface engineering by employing zeolitic imidazolate framework-8 (ZIF-8) as the only scaffold in the architecture of perovskite solar cells. ACS Applied Energy Materials. 2020;**3**:3134-3143. DOI: 10.1021/acsaem.9b02115

[154] Zhou X, Qiu L, Fan R, Ye H, Tian C, Hao S, et al. Toward high-efficiency

and thermally-stable perovskite solar cells: A novel metal-organic framework with active pyridyl sites replacing 4-tert-butylpyridine. Journal of Power Sources. 2020;**473**:228556. DOI: 10.1016/j.jpowsour.2020.228556

[155] Hazeghi F, Mozaffari S, Ghorashi SMB. Metal organic framework–derived core-shell CuO@NiO nanosphares as hole transport material in perovskite solar cell. Journal of Solid State Electrochemistry. 2020;**24**:1427-1438. DOI: 10.1007/s10008-020-04643-w

[156] Zhou X, Qiu L, Fan R, Zhang J, Hao S, Yang Y. Heterojunction incorporating perovskite and microporous metal–organic framework nanocrystals for efficient and stable solar cells. Nano-Micro Letters. 2020;**12**:80. DOI: 10.1007/s40820-020-00417-1

[157] Zhang J, Hu B, Guo Z, Luo H, Wang R, Li M, et al. Zr-metal–organic framework based cross-layer-connection additive for stable halide perovskite solar cells. Applied Surface Science. 2023;**628**:157339. DOI: 10.1016/j.apsusc.2023.157339

[158] Zhang M, Zhang A-M, Wang X-X, Huang Q, Zhu X, Wang X-L, et al. Encapsulating ionic liquids into POM-based MOFs to improve their conductivity for superior lithium storage. Journal of Materials Chemistry A. 2018;**6**:8735-8741. DOI: 10.1039/C8TA01062E

[159] Salcedo-Abraira P, Santiago-Portillo A, Atienzar P, Bordet P, Salles F, Guillou N, et al. A highly conductive nanostructured PEDOT polymer confined into the mesoporous MIL-100(Fe). Dalton Transactions. 2019;**48**:9807-9817. DOI: 10.1039/C9DT00917E

[160] Selvaraj B, Shanmugam G, Kamaraj S, Thirugnanasambandam E, Gunasekeran A, Sambandam A. Effect of an aqueous copper gel electrolyte with cobalt metal organic framework based additive on performance of aqueous-dye-sensitized solar cells. Solar Energy. 2022;**236**:586-598. DOI: 10.1016/j.solener.2022.03.034

[161] Tang B, He Y, Zhang Z, Wang Z, Ji L, Ma T, et al. Influence of N doping and the functional groups of graphene on a RGO/TiO_2 composite photocatalyst. Science China Technological Sciences. 2020;**63**:1045-1054. DOI: 10.1007/s11431-019-9749-1

[162] Ismail FM, O'Neil D, Youssef T, Elfeky SA. Ultrafast laser dynamics of metal organic frameworks/TiO_2 nano-arrays hybrid composites for energy conversion applications. Journal of Energy Chemistry. 2019;**29**:88-94. DOI: 10.1016/j.jechem.2018.02.017

[163] Gómez Andrade VA, Herrera Martínez WO, Redondo F, Correa Guerrero NB, Roncaroli F, Perez MD. Fe and Ti metal-organic frameworks: Towards tailored materials for photovoltaic applications. Applied Materials Today. 2021;**22**:100915. DOI: 10.1016/j.apmt.2020.100915

[164] Lee C, Chen C, Liao Y, Wu KC-W, Chueh C. Enhancing efficiency and stability of photovoltaic cells by using perovskite/Zr-MOF heterojunction including bilayer and hybrid structures. Advanced Science. 2019;**6**:1801715. DOI: 10.1002/advs.201801715

[165] Yuan S, Xian Y, Long Y, Cabot A, Li W, Fan J. Chromium-based metal–organic framework as A-site cation in $CsPbI_2$ Br perovskite solar cells. Advanced Functional Materials. 2021;**31**:2106233. DOI: 10.1002/adfm.202106233

Chapter 3

Graphitic Carbon Nitride as Photocatalysts for the Degradation of Organic Pollutants in Aqueous Suspension: A Review

Mohammad Muneer, Ziyaur Rasool, Shreya Parashar, Abdulaziz M. Almohyawi, Ziad Moussa and Saleh A. Ahmed

Abstract

The semiconductor photocatalysis has been the subject of much research interest in the recent decades due to its potential future technology for energy conversion and environmental remediation. Graphitic carbon nitride ($g\text{-}C_3N_4$) is a promising semiconductor photocatalyst due to its metal-free structure and responsiveness to visible light. However, its pristine form exhibits limitations such as poor solar absorption and rapid charge carrier recombination, which reduces photocatalytic efficiency. Various modification strategies, like doping and heterojunction formation, have significantly improved its optical and electronic properties, paving the way for advanced applications in environmental remediation and energy conversion. This review has emphasized the synthesis, advantages/disadvantages, and mechanistic aspects of $g\text{-}C_3N_4$ for the degradation of organic pollutants in aqueous suspension. In addition, the review discusses the use of different strategies such as doping of metals or non-metals, co-doping, and generation of heterojunction for modified $g\text{-}C_3N_4$ photocatalyst. These strategies alter the electronic and structural properties by greatly increasing the optical absorption behavior, enhancing charge separation, and prolonging the lifetime of charge carriers.

Keywords: graphitic carbon nitride, $g\text{-}C_3N_4$, modified photocatalyst, photocatalysis, visible light active catalyst, organic photodegradation

1. Introduction

One of the most alarming problems that our planet has recently faced is the challenge of environmental deterioration and the rapid depletion of non-renewable energy resources. Fossil fuels usage eventually contribute to the introduction of toxic compounds into the environment. To address these issues, researchers are increasingly turning to sustainable and inexhaustible energy resources, such as solar energy, for environmental remediation. The rapid industrialization and population growth have led to the direct or indirect discharge of hazardous pollutants into natural water systems, threatening aquatic ecosystems and human health. Among various water purification techniques, semiconductor-based photocatalysis stands out for its ability to utilize solar energy to degrade pollutants into non-toxic products, offering an eco-friendly and efficient alternative to conventional methods.

The priority contaminants include a wide range of organic pollutants, including pesticides, insecticides, phenol, chlorophenol, bisphenol A (BPA, a microplastic constituent), nitrophenol (NP), aniline, pharmaceuticals like paracetamol (PCM), tetracycline hydrochloride (TC), ciprofloxacin (CF), sulfamethoxazole, and numerous dyes, including methyl orange (MO), rhodamine B (RhB), methylene blue (MB), malachite gray (MG), and sulfonamides [1].

Consequently, researchers are now concentrating on creating effective techniques for purifying natural water sources. In the past, a variety of conventional techniques have been used to remove organic contaminants from wastewater, including adsorption, sedimentation, filtration, reverse osmosis, coagulation, precipitation, and biodegradation methods. Among the many drawbacks of such traditional methods are their high engineering cost and inability to completely remove hazardous chemicals from wastewater. These require a significant amount of time and work [1]. Therefore, it is imperative to switch to a more efficient and ecologically friendly method of removing harmful contaminants from the water supply.

In recent years, semiconductor-based photocatalysis, which transforms solar energy into chemical energy, has emerged as the most reliable technology for fostering sustainable development and effectively addressing the issue of pollution. Semiconductor-based photocatalysts are both cost-effective and energy-efficient. Photocatalysis uses light to generate catalytically active species. Semiconductors interact with the light of a particular wavelength, causing electron excitation from the valence band (VB) to the conduction band (CB), generating reactive oxidizing species (ROS), which are responsible for the photocatalytic transformation of toxic pollutant to non-toxic components [2]. Semiconductor-based photocatalysis, also known as heterogeneous photocatalysis, the photocatalyst used here has a range of advantages over conventional methods and is therefore a strong field of concern in the near future.

1.1 Advantages of semiconductor-based photocatalyst

- High photocatalytic stability
- Low cost
- Thermal and chemical stability
- Catalyze reaction effectively, leaving no harmful by-products
- Completely mineralize organic pollutants into non-hazardous substances

As a result, it paves the way for an environmentally sustainable process of eliminating toxins from water bodies.

In recent years, a large number of semiconductor photocatalysts, to eliminate the pollutants in water, have been synthesized using various methods involving TiO_2, ZnO, CdS, BiO_3, WO_3, and other materials [3]. The broad bandgap and low solar energy consumption, on the other hand, were roadblocks that necessitated the creation of a robust and visible-light sensitive photocatalyst. Furthermore, the majority of these semiconductors exhibit photocatalytic properties corresponding to the UV range, and the methods used in their synthesis to make them function in visible-light irradiation are time-consuming and costly.

For its unique advantages, g-C_3N_4 has recently engrossed the researchers to establish it as a novel photocatalyst that works well with visible region of solar light. Because of the structure of heptazine ring as well as high degree of condensation, g-C_3N_4 has a range of advantages [4]. As tectonic units, g-C_3N_4 has two allotropic forms: (1) triazine and (2) tri-s-triazine, having different stability as not only the environment of electrons around N atom varies but also the nitride pores' sizes vary, reported by Kroke et al. who determined that the allotrope that is more stable is the one, has its structure based on tris-s-triazine and is stable by about 30 kJ mol^{-1}. This was done using density-functional theory (DFT) [5].

1.2 Synthesis of g-C_3N_4

C-N core structures that are pre-bonded (derivatives of triazine and heptazine) such as cyanamide, melamine, dicyanamide, guanidine chloride, thiourea, urea, were used to make g-C_3N_4 *via* polymerization of such precursors that were rich in nitrogen and free from oxygen components. The reaction path involves polyaddition and polycondensation reactions that first condense into melamine, followed by condensation and removal of ammonia to form tris-s-triazine-based products. At around 520°C, tris-s-triazine further condenses into polymeric C_3N_4, which becomes unstable above 600°C and disappears at 700°C. Dicyanamide is recommended as a precursor to prevent sublimation of melamine at high temperatures [3, 5].

1.3 Advantages of g-C_3N_4

- Non-metal semiconductor [6–8]
- Good thermal as well as chemical stability
- Moderate bandgap (2.7 eV) (absorption peak at about 460 nm)
- Visible-light-sensitive photocatalyst
- Collected from inexpensive and earth-abundant components (carbon and nitrogen)
- High quantum efficiency
- Non-toxic nature
- Economically feasible
- Easy to handle
- Long service life
- π-π conjugated framework providing an anchorage to organic compounds

1.4 Disadvantages of pristine g-C_3N_4

- Photo-generated electron-hole (e^-/h^+) pairs recombine quickly [6–8]
- Less-positive VB potential
- Insufficient absorption of visible light (restricted to solar blue light)
- Poor electronic conductivity
- Low specific surface area($<10\ m^2g^{-1}$)
- The particle boundary interrupts electron delocalization

1.5 How to overcome the drawbacks of pure g-C_3N_4

To overcome the drawbacks related to pure g-C_3N_4 several modifications have been made to increase its photocatalytic activity by expanding the optical response in visible light, enhancing the physicochemical response, increasing the surface area of catalyst, and lowering the recombination rate of e^-/h^+ pairs formed by photons.

It has been noted that the two-dimensional layered structure of g-C_3N_4 favors its hybridization with other components [5]. There are few different approaches to develop an effective and modified g-C_3N_4 photocatalyst such as by controlling morphological features of g-C_3N_4, polymerization, doping with a metal (Fe, Cu, Mo, Ag, Pd) or non-metal (B, C, N, O, S, P), co-doping, formation of heterojunction (binary and ternary nanocomposites), etc.

Thus, the modified g-C_3N_4 can be further explored to utilize the process of heterogeneous photocatalysis in a much more productive way and to achieve desirable outcomes from decomposition of numerous pollutants existing in aquatic ecosystem.

Apart from pollutant degradation, the g-C_3N_4 based photocatalyst has other potential applications such as carbon dioxide reduction, splitting of water photocatalytically into hydrogen and oxygen (as depicted in **Figure 1**) [9].

2. Mechanism of photodegradation

Figure 2 illustrates the charge carrier generation, separation, and migration mechanisms in pure g-C_3N_4 and their role in pollutant degradation.

The bandgap of pure g-C_3N_4 is 2.7 eV, while the valence band (VB) and conduction band (CB) potentials are 1.56 and −1.09 eV, respectively [10].

The mechanism constituting these photocatalyzed (predominantly) oxidative processes has been discussed extensively in the literature and can be summarized as follows. Under illumination with photons of energy equal to or greater than its bandgap, g-C_3N_4 absorbs light, exciting electrons from the valence band (VB) to the conduction band (CB), thereby generating electron-hole pairs (e_{CB}^- and h_{VB}^+). These charge carriers migrate to the catalyst surface, where they participate in redox reactions, leading to the degradation of pollutants. These charge carriers may then either recombine (which occurs on the picosecond timescale and corresponds to a short-

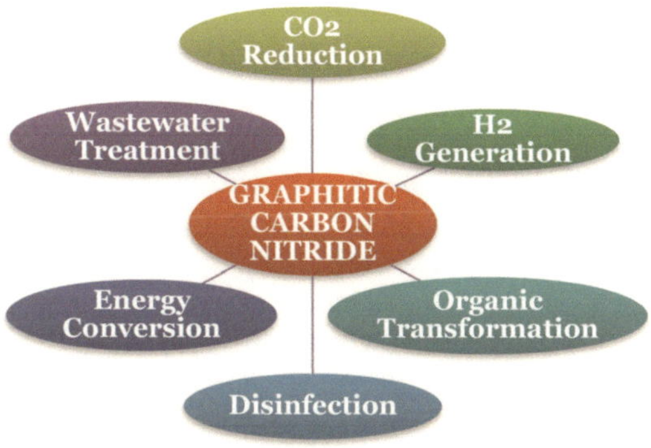

Figure 1.
Applications of materials based on g-C_3N_4.

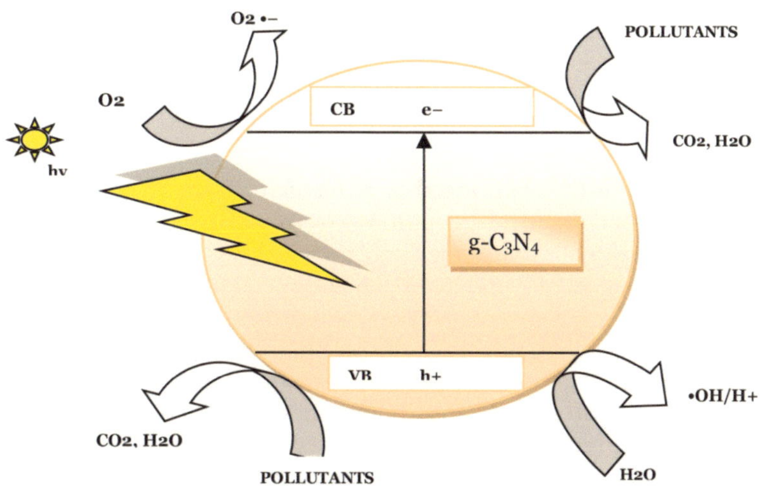

Figure 2.
Schematic diagram representing a mechanism of photodegradation with separation and migration of photoinduced charges in pure g-C_3N_4.

circuiting of the photocatalysis) or migrate to the catalyst surface where they participate in redox reactions with species in free solution (the so-called dynamic quenching processes) or adsorbed at the catalyst surface (the so-called static quenching processes). In particular, h_{VB}^+ may react with surface-bound H_2O or OH^- to produce the hydroxyl radical, while e_{CB}^- may reduce oxygen to generate the superoxide radical anion ($O_2^{\cdot-}$). The hydroxyl radicals ($^{\cdot}OH$) and superoxide anions ($O2^{\cdot-}$) formed through these pathways are highly reactive and attack the organic pollutants, breaking them down into non-toxic end products such as CO_2 and H_2O. The overall efficiency of the photocatalytic degradation process is dependent upon a number of factors including. The following equations define the degradation reaction process of a pollutant/contaminant: [11].

$g\text{-}C_3N_4 + h\upsilon \rightarrow h_{VB}^+ + e_{CB}^-$.

Contaminants + $h_{VB}^+ \rightarrow$ oxidation products.

Contaminants + $e_{CB}^- \rightarrow$ reduction products.

$H_2O + h_{VB}^+ \rightarrow {}^{\bullet}OH + H^+$.

$^{\cdot}OH$ + pollutant $\rightarrow$ intermediates $\rightarrow H_2O + CO_2$ (degradation products).

$O_2 + e_{CB}^- \rightarrow O_2^{\cdot -}$.

$O_2^{\cdot -}$ + pollutant → intermediates → $CO_2 + H_2O$ (degradation products).

Recombination of e^-/h^+ pairs:

$h_{VB}^+ + e_{CB}^- \rightarrow$ Heat + Radiationless decay.

The most significant restriction that lowers photocatalytic efficiency is the recombining of electrons and holes, which is caused by the narrow bandgap. If the bandgap is increased in any way, the number of pairs of e^-/h^+ will increase, increasing the degradation efficacy. Many changes have been made to pure g-C3N4 in order to improve performance. It is determined that hydroxyl radicals and superoxide radical anion are the primary species in charge of the degradation of dyes and drug derivatives with the aid of trapping and quenching studies [3].

3. Modified g-C_3N_4

The following strategies have been used to synthesize the more efficient photocatalysts;

3.1 Non-metal doped g-C_3N_4

Since g-C_3N_4 is an n-type non-metal semiconductor, its metal-free nature may be preserved by non-metal doping. By acquiring electrons during the reaction phase, covalent bonds between non-metals and other compounds can be easily formed. The symmetry of g-C_3N_4 is usually broken by the non-metal introduction, allowing for faster separation of e^-/h^+ pairs [12]. Various non-metal dopants, including oxygen, phosphorus, sulfur, halogens, nitrogen, and boron, have been extensively studied for their role in enhancing g-C_3N_4's photocatalytic activity. These dopants improve charge separation and extend light absorption into the visible spectrum [13]. According to the most recent reports on non-metal doped g-C_3N_4, its photocatalytic activity is significantly impacted by phenyl doping in its lattices [3, 13–16].

3.1.1 Oxygen-doped g-C_3N_4

Among the most widely used methods employed to increase the photocatalytic potential of g-C_3N_4 is O-doping. According to studies, the photocatalysts' surface morphology and microstructures seem to be the most critical in determining light-harvesting efficiency. Zeng et al. designed and created one-dimensional porous g-C_3N_4 nanorods (O-g-C_3N_4 NR) from hydrous nanofibers of melamine isolated from aqueous melamine solution [17]. Strong hydrogen bonding among amino groups and van der Waals interactions between melamine units facilitate the formation of one-dimensional nanorods during the polymerization process. The characterized O-doped g-C_3N_4 gives insight into its effective electronic, morphological, and optical surface as compared to pristine g-C_3N_4, and under similar experimental conditions, the degradation rate for MB due to O-doped g-C_3N_4 (81%) is nearly four times as compared to pure g-C_3N_4 (24%) [17].

3.1.2 Phosphorus-doped g-C_3N_4

Guo et al. used melamine molecules as raw material in a sodium pyrophosphate-aided hydrothermal process to synthesize P-doped tubular g-C_3N_4 (P-TCN) with significant defects related to the surface [18]. P-TCN features a tubular hexagonal nanostructure with an increased surface area, providing additional active sites for catalytic reactions, facilitating reactant and product diffusion, and improving visible light absorption.

3.1.3 Sulfur-doped g-C_3N_4

Thiourea (TU) is the most convenient starting material in producing S-doped g-C_3N_4. An et al. employed urea and TU as raw materials to successfully produce S-doped g-C_3N_4 nanomaterial. Experimental results showed that S-doped g-C_3N_4 would degrade RhB at a rate of 0.232986 h^{-1} [19]. Zhou et al. synthesized g-C_3N_4 nanosheet having a porous network as well as doping with S using a template-free one-step process. This approach avoids assemblage and also provides charge carriers with more active surface sites and porous channels [20]. The S-doped g-C_3N_4 removes more toxic contaminants than g-C_3N_4, rendering it as an innovative technologies for removing pollutants from the ecosystem.

3.1.4 Phenyl-doped g-C_3N_4

The functionalization of g-C_3N_4 nanosheet has been obtained using benzamide and urea as a precursor through the simple co-polymerization and *in situ* calcination technique, represented by **Figure 3** [3]. Ph-g-C_3N_4 makes it highly efficient by minimizing the back electron transfer and increasing the lifetime of charge carrier to generate the ROS. In this way, the promising photocatalytic behavior of Ph-g-C_3N_4 has been obtained for the photo-removal of MO and PCM.

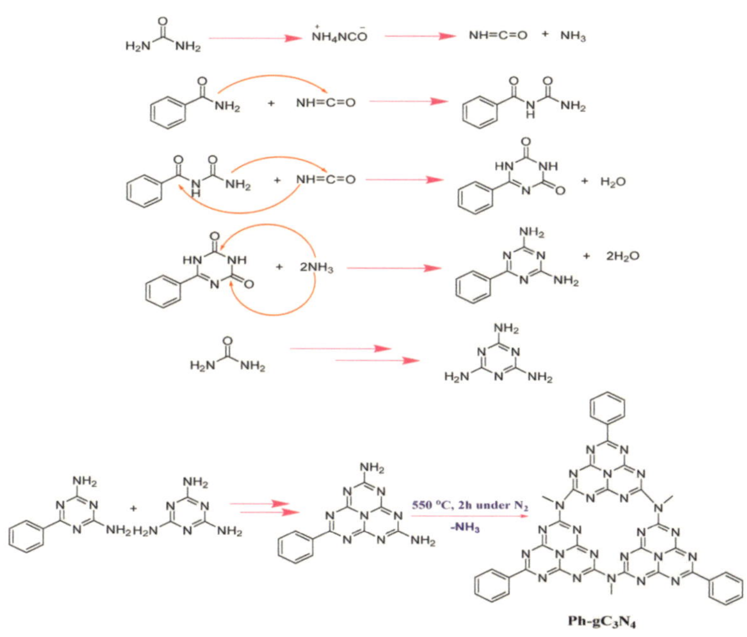

Figure 3.
Stepwise procedure for the preparation of Ph-gC_3N_4. Adapted with permission from Ref. [3], Copyright 2020, ACS Publications.

3.1.5 Boron-doped g-C_3N_4

Warshagha et al. proposed a hydrothermal-assisted synthesis of boron-doped phenyl functionalized g-C_3N_4 (B/PhCN) leading to a decrease in bandgap and slow charge carrier recombination due to the generation of impurity state of boron element [13]. Under visible light illumination, the recyclable and stable photocatalyst 0.2 B/PhCN shows proficient photocatalytic degradation toward MO and RhB.

3.2 Metal-doped g-C_3N_4

Metal doping imparts g-C_3N_4 with specific photocatalytic properties, such as narrowing the bandgap, increasing absorption of visible light, and improving catalytic efficiency. Transition metals (Fe, Cu, W) and alkali metals (Li, Na, K) are utilized as dopants. Usually, a metal-solubilized salt solution is uniformly combined with a precursor to thermal treatment. During the precursor's thermal condensation phase, metallic impurities will be introduced into the g-C_3N_4 structure at the same time [21].

3.2.1 Alkali metal-doped g-C_3N_4

The migration rate and separation rate of photo-induced e^-/h^+ pairs have been significantly enhanced *via* doping with alkali metals into the g-C_3N_4 system, and this leads to the induction of photocatalytic redox reactions in a more efficient way. Jiang, Yuan, et al. used calcination at high temperatures to dope Li^+, Na^+, and K^+ g-C_3N_4 (e.g., Li-CN, Na-CN, K-CN) [21]. The preparation of pristine g-C_3N_4 without the addition of NaBr and KBr is labeled as CN [22]. Hu et al. used precursors: dicyandiamide monomer and solution of potassium hydrate to make K-doped g-C_3N_4 with a tunable band distance [23]. Around the K^+ ion, a ring structure with C and N had been formed in K-doped g- C_3N_4. In addition, it was reported that by adjusting the quantity of doped K, the values of CB and VB in g-C_3N_4 can be regulated, and it could generate $OH^{\cdot}$ and $O_2^{\cdot-}$ simultaneously in the visible region and result in an improved degradation rate [23]. After doping Na on the g-C_3N_4 photocatalyst, the results obtained were similar. If Na concentration has been changed, then the potentials of CB and VB could be tuned. The RhB degradation rate for Na-doped g-C_3N_4 was lower in comparison with K-doped g-C_3N_4 in identical conditions as depicted in **Table 1**.

As shown in **Table 1**, K-doping exhibits superior performance due to its enhanced bandgap tuning and improved charge carrier mobility compared to Na-doping. This indicated that the effects of K and Na doping on g-C_3N_4 were comparable. However,

Precursor	Bandgap energy (eV)	Photocatalytic activity (min^{-1})	Activity compared to CN (min^{-1})	Reference
KOH(K) Dicyanamide (CN)	2.57	0.011	0.0017	[23]
NaOH(Na) Dicyanamide (CN)	2.58	0.0064	0.0018	[24]

Table 1.
Comparative degradation rates of RhB using alkali metal-doped g-C_3N_4 photocatalysts under visible light.

because of the extended conjugated systems, narrow bandgap, and positively-shifted VB location, the K doping in g-C_3N_4 results in increased visible light responsiveness, mitigates charge pair recombination, and has a strong oxidation capability. As a result, organic pollutants are efficiently degraded. Fe(III) metal-grafted K-doped g-C_3N_4/rGO exhibits effective photocatalytic activity, as evidenced by a lower charge carrier recombination rate due to interfacial charge transfer (IFCT) caused by the presence of Fe(III), according to additional research on the alkali metal-doped g-C_3N_4 reported by Azam et al. [25].

In addition to the fact that g-C_3N_4 doped with alkali metals is more effective at mineralizing organic pollutants through photocatalysis. Another important factor is the bandgap width, which can be controlled by altering the concentration of the doped alkali metal ions.

3.2.2 Transition metal-doped g-C_3N_4

Transition metals, such as Mo, Fe, Cu, W, and Zr, are commonly utilized as dopants for altering the electronic and optical characteristics of g-C_3N_4. Doping with metals significantly decreases the bandgap, promotes charge mobility, improves visible light absorption, and extends the lifetime of charge carriers, all the above-mentioned features are needed to increase the photocatalytic performance. In addition, the strong interactions between the negatively charged N atoms and metal cations, g-C_3N_4 can easily capture metal cations. Metal doping also boosts the activity of the g-C_3N_4 catalyst by mitigating the e^-/h^+ pair recombination rate.

Tonda et al. employed a simple and inexpensive way to make the Fe-doped g-C_3N_4 nanosheet. Doping with Fe poses a significant impact on both optical and electronic attributes of the g-C_3N_4 nanosheet. In the visible light spectrum, a red shift with greater absorption is observed in the Fe-doped g-C_3N_4 nanosheet. The Fe species were discovered to be in the +3 oxidation state, indicating that they could be used as a temporary site for trapping electrons and holes, when photogenerated. O_2 could be reduced to $O_2^{\cdot-}$ by these trapped electrons. Meanwhile, valance holes can use Fe^{3+} as a mediator to oxidize OH^- or H_2O molecules into $^{\cdot}OH$. Thus, the activity of photocatalysts for RhB degradation would be greatly increased [26].

In order to enhance the degradation of the dye RhB, g-C_3N_4 was doped with yttrium, a rare earth element, using a simple pyrolysis process with the precursor urea and yttrium nitrate as the Y source [27]. The effective Y doping ratio caused the bandgap of g-C_3N_4 to decrease from 2.70 to 2.50 eV. By limiting the recombination of photoinduced e^-/h^+ pairs, an appropriate doping ratio raises the photocatalyst's electronic conductivity and boosts photocatalysis's energy efficiency [27].

It is clear from the discussion above that metal ions have a wide range of potential applications as dopants in g-C_3N_4. The production of new energy levels by metal ions can intensify the response in visible light and frequently stop e^-/h^+ pairs from recombining. Metal doping, however, has also been shown to have certain drawbacks. The thermal stability of the doped ions is low. Furthermore, the newly created energy bands might act as recombination sites, which would lower quantum efficiency [28].

The g-C_3N_4 can also be functionalized with noble metals (Pt, & Pd, etc.) for maximizing carrier mobility, e^-/h^+ separation, bandgap narrowing, and encouraging increased absorption and activity of photocatalysts in the visible region. Doping with noble metals could boost catalytic performance in g-C_3N_4, but the high cost for noble metals prevents its widespread application.

3.3 Co-doping

It has been noted that non-metal/non-metal co-doping will lead to an enhancement in the efficiency of the g-C_3N_4 photocatalyst. For instance, Ma et al. synthesized P and O co-doped g-C_3N_4 which showed increased RhB mineralization in hypoxic conditions, which was caused by a smaller bandgap, increased specific area as well as improved e^-/h^+ pairs separation [29].

For enhanced catalytic efficiency, tridoped g-C_3N_4 has also been fabricated by integrating three separate heteroatoms into the g-C_3N_4 system. Ma et al. used hydrothermal post-treatment without using H_2O_2 to synthesize a new S—Co—O tridoped g-C_3N_4. The annealing of a mixture of thiourea (TU) and $Co(NO_3)_2 \cdot 6H_2O$, followed by a hydrothermal method, was employed to fabricate S—Co—O tridoped g-C_3N_4. The surface area, bandgap energy, and separation of photo-induced e^-/h^+ pairs were all altered. Furthermore, under hypoxic conditions, O doping increases the adsorption capacity of g-C_3N_4 by trapping photo-induced electrons and causing degradation of RhB by utilizing holes [30].

In general, element doping will lead to the mitigation of bandgap energy and also result in the improvement of the separation efficiency of photogenerated e^-/h^+ pairs, resulting in increased photocatalytic application. Excess non-metal and/or metal doping, on the other hand, is reported to show a negative impact on photocatalysis leading to the advent of further defects for e^-/h^+ pair recombination.

Chen et al. with the help of TU (S source) and $NaHCO_3$ (Na source) synthesized Na-S-co-doped g-C_3N_4 nanotubes ($Na_xSCNNTs$), where x indicates the mass ratio of $NaHCO_3$ to total of urea and TU (x = 0, 0.05, 0.1, 0.2, 0.5, 1.5 wt. %). Varying the quantity of $NaHCO_3$, different $Na_xSCNNTs$ could be yielded. If x = 0 it depicted the S-doped g-C_3N_4 NT synthesized under similar conditions without $NaHCO_3$ [31].

Among the various compositions, the photocatalytic operation of $Na_{0.1}SCNNT$ for degrading RhB was observed to be the highest, which is about 3.4 times more as compared to pristine g-C_3N_4. The broad specific surface area, narrow bandgap, improved absorption in visible light, and down-shift in VB of $Na_{0.1}SCNNT$ are responsible for the higher activity of the photocatalyst.

3.4 Heterojunction-based g-C_3N_4 photocatalysts

Since most photocatalysts consist of a single component and have a sufficient bandgap, they can efficiently conduct one of the two processes, namely oxidation or reduction, but not both. Heterojunction-based photocatalysts enable two or more semiconductors to be integrated to carry out photocatalytic reactions more efficiently.

The photo-excited electrons and photo-induced holes can be separated more effectively when a semiconductor having an appropriate band structure approaches g-C_3N_4. This is due to the effective charge transfer through the two semiconductors' interface.

According to its efficacy in upgrading photocatalytic efficiency, the fabrication of heterojunction-based g-C_3N_4 photocatalysts has been a substantial topic in research and technology. Due to the higher separation between the e^-/h^+ pairs, designing heterojunction-based g-C_3N_4 photocatalysts is a vital challenge among the ongoing innovations. Regarding the synergic impact involving g-C_3N_4 as well as the other components, the g-C_3N_4-based heterostructures result in preventing recombination of charge carriers that are generated *via* photocatalysis and also enrich the g-C_3N_4

photocatalyst with some innovative features, establishing g-C_3N_4 catalyst, as a valuable photocatalytic nanomaterial for numerous energy-efficient applications.

The heterojunction-based g-C_3N_4 photocatalysts can be classified as the following based on various transfer mechanisms of photo-induced charges occurring between g-C_3N_4 and components coupled with it: [32].

- g-C_3N_4-based type II heterojunction (Conventional)
- g-C_3N_4-based Z-scheme heterojunction
- g-C_3N_4-based p–n heterojunction
- g-C_3N_4/metal heterostructure
- g-C_3N_4/carbon heterostructure

A higher positive VB and a higher negative CB will increase the photocatalyst's activity. In addition, only a few e^-/h^+ pairs are generated because the bandgap of a heterojunction-based semiconductor is too large for effective optical absorption. Because heterojunction-based g-C_3N_4 photocatalysts effectively separate the semiconductor components that make up the heterojunction across the interface, they can significantly reduce the recombination of photo-generated e^-/h^+ pairs. Additional advantages that can result from the addition of the second component include enhanced light absorption, better band placement, and higher surface reaction activity —all of which are conducive to higher catalytic activity.

3.4.1 Binary heterojunction-based g-C_3N_4 nanocomposites

For the heterojunction-based g-C_3N_4 composites based on type II, the choice of the second semiconductor for constructing heterojunctions relies on its band structure. For designing and building type II heterojunction composites of g-C_3N_4, different metal oxides, sulfides, and halides were employed for modification purposes.

Using g-C_3N_4/ZnO binary composites, Sun et al. [33] documented the photocatalytic degradation of MO in visible light. When compared to pure g-C_3N_4 and ZnO, the photocatalytic performance of g-C_3N_4/ZnO binary composites was found to be higher. Visible light could not excite ZnO due to its large bandgap; however, in composite g-C_3N_4/ZnO, the g-C_3N_4 could absorb visible light and generate an e^-/h^+ pair. The photo-excited electrons in the CB of g-C_3N_4 will transfer to the CB of ZnO due to a suitable band match.

The collection of photoexcited electrons and holes in ZnO's CB and g-C_3N_4's VB, respectively, allows for the separation of charge carriers that were produced photocatalytically. Photo-induced holes on VB of g-C_3N_4 can oxidize MO directly, whereas photo-induced electrons in ZnO's CB could reduce O_2 to $O_2^{\cdot -}$ with high oxidation potential. The primary active species in MO degradation was $O_2^{\cdot -}$, according to experiments involving the quenching of radicals.

The migration of electrons produced during photocatalysis from g-C_3N_4 to ZnO was the primary factor in preventing charge carrier recombination and in producing more $O_2^{\cdot -}$, demonstrating that the g-C_3N_4/ZnO composite is effective in boosting the photocatalyst's efficiency [33].

Binary heterojunction composites	Pollutant degraded	Fold increment in degradation compared to pure g-C_3N_4	References
ZnO/g-C_3N_4	MO	3.0	[33]
WO_3/g-C_3N_4	MB	2.9	[34]
ZnS/g-C_3N_4	RhB	2.8	[35]
CdS/g-C_3N_4	MB	3.0	[36]
AgI/g-C_3N_4	MB	2.3	[37]

Table 2.
Binary heterojunction composites of conventional type II.

Based on a similar mechanism a number of other binary heterojunction type II photocatalysts are listed in **Table 2**, according to their role in degrading various dyes.

Other binary heterojunction composites based on the direct Z-scheme electron transfer mechanism include g-C_3N_4. The best illustration of this kind of system can be found in nature itself, and it is connected to the most beneficial process that green plants undergo, called photosynthesis. With the help of electron mediators, charge transfer in the plant under visible light irradiation takes a different route than that of type II conventional heterojunctions. Photocatalytically generated charge carriers can be effectively separated spatially while maintaining their potent redox capabilities attributed to the Z-scheme method of charge transfer [38]. Moreover, a cost-effective method of increasing the efficiency of g-C_3N_4 photocatalysts was the development of Z-scheme heterojunction-based g-C_3N_4 photocatalysts.

Azam et al. recently reported the synthesis of a photocatalytically active 3D-flower-like CuS/g-C_3N_4 nanocomposite *via* a hydrothermal method that degraded MB dye in aqueous suspension under visible light [39]. A strong interfacial interaction exists between CuS and g-C_3N_4, according to SEM analysis.

One advantage of this binary composition is that the catalyst exhibits remarkable stability even after five-run cycles on irradiation of an aqueous suspension of MB in the presence of 10% CuS/g-C_3N_4 composite, in 90 min. **Figure 4** represents the mechanism of MB degradation with a binary heterojunction system of CuS/g-C_3N_4

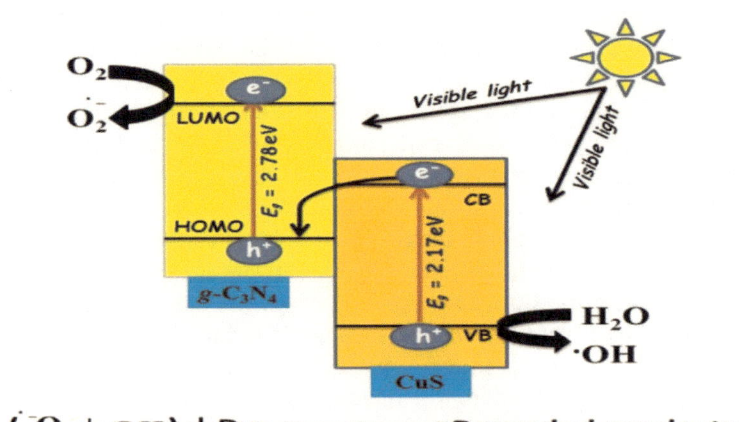

Figure 4.
Scheme representing the degradation of MB via *Z-scheme-based CuS/g-C_3N_4 heterojunction under visible light. Adapted with permission from Ref. [39]. Copyright 2018, Elsevier.*

Binary composite	Pollutants degraded	References
$CuS/g\text{-}C_3N_4$	MB	[39]
$Ag_2WO_4/g\text{-}C_3N_4$	MO	[40]
$WO_3/g\text{-}C_3N_4$	MB	[41]
$Fe_2O_3/g\text{-}C_3N_4$	RhB	[42]
$V_2O_5/g\text{-}C_3N_4$	RhB, MO, MB	[43]
$Bi_2WO_6/g\text{-}C_3N_4$	RhB	[44]

Table 3.
Direct Z-scheme-based binary heterojunctions composites of g-C_3N_4.

under visible light leading to the excitation of both CuS and g-C_3N_4 catalysts simultaneously due to the formation of heterojunction resulting in the generation of electrons in CB and hole in VB.

The electron from the CB of CuS migrates to the hole in g-C_3N_4, leading to an accumulation of electrons in the CB of g-C_3N_4 and the accumulation of holes in the VB of CuS. This transfer pathway is responsible for the separation of charge carriers and a longer lifetime of the photoinduced e^-/h^+ pairs, which increases the photocatalytic activity of the composite. The electrons collected in the CB of g-C_3N_4 react with dissolved O_2 to generate $O_2^{\cdot -}$, while the potential of the holes accumulated in the VB of CuS is more positive to oxidize H_2O to generate $^{\cdot}OH$. This is considered to be the key species for the Z-scheme CuS/g-C_3N_4 photocatalyst, and such a heterojunction offers promising potential in the visible region. The entire Z-scheme reaction series is described by the following equations (**Table 3**) [39].

$$CuS/g\text{-}C_3N_4 \rightarrow g\text{-}C_3N_4\ (e^- + h^+) + CuS\ (e^- + h^+)$$
$$g\text{-}C_3N_4\ (e^- + h^+) + CuS\ (e^- + h^+) \rightarrow g\text{-}C_3N_4\ (e^-) + CuS\ (h^+)$$
$$O_2 + g\text{-}C_3N_4\ (e^-) \rightarrow {}^{\cdot}O_2^-$$
$$H_2O + CuS\ (h^+) \rightarrow H^+ + {}^{\cdot}OH$$
$$\text{Dye} + ({}^{\cdot}O_2^-, {}^{\cdot}OH) \rightarrow \text{Degraded Products}$$

Several other binary heterojunctions have been synthesized by various researchers based on p-n heterojunctions and numerous catalysts have been prepared using co-doping in heterojunctions. For example, Azam et al. synthesized a Fe(III)-grafted, K-doped binary g-C_3N_4/rGO composite that exhibited strong photocatalytic activity in the degradation of MO, RhB, and MB dyes under visible light [25]. Danish et al. also reported a binary composite based on g-C_3N_4 and Ni-ZnS/g-C_3N_4, which increased the photocatalytic efficiency of RhB, MB, and PCM [45]. Few researchers have also investigated the g-C_3N_4-based binary composites for the efficient degradation of the priority colorless pollutants such as BPA and TC [15, 16].

3.4.2 Ternary heterojunction-based g-C_3N_4 nanocomposites

Compared to binary composites and pure g-C_3N_4, the synergistic effect of a multicomponent system in ternary photocatalysts with heterojunction leads to stronger segregation of the photoinduced e^-/h^+ pairs, resulting in a more efficient mineralization of organic pollutants. Akhundi and Habibi-Yangjeh prepared a ternary heterojunction photocatalyst of g-C_3N_4/ZnO/AgCl that effectively degraded RhB upon irradiation with visible light [46].

The ternary heterojunction photocatalyst therefore has an improved photocatalytic effect compared to a binary heterojunction or pure g-C_3N_4. The electrons generated by the photocatalytic process migrate from g-C_3N_4 to ZnO and AgCl due to suitable band structures, which leads to a synergistic improvement in the removal efficiency of charge carriers from g-C_3N_4 [46].

Likewise, we have also fabricated and reported several g-C_3N_4 ternary composites for photo-removal of many priority organic contaminants such as Co@ZnSQDs/g-C_3N_4/MWCNT, α-Fe_2O_3/CdS/g-C_3N_4, Fe_2O_3/g-C_3N_4/GO, and $LaNiO_3$/g-C_3N_4/MWCNT [47–51]. These materials have proven to be quite effective ternary nanocomposites for the removal of dyes, pharmaceutical drugs, and other organic contaminants from the water bodies efficiently leading to the purification of water. Among the aforementioned reports, α-Fe_2O_3/CdS/g-C_3N_4 has notably followed the double Z-scheme electron transfer mechanism between its components α-Fe_2O_3, CdS, and g-C_3N_4, degrading RhB and MB effectively [48]. These three photocatalysts formed the heterojunction in such a way that under illumination with visible light, the transfer of an electron from the conduction band of Fe_2O_3 to the valence band of CdS takes place, followed by a further transfer of an excited electron from the conduction band of CdS to the valence band of g-C_3N_4. A further excitation of an electron to the conduction band of g-C_3N_4, and finally, the formation of a superoxide radical in the conduction band of g-C_3N_4 and a hydroxyl radical in the valence band of Fe_2O_3 is expected as shown in **Figure 5(b)**, in contrast to the type II photocatalytic mechanism, which is not possible due to an insufficient band edge potential required for the generation of ROS species represented in **Figure 5(a)** [48].

The construction of a ternary heterojunction provides an excellent enhancement of the photocatalytic performance of pure g-C_3N_4 by improving the separation of e^-/h^+ pairs to a greater extent than binary nanocomposites and pure g-C_3N_4. **Table 4** provides a list of the various ternary heterojunction-based composites synthesized by different methods to degrade various dyes.

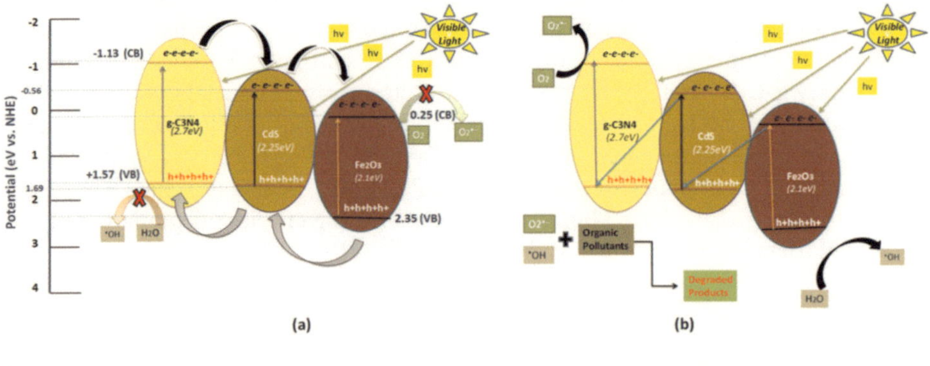

Figure 5.
(a) Type-II heterojunction photocatalytic mechanism and (b) Z-scheme mechanism. Adapted with permission from Ref. [48]. Copyright 2021, Elsevier.

Ternary composite	Dye degraded	How much fold increment in degradation compared to pure g-C_3N_4	References
YVO_4/g-C_3N_4/Ag	MO	3	[52]
g-C_3N_4/ZnS/SnS_2	MB	8.74	[53]
g-C_3N_4/Ag_3VO_4/AgBr	RhB	116	[54]
g-C_3N_4/Fe_3O_4/BiOI	RhB MO MB	21	[55]
g-C_3N_4/Fe_3O_4/$CuWO_4$	MO MB RhB	12.5 17 10.5	[56]

Table 4.
Ternary heterojunction-based g-C_3N_4 photocatalysts.

4. Conclusion and future perspectives

Pristine g-C_3N_4's photocatalytic efficiency is limited by its narrow light absorption range and rapid recombination of photoinduced charge carriers. However, apart from that, the rising enthusiasm of researchers in employing g-C_3N_4 as a photo-catalyst indicates that it has become a widely used material in a variety of technological and research fields, including removal of pollutants, CO_2 reduction, H_2O splitting, and organic synthesis. Owing to its distinctive attributes, such as metal-free design, visible light responsiveness, ease of fabrication, sufficient bandgap, the effective potential to cause redox reactions, good physical, and chemical stability, ecologically friendly behavior, polymeric framework, that further aids modifications in pure g-C_3N_4, make g-C_3N_4 efficient enough for being used in a variety of applications.

Significant innovations related to the production of reliable and non-expensive modified g-C_3N_4 photocatalysts are outlined in this article that are carried out *via* doping with non-metal or metal, co-doping, and creating heterojunctions based g-C_3N_4 semiconductor photocatalysts. The modifications done are responsible for enhanced catalytic efficiency by boosting light absorption, extending charge carrier lifetime, and increasing the e^-/h^+ pair separation.

Elemental-doped g-C_3N_4 photocatalysts are expected to receive increasing research emphasis in the coming years. Elemental doping has emerged as an effective method that can alter the bandgap of g-C_3N_4 resulting in the separation of e^-/h^+ pairs. In a nutshell, doping with metal contributes to the emergence of new levels of energy in the bandgap as well as in extension of the spectral response. Also, the doped ions do not exhibit good thermal stability. In addition, the new energy bands formed may lead to a reduction in quantum efficiency as these may provide sites for recombination.

If we need to combine the strengths of the individual dopants, co-doping is desirable, contributing to increased photocatalytic performance, as well as optimized structural and optical properties. The heterojunction photocatalysts are designed to achieve a high separation rate of the e^-/h^+ pair, which will be effective enough to carry out visible-light photocatalysis efficiently.

The utilization of g-C_3N_4 in various applications will undoubtedly lead to significant developments in future sustainable development and the creation of new green technology, despite such promising results, g-C_3N_4 still encounters some challenges that are listed below:

- Since the absorption of g-C_3N_4 is primarily restricted to the blue-violet region of visible light, the efficacy of visible light is reduced and we obtain a low photocatalytic output.
- The prediction of the surface morphology of g-C_3N_4 is difficult to monitor precisely if any modifications are done.
- The position of dopant atoms or heteroatoms in modified g-C_3N_4 is not easy to analyze.
- The mechanisms involved in photocatalysis done by an element-doped g-C_3N_4 system need further explanation to have a better understanding of the working related to g-C_3N_4.
- The g-C_3N_4 synthesized in laboratories sometimes is found to have limited efficiency because of its low crystallinity and defects present in large numbers.
- It is not well understood what impact the defects and doping pose on the electronic and structural characteristics of g-C_3N_4.

Therefore, the construction as well as production of effective photocatalysts based on g-C_3N_4 with superior capacity to degrade the organic contaminants by irradiating the catalyst in visible light is critical in carrying out the purification of water resources. As technology has progressed, it has become easier to prepare g-C_3N_4-based photocatalysts. Considering the success of photocatalysts based on g-C_3N_4, the strategies that need to be further explored in future research studies are as follows:

- When modifying g-C_3N_4, theoretical analysis (DFT modeling) and practical implementations may be employed for optimizing the doping effect and heterojunction formation which can modify the bandgap, and charge carrier separation, and structural stability.
- Probable structure and morphology can be estimated by integrating theoretical calculations into the catalyst's structural system.
- A novel doping strategy for adjusting the HOMO and LUMO with high oxidation and reduction potential should be established because this tends to mitigate the bandgap and improves the segregation between photo-generated electrons and holes giving rise to improved photocatalytic performance.
- Modifications done should be eco-friendly, cost-effective, scalable, and non-hazardous.
- Examining the molecular mechanism of photocatalysis and witnessing the entire photocatalytic process should be emphasized.
- Doped g-C_3N_4-based materials should have to be thermally and chemically stable.
- Kinetic and thermodynamic monitoring of the mechanism should be used in alterations.

- Further analysis of g-C_3N_4 surface/textural properties and tuning of the bandgap is needed to acquire optimal optical and photocatalytic properties.
- The quest for low-cost noble metal substitutes should be strongly encouraged.
- A clear picture of the reaction mechanism involved in photocatalysis is needed to upgrade the efficiency and development of heterojunction-based g-C_3N_4 photocatalysts.
- Charge transfer pathways in various forms of heterojunction-based g-C_3N_4 photocatalysts, and even the photocatalytic process, need to be explored further.

In the coming years, we believe there are many routes worthy of consideration for modifying the g-C_3N_4 photocatalyst and utilizing its full potential in improving the efficiency of degradation or mineralization of emerging organic pollutants such as pharmaceuticals and microplastics that are persistent in aquatic reservoirs or are incorporated into it *via* some other industrial sources.

Acknowledgements

Dr. Ziad Moussa is grateful to the United Arab Emirates University (UAEU) and to the Research Office for supporting the research developed in his laboratory and reported herein (UPAR grant code G00004605). The authors are thankful to the DST, UGC for DRS II, PURSE & FIST support to the Department of Chemistry, AMU Aligarh. Ziyaur Rasool thanks the DST, New Delhi, India, for the grant of the DST-Inspire fellowship (IF200082).

Author details

Mohammad Muneer[1*], Ziyaur Rasool[1], Shreya Parashar[1], Abdulaziz M. Almohyawi[2], Ziad Moussa[3] and Saleh A. Ahmed[2]

1 Department of Chemistry, Aligarh Muslim University, Aligarh, India

2 Faculty of Science, Department of Chemistry, Umm Al-Qura University, Makkah, Saudi Arabia

3 Department of Chemistry, College of Science, United Arab Emirates University, Al Ain, United Arab Emirates

*Address all correspondence to: m.muneer.ch@amu.ac.in; readermuneer@gmail.com and saahmed@uqu.edu.sa

IntechOpen

© 2025 The Author(s). Licensee IntechOpen. This chapter is distributed under the terms of the Creative Commons Attribution License (http://creativecommons.org/licenses/by/4.0), which permits unrestricted use, distribution, and reproduction in any medium, provided the original work is properly cited.

References

[1] Qi K, Yuan LS, Zada A. Graphitic carbon nitride, a polymer photocatalyst. Journal of the Taiwan Institute of Chemical Engineers. 2020;**109**: 111-123

[2] Pelaez M, Nolan NT, Pillai SC, Seery MK, Falaras P, Kontos AG, et al. A review on the visible light active titanium dioxide photocatalysts for environmental applications. Applied Catalysis B: Environmental. 2012;**125**: 331-349

[3] Warshagha MZA, Muneer M. Synthesis of Ph-modified graphitic carbon nitride for degradation of different chromophoric organic pollutants in aqueous suspension under visible light. Langmuir. 2020;**36**(33): 9719-9727

[4] Zhao Z, Sun Y, Dong F. Graphitic carbon nitride based nanocomposites: A review. Nanoscale. 2015;7:15-37

[5] Thomas A, Fischer A, Goettmann F, Antonietti M, Müller JO, Schlögl R, et al. Graphitic carbon nitride materials: Variation of structure and morphology and their use as metal-free catalysts. Journal of Materials Chemistry. 2008; **18**(41):4893-4908

[6] Jourshabani M, Dominic JA, Achari G, Shariatinia Z. Synergetic photocatalytic ozonation using modified graphitic carbon nitride for treatment of emerging contaminants under UVC, UVA and visible irradiation. Chemical Engineering Science. 2019;**209**:115181

[7] Akhundi A, Habibi-Yangjeh A, Abitorabi M, Rahim PS. Review on photocatalytic conversion of carbon dioxide to value-added compounds and renewable fuels by graphitic carbon nitride-based photocatalysts. Catalysis Reviews - Science and Engineering. 2019;**61**(4):595-628

[8] Zhang C, Li Y, Shuai D, Shen Y, Xiong W, Wang L. Graphitic carbon nitride (g-C_3N_4)-based photocatalysts for water disinfection and microbial control: A review. Chemosphere. 2019; **214**:462-479

[9] Akhundi A, Badiei A, Ziarani GM, Habibi-Yangjeh A, Muñoz-Batista MJ, Luque R. Graphitic carbon nitride-based photocatalysts: Toward efficient organic transformation for value-added chemicals production. Molecular Catalysis. 2020;**488**:110902

[10] Xu B, Ahmed MB, Zhou JL, Altaee A, Xu G, Wu M. Graphitic carbon nitride based nanocomposites for the photocatalysis of organic contaminants under visible irradiation: Progress, limitations and future directions. Science of the Total Environment. 2018;**633**: 546-559

[11] Sharma S, Dutta V, Singh P, Raizada P, Rahmani-Sani A, Hosseini-Bandegharaei A, et al. Carbon quantum dot supported semiconductor photocatalysts for efficient degradation of organic pollutants in water: A review. Journal of Cleaner Production. 2019;**228**: 755-769

[12] Huang Z-F, Song J, Lun Pan Z, Wang XZ, Ji-Jun Zou WM, Xiangwen Zhang LW. Carbon nitride with simultaneous porous network and O-doping for efficient solar-energy-driven hydrogen evolution. Nano Energy. 2015; **12**:646-656

[13] Warshagha MZA, Muneer M. Visible light active boron doped phenyl-g-C_3N_4 nanocomposites for decomposition of

dyes. Surfaces and Interfaces. 2021;**26**: 101394

[14] Warshagha MZA, Muneer M. Synthesis of ZnO co-doped Ph-g-C_3N_4 for enhanced photocatalytic organic pollutants removal under visible light. International Journal of Environmental Analytical Chemistry. 2022;**102**(18): 6339-6358

[15] Rasool Z, Athar MS, Muneer M. Construction of flake ball-shaped Bi_2WO_6 embedded on phenyl functionalized g-C_3N_4 nanosheet for efficient degradation insight of colorless pollutants and its biological application. Environmental Science and Pollution Research. 2024;**31**(21):31259-31272

[16] Athar MS, Rasool Z, Muneer M, Altass HM, Althagafi II, Ahmed SA. Fabrication of direct Z-scheme $CoNiWO_4$/Ph-gC_3N_4 heterocomposites: Enhanced photodegradation of bisphenol A and anticancer activity. ACS Omega. 2023;**8**(41):38272-38287

[17] Zeng Y. Scalable one-step production of porous oxygen-doped g-C_3N_4 nanorods with effective electron separation for excellent visible-light photocatalytic activity. Applied Catalysis B: Environment and Energy. 2018;**224**: 1-9

[18] Guo S, Tang Y, Xie Y, Tian C, Feng Q, Zhou W, et al. P-doped tubular g-C_3N_4 with surface carbon defects: Universal synthesis and enhanced visible-light photocatalytic hydrogen production. Applied Catalysis B: Environmental. 2017;**218**:664-671

[19] An TD, Van PN, Tri NN, Phu HT, Hung NP, Vo V. Sulfur-doped g-C_3N_4 with enhanced visible-light photocatalytic activity. Applied Mechanics and Materials. 2019;**889**: 43-50

[20] Zhou Y, Lv W, Zhu B, Tong F, Pan J, Bai J, et al. Template-free one-step synthesis of g-C_3N_4 nanosheets with simultaneous porous network and S-doping for remarkable visible-light-driven hydrogen evolution. ACS Sustainable Chemistry & Engineering. 2019;**7**(6):5801-5807

[21] Jiang L, Yuan X, Pan Y, Liang J, Zeng G, Wu Z, et al. Doping of graphitic carbon nitride for photocatalysis: A review. Applied Catalysis B: Environmental. 2017;**217**:388-406

[22] Xiong T, Cen W, Zhang Y, Dong F. Bridging the g-C_3N_4 interlayers for enhanced photocatalysis. ACS Catalysis. 2016;**6**(4):2462-2472

[23] Hu S, Li F, Fan Z, Wang F, Zhao Y, Lv Z. Band gap-tunable potassium doped graphitic carbon nitride with enhanced mineralization ability. Dalton Transactions. 2014;**44**(3):1084-1092

[24] Zhang J, Hu S, Wang Y. A convenient method to prepare a novel alkali metal sodium doped carbon nitride photocatalyst with a tunable band structure. RSC Advances. 2014;**4**(108): 62912-62919

[25] Khan A, Alam U, Zafar S, Muneer M. Fe (III) -grafted K-doped g-C_3N_4/rGO composite photocatalyst pollutants. Journal of Chemical Sciences. 2018; **130**(10):1-9

[26] Tonda S, Kumar S, Kandula S, Shanker V. Fe-doped and -mediated graphitic carbon nitride nanosheets for enhanced photocatalytic performance under natural sunlight. Journal of Materials Chemistry A. 2014;**2**(19): 6772-6780

[27] Wang Y, Li Y, Bai X, Cai Q, Liu C, Zuo Y, et al. Facile synthesis of Y-doped graphitic carbon nitride with enhanced

photocatalytic performance. Catalysis Communications. 2016;**84**:179-182

[28] Zhou L, Zhang H, Sun H, Liu S, Tade MO, Wang S, et al. Recent advances in non-metal modification of graphitic carbon nitride for photocatalysis: A historic review. Catalysis Science & Technology. 2016;**6**(19):7002-7023

[29] Ma H, Li Y, Li S, Liu N. Novel P-O codoped g-C_3N_4 with large specific surface area: Hydrothermal synthesis assisted by dissolution-precipitation process and their visible light activity under anoxic conditions. Applied Surface Science. 2015;**357**:131-138

[30] Ma H, Zhao S, Li S, Liu N. A facile approach to synthesizing S-Co-O tridoped g-C_3N_4 with enhanced oxygen-free photocatalytic performance via a hydrothermal post-treatment. RSC Advances. 2015;**5**(97):79585-79592

[31] Chen KL, Zhang SS, Yan JQ, Peng W, Lei DP, Huang JH. Excellent visible light photocatalytic efficiency of Na and S co-doped g-C_3N_4 nanotubes for H_2 production and organic pollutant degradation. International Journal of Hydrogen Energy. 2019;**44**(60): 31916-31929

[32] Fu J, Yu J, Jiang C, Cheng B. G-C_3N_4-based heterostructured photocatalysts. Advanced Energy Materials. 2018;**8**(3): 1-31

[33] Sun JX, Yuan YP, Qiu LG, Jiang X, Xie AJ, Shen YH, et al. Fabrication of composite photocatalyst g-C_3N_4-ZnO and enhancement of photocatalytic activity under visible light. Dalton Transactions. 2012;**41**(22):6756-6763

[34] Huang L, Xu H, Li Y, Li H, Cheng X, Xia J, et al. Visible-light-induced WO_3/g-C_3N_4 composites with enhanced photocatalytic activity. Dalton Transactions. 2013;**42**(24):8606-8616

[35] Wang J, Guo P, Guo Q, Jönsson PG, Zhao Z. Fabrication of novel g-C_3N_4/nanocage ZnS composites with enhanced photocatalytic activities under visible light irradiation. CrystEngComm. 2014; **16**(21):4485-4492

[36] Jiang F, Yan T, Chen H, Sun A, Xu C, Wang X. Ag-C_3N_4-CdS composite catalyst with high visible-light-driven catalytic activity and photostability for methylene blue degradation. Applied Surface Science. 2014;**295**:164-172

[37] Liua L, Qi Y, Yang J, Cui W, Xingang Li ZZ. An AgI@g-C_3N_4 hybrid core@shell structure: Stable and enhanced photocatalytic degradation. Applied Surface Science. 2015;**358**: 319-327

[38] Zhou P, Yu J, Jaroniec M. All-solid-state Z-scheme photocatalytic systems. Advanced Materials. 2014;**26**(29): 4920-4935

[39] Khan A, Alam U, Raza W, Bahnemann D, Muneer M, Alam U. One-pot, self-assembled hydrothermal synthesis of 3D flower-like CuS/g-C_3N_4 composite with enhanced photocatalytic activity under visible-light irradiation. Journal of Physics and Chemistry of Solids. 2017;**115**:59-68

[40] Zhu B, Xia P, Li Y, Ho W, Yu J. Fabrication and photocatalytic activity enhanced mechanism of direct Z-scheme g-C_3N_4/Ag_2WO_4 photocatalyst applied surface science fabrication and photocatalytic activity enhanced mechanism of direct. Applied Surface Science. 2016;**391**:175-183

[41] Cui L, Ding X, Wang Y, Shi H, Huang L, Zuo Y, et al. Facile preparation of Z-scheme WO_3/g-C_3N_4 composite

photocatalyst with enhanced photocatalytic performance under visible light. Applied Surface Science. 2017;**391**:202-210

[42] Wang J, Li C, Cong J, Liu Z, Zhang H, Liang M, Gao J, et al. Facile synthesis of nanorod-type graphitic carbon nitride/Fe_2O_3 composite with enhanced photocatalytic performance. Journal of Solid State Chemistry. 2016; **238**:246-251

[43] Hong Y, Jiang Y, Li C, Fan W, Yan X, Yan M, et al. In-situ synthesis of direct solid-state Z-scheme V_2O_5/g-C_3N_4 heterojunctions with enhanced visible light efficiency in photocatalytic degradation of pollutants. Applied Catalysis B: Environmental. 2016;**180**: 663-673

[44] Chen W, Liu TY, Huang T, Liu XH, Zhu JW, Duan GR, et al. In situ fabrication of novel Z-scheme Bi_2WO_6 quantum dots/g-C_3N_4 ultrathin nanosheets heterostructures with improved photocatalytic activity. Applied Surface Science. 2015;**355**:379-387

[45] Danish M, Muneer M. Excellent visible-light-driven Ni-ZnS/g-C_3N_4 photocatalyst for enhanced pollutants degradation performance: Insight into the photocatalytic mechanism and adsorption isotherm. Applied Surface Science. 2021;**563**:150262

[46] Akhundi A, Habibi-Yangjeh A. Ternary g-C_3N_4/ZnO/AgCl nanocomposites: Synergistic collaboration on visible-light-driven activity in photodegradation of an organic pollutant. Applied Surface Science. 2015;**358**:261-269

[47] Danish M, Muneer M. Facile synthesis of highly efficient Co@ZnSQDs/g-C_3N_4/MWCNT nanocomposites and their photocatalytic potential for the degradation of RhB dye: Efficiency, degradation kinetics, and mechanism pathway. Ceramics International. 2021;**47**(9):13043-13056

[48] Athar MS, Danish M, Muneer M. Fabrication of visible light-responsive dual Z-scheme (α-Fe_2O_3/CdS/g-C_3N_4) ternary nanocomposites for enhanced photocatalytic performance and adsorption study in aqueous suspension. Journal of Environmental Chemical Engineering. 2021;**9**(4):105754

[49] Danish M, Athar MS, Ahmad I, Warshagha MZA, Rasool Z, Muneer M. Highly efficient and stable Fe_2O_3/g-C_3N_4/GO nanocomposite with Z-scheme electron transfer pathway: Role of photocatalytic activity and adsorption isotherm of organic pollutants in wastewater. Applied Surface Science. 2022;**604**:154604

[50] Ahmad I, Danish M, Khan A, Muneer M. One-pot hydrothermal synthesis of a double Z-scheme g-C_3N_4/AgI/β-$AgVO_3$ ternary nanocomposite for efficient degradation of organic pollutants and DPC–Cr(VI) complex under visible-light irradiation. Photochemical & Photobiological Sciences. 2022;**21**(8):1371-1386

[51] Athar MS, Saleem N, Ahmad I, Fazil M, Ahmad T, Muneer M. Z-scheme designed $LaNiO_3$/g-C_3N_4/MWCNT nanohybrid with bifunctional photocatalytic applications under visible light. Materials Today Sustainability. 2024;**1**(26):100779

[52] Gao Y, Lin J, Zhang Q, He YF, Xu DB, Yaguang Sun ZX. Facile synthesis of heterostructured YVO_4/g-C_3N_4/Ag photocatalysts with enhanced visible-light photocatalytic performance. Applied Catalysis B: Environmental. 2018;**224**:586-593

[53] Dai K, Lv J, Zhang J, Liang C, Zhu G. Band structure engineering design of g-C_3N_4/ZnS/SnS_2 ternary heterojunction visible-light photocatalyst with ZnS as electron transport buffer material. Journal of Alloys and Compounds. 2019; **778**:215-223

[54] Barzegar J, Habibi-Yangjeh A, Anise Akhundi SV. Novel ternary g-C_3N_4/Ag_3VO_4/AgBr nanocomposites with excellent visible-light-driven photocatalytic performance for environmental applications. Solid State Sciences. 2018;**78**:133-143

[55] Mousavi M, Habibi-Yangjeh A. Magnetically separable ternary g-C_3N_4/Fe_3O_4/BiOI nanocomposites: Novel visible-light-driven photocatalysts based on graphitic carbon nitride. Journal of Colloid and Interface Science. 2016;**465**: 83-92

[56] Habibi-Yangjeh A, Mousavi M. Deposition of $CuWO_4$ nanoparticles over g-C_3N_4/Fe_3O_4 nanocomposite: Novel magnetic photocatalysts with drastically enhanced performance under visible light. Advanced Powder Technology. 2018;**29**(6):1379-1392

Chapter 4

Research and Application of Sandwiched Dissimilar Metals

Huisheng Cai, Siqi Yang and Huaqing Li

Abstract

Metal matrix composites are widely used. The demand for laminated composite metal materials is increasing rapidly. Copper steel, copper aluminum, titanium aluminum, and other layered composite metals (LCMs) materials have become a variety of metals with excellent performance. As a kind of cladding composite material with excellent comprehensive properties, sandwiched dissimilar metals (SDMs) have broad application prospects in many fields. The electric connection materials of electric vehicles, the shell materials of power batteries, and the heat dissipation materials of AI servers all put forward new requirements for LCMs. SDMs guided by the needs of industrial applications will usher in rapid development. In this paper, the characteristics, preparative technique, classification, interface bonding mechanism, methods and approaches to improve the interface bonding strength, and the application of SDMs are systematically summarized and introduced. It also proposes targeted problems and solutions encountered in the research process of SDMs and finally prospects for the development trend of them.

Keywords: SDMs (sandwiched dissimilar metals), CMS (cladding metals), MMCs (metal matrix composites), LCMs (layered composite metals), CAC (copper-aluminum composite), CRC (cast-rolling composites)

1. Introduction

The origin of SDMs can be traced back to several hundred years BC. The Achilles Shield in 800 BC is one of the typical SDMs. The shield consists of two layers of bronze, two layers of tin, and one layer of gold, in the order of bronze/tin/gold/tin/bronze [1]. In addition, composite material technology was used in the preparation of King Yue's Gujian sword and the Damascus sword. Composite metal preparation technology makes sword performance more excellent.

As one of the metal composite preparation technologies, there are two main reasons for the emergence of SDMs. Firstly, the single material cannot meet the comprehensive performance requirements in specific scenarios. Secondly, precious metals are cladded on the surface of low-cost metals to reduce economic costs.

With the development of the economy and the emergence of various new technologies and new industries, the demand for engineering materials with various properties is becoming more and more urgent. The application fields of single metal

IntechOpen

materials are greatly challenged due to the limitations of natural resources or the insufficient comprehensive performance.

The positive and negative collectors of the power battery are aluminum foil and copper foil, respectively. When connecting to an external busbar, there is a part where copper and aluminum materials overlap. Because of the potential difference between copper and aluminum, the lap part is prone to corrosion. In order to avoid failure due to corrosion, conversion components made of copper-aluminum bimetallic materials are needed to link the battery to the external material.

For vehicle batteries, the battery cover plate should be of high strength, and the aluminum layer should be retained for welding with the shell, so it is necessary to develop a steel-aluminum bimetal.

AI computing server, heat dissipation requirements, cabinet shell, and chip parts of the heat sink material, need to improve strength and heat dissipation. The heat dissipation capacity of titanium is insufficient, and the heat conduction capacity of aluminum and copper is strong, but the strength is low. The rapid development of AI has led to the demand for titanium-aluminum and copper-aluminum bimetals, which is not only urgent but also a huge demand.

In this case, the importance of composite materials becomes more and more prominent. Composite materials can be generally divided into SDMs, cladding metal, particle-reinforced composites, and fiber-reinforced composites. Because of its high strength, good corrosion resistance, excellent thermal conductivity, and electrical conductivity, SDMs have been widely used in aerospace, petroleum, machinery, chemical industry, automobiles, ships, construction, electric power, electronics, home appliances, and other fields. It has great application potential in electric vehicles, energy storage batteries, 5G communication, artificial intelligence, and so on.

2. Characteristics of SDMs

SDMs are laminar composites that combine two or more metals with different physical, chemical, and mechanical properties and electrical conductivity through a preparation process; the microstructure and properties are designable. SDMs fully utilize the excellent properties of each layer and have better comprehensive performance than the single matrix materials. For example, Cu-Al cladding sheets comprehensively utilize the suitable corrosion resistance, conductivity, and heat dissipation performance of copper, as well as the advantages of low density, large heat capacity, and low cost of base metals.

SDMs are currently one of the most widely used cladding metals, and the interface bonding is closely related to the surface oxidation of the metal during the preparation process, especially in the high-temperature preparation process, such as hot rolling. Effectively reducing the surface oxidation during the productive process is crucial for preparing metal SDMs with good metallurgical bonding interfaces. In addition, the synergistic deformation during the preparation process and the subsequent heat treatment are also important aspects that need to be paid more attention in the preparation of SDMs. The significant difference in work hardening between the cladding metals usually reduces the interface bonding strength; sometimes even cracking and delamination may occur.

When the melting point temperature is close, it is favorable for laminar recombination. Otherwise, when the melting point difference between the base metal and the cladding metal is large, there are many negative effects on layered metals. Stress may occur at the interface, or coarse microstructure may be created, and even the deformation may occur during subsequent heat treatment.

Therefore, when designing and producing SDMs, it is necessary to fully consider the characteristics of cladding metals. Proper selection of materials is conducive to the production of layered metals. To obtain an extraordinary combination of properties, materials with vastly different intrinsic properties are often selected. It is a significant and challenging work to develop production processes for materials with widely varying properties.

3. Preparation technology of SDMs

The kinds of SDMs are plentiful, the application range is wide, and the preparation technology is various. According to the state of the matrix metal and cladding metal, the preparation technology of SDMs is mainly divided into solid-solid cladding forming, solid-liquid casting rolling forming, and liquid-liquid casting and/or rolling technology. The SDMs fabricated by adhesive technology and mechanical meshing technology also have wide engineering applications, but they are not introduced in this paper because no metallurgical bonding interface was formed.

With the progress of equipment technology, the requirements for material properties are constantly improving. Higher requirements have been put forward for SDMs. Integrating multiple fabrication technologies is essential for the flexible and efficient preparation of high-performance SDMs. For example, the Cu/Al SDMs used for the filter cover plate in 5G communication have high requirements for the surface quality and thickness percentage of copper. A single rolling preparation process cannot meet the requirements, and there is an urgent need to develop new composite technologies to meet application requirements.

3.1 Solid-solid cladding forming

Solid-solid cladding forming is one of the important production processes for SDMs, which can fabricate SDMs with wide widths but it is unsuitable for fabricating special-shaped parts with complex structures. The interface bonding strength of SDMs fabricated by solid-solid cladding technology is usually lower than that of roll casting items. There exists a troublesome residual stress between the interfaces. Annealing treatment is required to transform the mechanical bonding interface into a metallurgical layer through atomic diffusion, improve the microstructure of the interface, enhance the interface bonding strength, and eliminate processing stress. The solid-solid cladding technology mainly includes: roll cladding, extrusion cladding, welding cladding, explosive cladding, and other preparation technologies.

3.1.1 Roll cladding

The roll cladding is an important preparation technology that combines two or more pretreated metal plates by rolling them in a mill to make different materials combine with each other. Generally, the roll cladding technology is divided into cold rolling and hot rolling techniques according to the different roll cladding temperatures.

Cold rolling is the process of rolling metal plates through a rolling mill at room temperature. Under the strong pressure of the rolling mill, the laminated metals are bonded together to achieve a certain bonding strength at the cladding interface. The products after cold rolling need to be heat treated to optimize the interface microstructure and improve the interface bonding strength of the composites. The schematic diagram of cold rolling composite technology is shown in **Figure 1**. Different from the high temperature and high pressure of hot rolling technology, cold rolling technology is completed at room temperature, which can effectively avoid surface oxidation of the matrix metal and the cladding metal. However, in order to achieve a better cladding effect, the cold rolling forming often needs to increase the deformation of the first pass, and the reduction rate of the first pass is as high as about 70%. Such deformation needs high requirements for the rolling mill, and the difference in work hardening tendency should not be too large between the cladding metals. Therefore, the application range of cold roll cladding products is limited. Li et al. [2] summarized that the metal cladding that can be successfully prepared into SDMs through cold bonding technology such as cold rolling or accumulative roll bonding (ARB). Metals with face-centered cubic (FCC) lattice structures are suitable for cold bonding due to their lower work hardening rate, such as aluminum, copper, gold, silver, and platinum. Metals with hexagonal close packing (HCP) lattice structures are difficult to be clad by cold compounding, such as magnesium, cadmium, and zirconium, as shown in **Figure 2**.

During the hot rolling process, the matrix metal and the cladding metal are heated to a suitable temperature range (above the recrystallization temperature), and then rolling deformation is performed with a large reduction. The contact interfaces between cladding metals are subjected to the combined action of heat and force to form interfaces with good metallurgical bonding. The schematic diagram of the hot rolling process is shown in **Figure 3**. Hot rolling technique effectively solves the problem of limited application range of cold rolling technology. Hot rolling produces microstructure rheology and dynamic atomic diffusion due to the high energy of metal atoms, thus the SDMs can be reduced or exempted from subsequent annealing treatment. Because of the good plastic deformation of metal at high temperatures, hot rolling preparation technology can also be used for hard metals cladding. Hot rolling technique has been used in industrial production since the 1950s. However, due to high temperatures, two issues need to be noted during the hot rolling preparation process. (1) During the high-temperature heating process, the surface oxidation, especially for easily oxidizable metals, will affect the interface bonding strength. In severe cases, interface delamination will occur, leading to product scrap; (2) The final

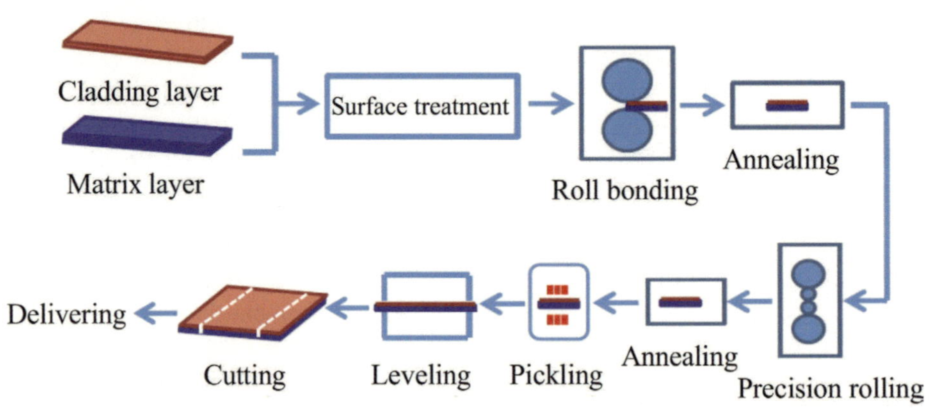

Figure 1.
The schematic diagram of cold rolling composite technology.

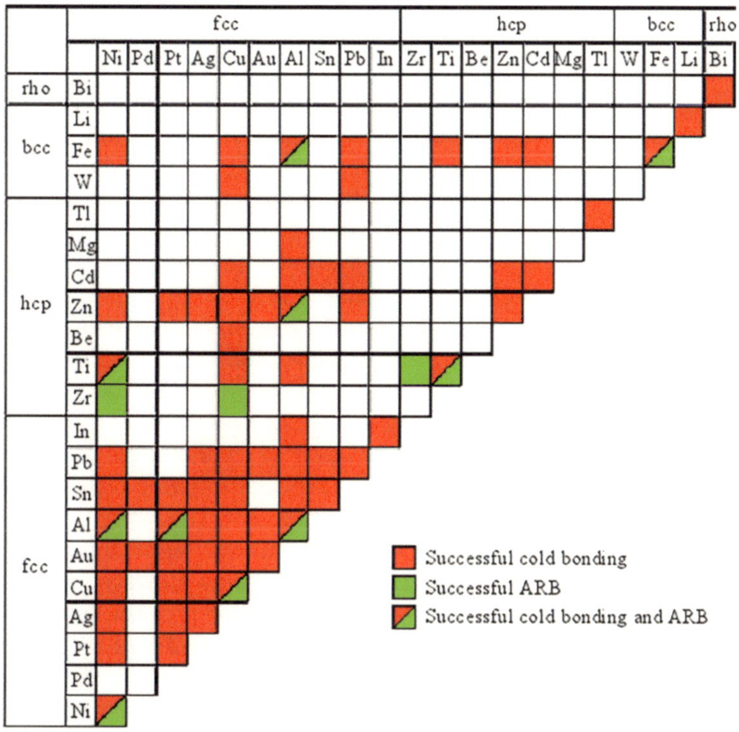

Figure 2.
Chart of metals suitable for cold bonding by rolling and/or by applying pressure and ARB, according to lattice structure and hardness of metals [2].

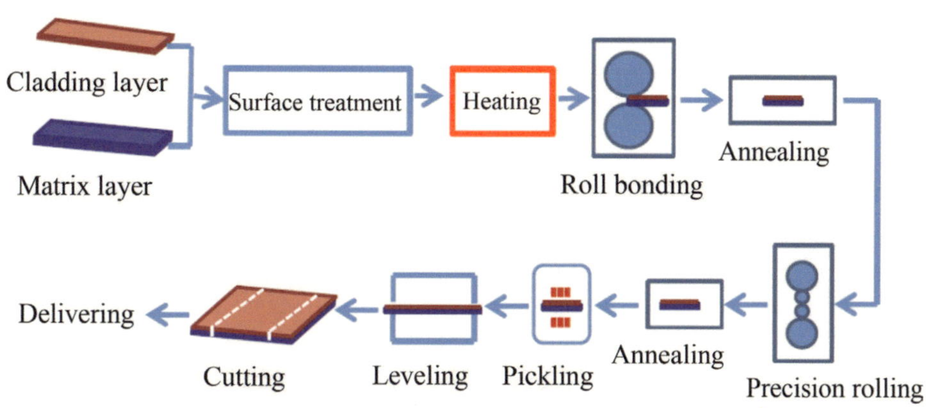

Figure 3.
The schematic diagram of hot rolling cladding technique.

rolling temperature of SDMs is usually strictly limited to ensure the interface bonding strength. Thus, continuous heating and insulation devices are necessary; otherwise, the length of continuous rolling of the metal plates is limited.

Soviet scientists carried out systematic research on the vacuum rolling process between easily oxidized metals and made a major breakthrough. JFE Corporation (Japan) has developed a vacuum rolling technology by combining vacuum electron beam welding technology with hot rolling. Although vacuum rolling can effectively alleviate the oxidation problem on the surface of metal plates, the rolling process is complex and cumbersome. In recent years, a new type of high-temperature, oxygen-free hot rolling composite technology has been gradually developed. This method greatly reduces the oxidation problem of the interface during the hot rolling preparation process by controlling the atmosphere, so as to prepare SDMs with a better metallurgical bonding interface. However, the equipment and process are also more complex, which limits the application and promotion.

In addition, Saito et al. [3] proposed a more cost-effective accumulative roll bonding (ARB) technology based on rolling composite technology for preparing SDMs. In this preparation process, the plates after surface pretreatment are accumulated in a certain order, and one end is punched and riveted, followed by rolling with a large pass reduction. The first rolled plates are cut, and the process of "surface pretreatment, accumulation, riveting, and rolling" is repeated many times to prepare SDMs. The schematic diagram of accumulative roll bonding technology is shown in **Figure 4**. ARB composite technology is a kind of large plastic deformation technology with the advantages of simple operation and low production cost. The strength of SDMs prepared by ARB can be improved by continuously accumulating strain, crushing grains, and forming a large number of fine-grain microstructures. However, the metal plates will undergo severe plastic deformation during the rolling process, and the increase in strength is often accompanied by a decrease in plasticity. Adding hard plates during the preparation of ARB can change the stress state of the metal plates during rolling, significantly affecting the interface bonding strength, mechanical properties, microstructure, and interface morphology of SDMs. The hard plates can convert the shear stress in the rolling direction into normal compressive stress, effectively increasing the interface bonding strength [5, 6].

The popularity of artificial intelligence, especially big data computing, has greatly improved the server's requirements for material heat dissipation performance. Traditional server shell materials require higher strength and excellent heat dissipation performance. Therefore, titanium-aluminum laminate composite materials

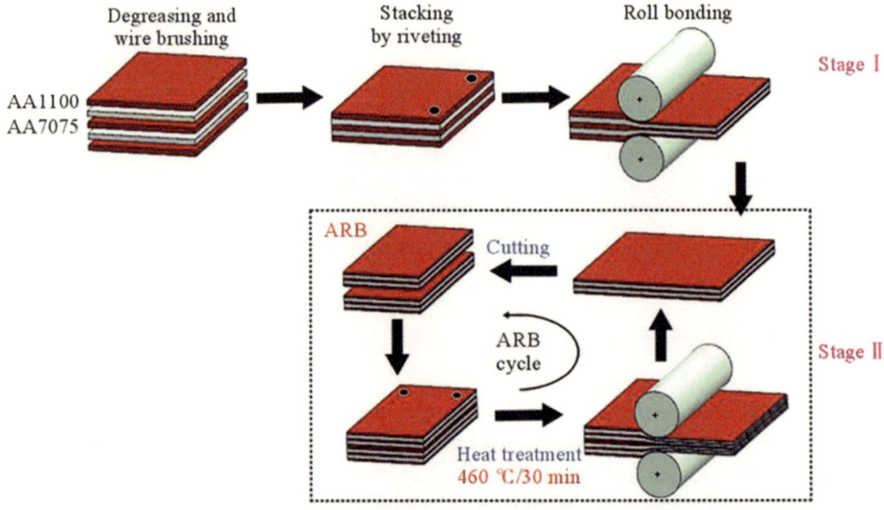

Figure 4.
The schematic diagram of accumulative roll bonding composite technology [4].

began to be applied to server hardware. Due to the rapid development of artificial intelligence and a large number of equipment updates, the demand for this material will explode. Equipment and process technology need to be updated as soon as possible to keep up with the pace of big data developments.

Rolling equipment is very popular in the steel and non-ferrous industries. The level of equipment is increasing day by day. The strip has high rolling efficiency, high precision, and excellent product performance. The rolling process is an ideal technical route to develop bimetal and multilayer metal production.

3.1.2 Extrusion cladding forming

Extrusion cladding forming is an important solid-solid composite technology. Firstly, the pretreated heterogeneous metal materials were prepared into extrusion blanks. Then the extrusion blank is subjected to severe plastic deformation under high temperature and high pressure so that the heterogeneous metal interface forms a strong metallurgical bond. When the extrusion composite technology is used to prepare the sandwich dissimilar material, the metal will have severe plastic deformation, which can significantly reduce the porosity and porosity defects, so as to improve the performance of SDMs.

There are a large number of banded deformation microstructures and fine recrystallized grains in SDMs prepared by extrusion cladding technology, and the interface with good metallurgical bonding will be formed, which can significantly improve the mechanical properties and interface bonding strength of the composites.

The extrusion cladding forming is mainly divided into horizontal extrusion, hydrostatic extrusion, and continuous extrusion processes. During the preparation, fresh contact surfaces will be formed at the interface, which will promote element diffusion under high temperature and high pressure, thus facilitating the formation of a metallurgical bonding interface. However, the preparation of SDMs by extrusion cladding has high requirements for the cladding metals. First of all, the material with close deformation resistance should be preferred. If the deformation resistance difference is great, the extension consistency between layers decreases, and the surface evenness of SDMs is difficult to control. Secondly, it is necessary to ensure that the two metals are in a suitable plastic deformation state during the preparation process. When the amount of deformation exceeds the elongation of one layer, fracture occurs. It also causes defects such as uneven wall thickness and macroscopic performance fluctuation of the composites. The hydrostatic extrusion transfers the pressure to the billets through the liquid, which can effectively ensure the plastic deformation uniformity, so as to obtain an excellent cladding effect. Continuous extrusion is an important preparation technology for the wires. The production process is simple and can achieve continuous production on the premise of ensuring the cladding with a good metallurgical bonding interface.

During the extrusion process, the material is subjected to three compressive stresses. So, the material is dense inside. The SDMs produced by the extrusion process have high interfacial binding strength. It is one of the ideal lamellar metal production processes and is more suitable for small-section SDM products.

3.1.3 Welding cladding technology

Welding cladding technology is an important solid-solid cladding technology. The interface of the cladding metals prepared by welding technology is a metallurgical interface, and the metals can achieve a good bonding effect. Welding cladding technology mainly includes seam welding, diffusion welding, friction stir welding, and so on.

Seam welding cladding uses a rotating disk-shaped electrode for continuous resistance welding to prepare SDMs. Welding cladding can obtain SDMs with large roll weight, which is an efficient production technology. The characteristics of seam welding cladding technology are that the higher interface strength between the metals, automated production, and low cost.

The diffusion welding cladding technology is an important welding technology in which the pretreated matrix metal and cladded metal are placed at a temperature between 0.5 and 0.8 Tm, and a certain pressure is applied to form a metallurgical interface through element diffusion. The applied pressure should be less than the yield strength of the matrix metal and the cladded metal to ensure that the metals do not deform during the process of cladding. The diffusion welding technology is mainly suitable for the production of titanium/stainless steel, titanium/steel, titanium/titanium, titanium/copper, etc., which are difficult to be prepared by the seam welding technology. Diffusion welding is suitable for a variety of materials; almost all metals can be welded, and defects such as cracks and segregation can be effectively avoided. However, the mechanical properties of the welding cladding products are poor, the production efficiency is low, and continuous production cannot be achieved.

The friction stir welding cladding technology has shown great advantages in the connection of heterogeneous materials and has broad market application prospects. During the friction stir welding process, the matrix metal remains in a plastic state, which can effectively avoid the defects caused by fusion welding. The schematic diagram of friction stir welding composite technology is shown in **Figure 5**.

Intermetallic will be formed at the interface of the SDMs prepared by the friction stir welding technology, which improves the cladding strength. The parameters such as the rotation speed, the welding speed, the position, and the thickness of the matrix metal and cladded metal directly affect the product quality.

Due to the easy formation of brittle intermetallic compounds, the welding process parameters should be strictly controlled. Avoid forming brittle intermetallic compound particles or regions at the interface so as not to affect the cladding effect.

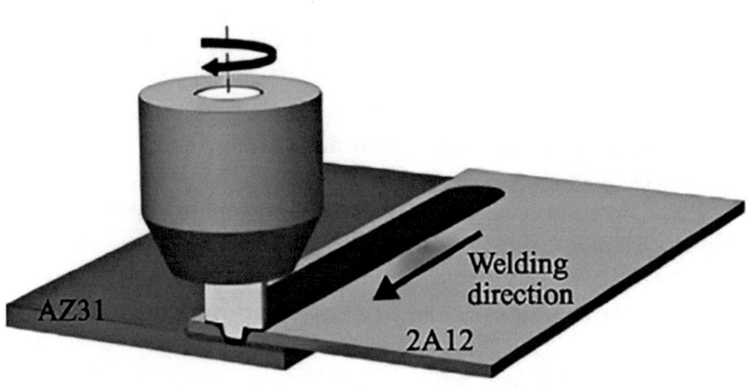

Figure 5.
The schematic diagram of friction stir welding composite technology.

The preparation of SDMs by friction stir welding composite technology has the characteristics of high efficiency, being environmentally friendly, and reliable quality, and will play an increasingly important role in the field of SDM preparation.

3.1.4 Explosive cladding

The explosive cladding technology utilizes the explosive force to generate a strain rate of up to 10^6–10^7 s^{-1} and a high pressure of 10^4 MPa near the collision point of two metal plates in a very short time, thereby achieving welding composite. The schematic diagram of explosive composite technology is shown in **Figure 6**. Explosive composite will form a very thin plastic deformation zone on the composite surface of two metal plates in a very short time, and there will be both metallurgical and mechanical bonding at the interface. The advantages of explosive composites are that the production process is simple, which can achieve a good composite of metals with large differences in thermal expansion coefficients.

The composite interface is formed at a lower temperature, which suppresses the formation of brittle intermetallic. In addition, the instantaneous high pressure generated by the explosion can effectively damage the oxide film on the surface of the metal and improve the interface bonding strength.

The process parameters, such as the space between metal plates, the thickness of explosives, and the explosion velocity during the preparation of SDMs by explosive composite technology, will affect the performance of composites, and the influence of the thickness of the explosive is the most significant. If the thickness of the explosive is too small, the pressure generated by the explosion is small, which is easy to cause a poor composite effect. If the thickness of the explosive is too large, it may cause the metal plate to melt and form a large number of intermetallics and impurities, which will also affect the composite effect. The interface of SDMs prepared by explosive welding is mainly divided into wave layer, flat layer, and melting layer. The formation of different interface layers mainly depends on the collision speed and collision angle. The interface of the wave layer can be obtained by reasonably adjusting the collision speed and collision angle. Due to the mechanical interlocking of the wave layer, the interface composite effect is better than that of the flat layer and the melting layer [7]. In the case of a lower explosive ratio, the composite interface is flat, while in the case of a higher explosive ratio, the composite interface is wavy. The grains near the interface are generally elongated along the direction parallel to the explosion, due to plastic deformation caused by the explosion wave.

However, explosive composite technology requires the use of explosives, and there are safety hazards, environmental pollution, noise pollution, and other issues during

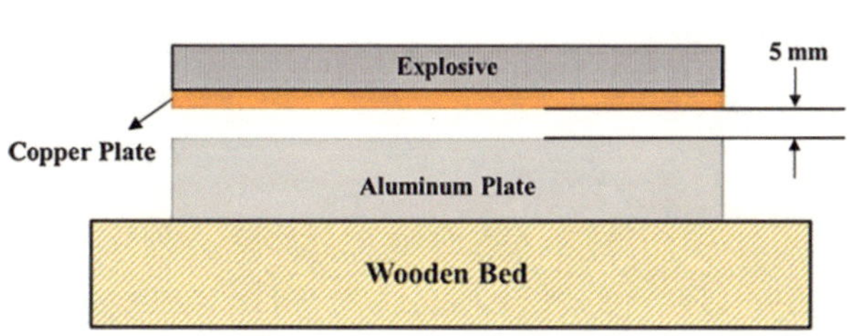

Figure 6.
The schematic diagram of explosive composite technology.

the preparation process, which limits the application and popularization of explosive composite technology in the preparation of SDMs.

3.2 Solid-liquid composite technology

Solid-liquid composite technology is an important composite technology for preparing SDMs by contacting solid metal and liquid metal. The interface of SDMs prepared by solid-liquid composite technology will be formed through fusion bonding and diffusion bonding. The interface bonding mode is mainly metallurgical bonding, and the interface bonding strength is high. The control of the interface temperature field is particularly crucial during the solid-liquid preparation process. Higher temperature can cause local melting of solid metals, reducing interface stability and uniformity, while lower temperature may limit the formation of metallurgical bonding interfaces. The solid-liquid composite technology mainly includes cast-rolling composite technology and immersion composite technology.

3.2.1 Cast-rolling composite technology

Cast-rolling composite technology is an important solid-liquid composite technology that continuously pours liquid metal onto the surface of a solid metal plate and then enters the rolling mill together to prepare SDMs. Liquid metal undergoes two processes of solidification and hot rolling between the two roller gaps. Therefore, the cast-rolling composite technology is a short-process and high-efficiency composite technology that combines casting and rolling technology to prepare SDMs. The schematic diagram of cast-rolling composite technology is shown in **Figure 7**. Cast-rolling composite technology utilizes the high pressure of the rolling mill and the high temperature of the liquid metal, and the interdiffusion of metal atoms can be fully carried out, thereby forming a metallurgical bonding interface with high interface bonding strength. The cast-rolling composite technology can be divided into horizontal cast-rolling, vertical cast-rolling, and inclined cast-rolling. The process parameters such as cast-rolling speed, solid metal preheating temperature, liquid metal pouring temperature, and roller gap space have important influences

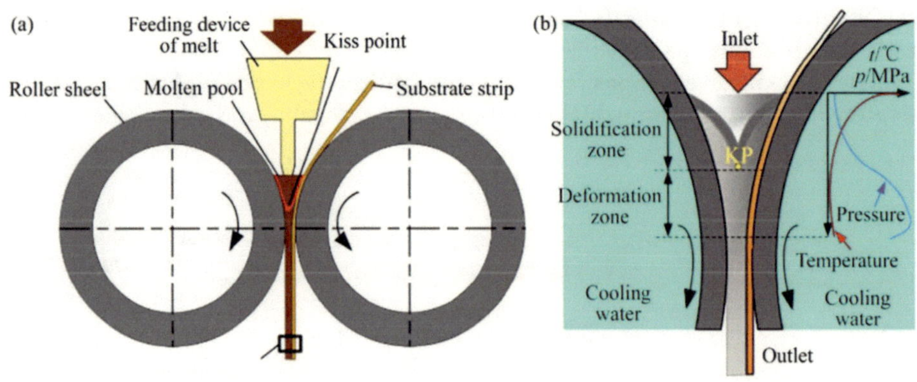

Figure 7.
Schematic diagram of cast-rolling composite technology. (a) Solid-liquid cast-rolling bonding; (b) cast-rolling zone [8].

on the final composite effect of the SDMs. The preparation of SDMs by cast-rolling composite technology has the advantages of a simple production process and high production efficiency.

The interface bonding mechanism of SDMs prepared by the cast-rolling composite technology is mainly to form a metallurgical bonding interface through atomic diffusion and reaction diffusion. The cast-rolling zone can be divided into the casting zone and the rolling zone. The driving force of diffusion depends on the heat and pressure of the casting zone and the rolling zone. When the liquid metal is poured onto the surface of the preheated solid metal plate, under the combined action of sensible heat and latent heat of solidification, the atoms on the surface of the solid metal obtain energy, and the local atoms will be violently active. Interdiffusion will occur between the solid-activated atoms and the liquid metal atoms, thereby forming a thin diffusion transition layer at the interface. As atomic diffusion progresses, the atomic concentration at some positions may exceed the solid solubility of atoms. At this time, chemical reactions may occur, and the intermetallics with the smallest Gibbs free energy will be preferentially formed. The growth process of intermetallics is mainly divided into four stages: physical contact, intermetallic nucleation, intermetallics connected along the interface, and the increase of intermetallics layer thickness. The interface of the SDMs prepared by cast-rolling is mainly a continuous metallurgical bonding interface, which has good interface bonding strength.

Because casting and rolling hair has matured in the field of aluminum alloy plates and strips, it has been popularized. The equipment conditions are provided for the development of copper-aluminum bimetallic belts.

The positive and negative electrode conversion material of an automotive battery is the copper-aluminum bimetal plate produced by this technology. Due to high production efficiency and excellent consistency, the technology has been rapidly popularized.

3.2.2 Immersion composite technology

The immersion composite technology refers to the process of immersing the plated workpiece into the molten metal liquid and then taking it out and cooling it to form a metal coating on the surface. It is mainly suitable for preparing the coatings of low-melting-point metals: zinc, tin, aluminum, lead, and their alloys. The most common preparation process is hot-dip, which is a simple and effective method for preparing SDMs. In order to improve the preparation effect of SDMs, the plated workpiece should be strictly pretreated, the surface of the plated workpiece should be cleaned, and the surface oxides should be removed to make the surface appear fresh matrix metal. The composites prepared by hot-dip can improve the corrosion resistance and oxidation resistance of the matrix material and have good interface bonding. However, the thickness of the coating metal is thinner, which is prone to uneven coating, and the application field is limited.

Hot-dip plating has high production efficiency and is suitable for continuous production. Due to the characteristics of the equipment, it is difficult to adjust the process parameters. Therefore, it is more suitable for products with stable processes and fixed models.

This technology has been widely used in the field of copper processing and the production process of steel strips for construction.

3.3 Liquid-liquid composite technology

Regardless of solid-solid composite technology or solid-liquid composite technology, the preparation process of SDMs has higher requirements on the surface of solid metal, which requires strict pretreatment, resulting in low production and processing efficiency. Based on this, liquid-liquid composite has gradually developed. This composite technology has lower requirements for the pretreatment process of materials and a shorter process. It is an efficient composite technology for preparing SDMs. Liquid-liquid composite technology mainly includes core-filling continuous casting composite technology and liquid-liquid cast-rolling composite technology.

3.3.1 Core-filling continuous casting composite technology

The core-filling continuous casting composite technology is a process for preparing high-melting-point metal-coated low-melting-point metal composites. The core metal liquid is filled into the outer metal shell of the continuous casting and solidified to achieve the combination of the two metals. It is mainly divided into vertical core-filling continuous casting and horizontal core-filling continuous casting. The schematic diagram of core-filling continuous casting composite technology is shown in **Figure 8**. The core metal and the outer metal are heated to different temperatures in the two heaters, respectively, for melting. The outer metal liquid enters the crystallizer to form a solidified shell of a certain thickness. The core metal liquid enters the crystallizer through the mandrel tube and fills the solidified shell of the outer metal liquid. The outer metal billet is pulled down at a certain speed under the traction of the dummy bar to ensure that the outer metal billet does not leak, thereby realizing the composite of the core metal and the outer metal.

The process parameters, such as the temperature of the two metal liquids, the casting speed, the cooling strength of the crystallizer, the length, and the taper of the mandrel tube, directly affect the composite effect of the composites. The core-filling continuous casting composite technology has the advantages of a simple process, high production efficiency, and easy realization of continuous and automated processes. However, due to the particularity of the composite technology, the prepared composites are relatively simple and mainly used for the preparation of coated composites, such as copper-coated aluminum composites. The bonding mechanism of the composites prepared by the core-filling continuous casting composite technology is essentially a solid-liquid composite. The formation process of the interface intermetallics is closely related to the interface thermal history. The cooling intensity of the crystallizer and the secondary cooling process have a great influence on the interface bonding.

The characteristics of the core-filling continuous casting composite technology are that the outer metal with a high melting point first solidifies into a tubular billet in the crystallizer, and then the tubular billet is pulled down, and the core metal liquid is filled into the inner cavity of the tubular billet and gradually solidified in the tubular billet, so as to realize the bonding of the two metals. With the continuous development of composite technology, various new core-filling continuous casting composite technologies have been developed. For example, in order to solve the oxidation during the preparation process, a core-filling continuous casting technology that can achieve bonding in an oxygen-free environment has been developed, which greatly improves the interface bonding strength of the composites. In addition,

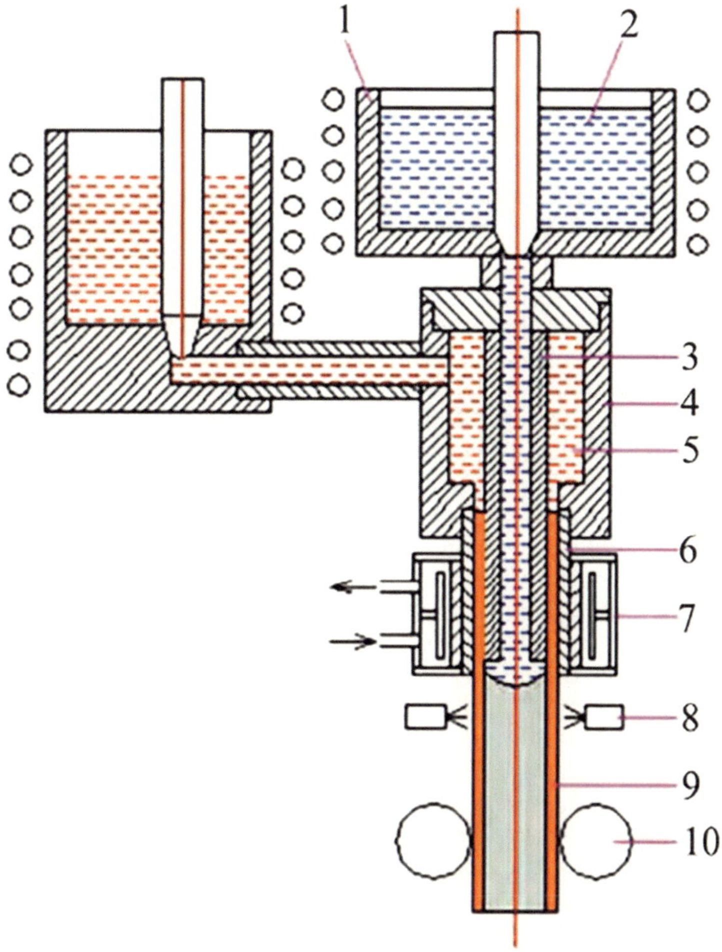

Figure 8.
Schematic diagram of core-filling continuous casting composite technology [9]. 1—core metal liquid holding furnace, 2—core metal liquid, 3—graphite mandrel tube, 4—outer metal liquid holding furnace, 5—outer metal liquid, 6—graphite casting mold, 7—crystallizer, 8—secondary cooling water, 9—composite casting rod, 10—drawing machine.

the core-filling continuous casting composite technology with double crystallizers has been developed, and the composites have good surface quality and an excellent interface composite effect.

The casting process changes metal from liquid to solid. The interfacial atoms are more active, which makes it easy to form metallurgical bonding, and the thickness of the bimetallic transition layer is higher by then. For materials that are easy to form brittle phases, attention should be paid to composition design, and appropriate process parameters should be selected. In order to obtain a comprehensive performance of excellent products.

3.3.2 Liquid-liquid cast-rolling composite technology

Liquid-liquid cast-rolling composite technology combines the advantages of solid-liquid cast-rolling composite technology and core-filling continuous casting composite technology. Unlike solid-liquid cast-rolling composites, there is no solid metal plate during the preparation process, and both the matrix metal and the cladding metal are molten. Different from the core-filling continuous casting composite technology, the preparation process needs to be rolled by the rolling mill, and the composite interface is formed under large pressure. Two-layer, three-layer, or even multilayer composites can be prepared by liquid-liquid cast-rolling composite technology. The schematic diagram of liquid-liquid cast-rolling composite technology is shown in **Figure 9**. The interface bonding effect of composites prepared by liquid-liquid cast-rolling is usually good and has excellent interface bonding strength. However, it is necessary to reasonably control the process parameters such as metal liquid temperature, rolling speed, and melt flow rate during the cast-rolling process to prepare composites with a better composite effect. In order to achieve a better composite effect, the solidification process of the metal liquid should be reasonably controlled. If the matrix metal liquid has already solidified before entering the flow channel, it will block the flow channel and prevent the subsequent preparation process. If the matrix metal liquid still does not solidify after entering the channel, it may mix with the cladding metal liquid, leading to the failure of the cast-rolling composite. If the cladding metal liquid is not completely solidified in the composite zone, it will flow out of the roller gap, which can also cause composite failure. The liquid-liquid cast-rolling composite technology eliminates the pretreatment of alloy plates, simplifies the process, and effectively reduces the occurrence of interface oxidation, and the prepared composites have a good interface bonding effect. However, the process is complex and limits its application.

The interface bonding of composite materials prepared by the liquid-liquid casting and rolling process is similar to solid-liquid interface bonding. The interface formation process is as follows. After the metal flowing through the front casting roll solidifies, the metal plate comes into contact with the metal liquid, and inter-atomic diffusion occurs, forming a diffusion layer with an atomic concentration gradient. When the atomic concentration exceeds the solid solubility, effective diffusion occurs at the interface, forming an intermetallic compound layer. Temperature, residence time, and rolling force affect the diffusion effect, which indirectly affects the thickness of the intermetallic compound layer.

3.4 Other composite technology

In addition to the above-mentioned composite technologies of SDMs, the composites can also be fabricated by surface treatment techniques such as vapor deposition,

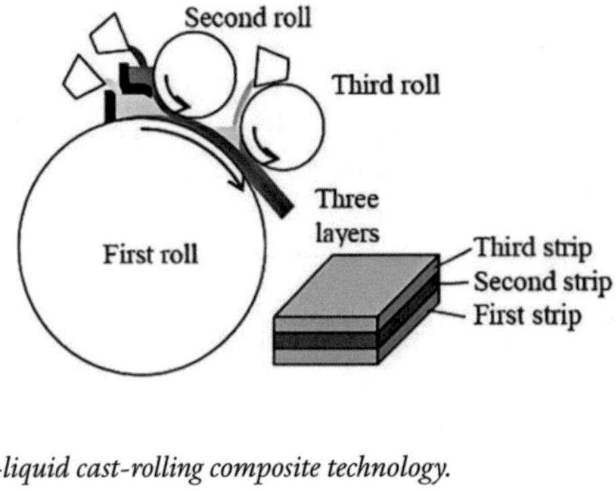

Figure 9.
Schematic diagram of liquid-liquid cast-rolling composite technology.

electroplating, anodic oxidation, 3D printing (additive manufacturing), surfacing welding, and spraying.

Evaporation coating is a kind of vapor deposition technology. The film material is heated to evaporate under vacuum conditions and then vaporized and deposited on the surface of the substrate to form a clean and dense film of metal, alloy, or compound. Sputtering coating is also a kind of vapor deposition technology. It uses particles with certain kinetic energy to bombard the target. The atoms on the surface of the target will be sputtered out. The moving direction and path of the sputtered atoms are controlled by electric fields or magnetic fields so that they can be deposited into a film to prepare SDMs. Multilayer metal or non-metal films with alternating compositions can be prepared by using different targets or intermittently introducing reactive gas during sputtering. In addition, molecular beam epitaxy can also prepare film. Although the above composite technology can be used to prepare SDMs, the prepared film is thin, and the film growth rate is low. However, these composite technologies have great advantages in the preparation of functional films and the preparation of functional SDMs.

Electroplating is a process in which a dense and uniform metal layer is formed on the surface of metal or non-metal by an electrochemical reaction at the interface between the electrode and the solution after direct current is introduced into the electrolyte. In industry, zinc plating, copper plating, chromium plating, and alloy plating are usually used to prepare SDMs with certain surface properties. In addition, SDMs can also be prepared by anodic oxidation. The anodic oxide film has good mechanical properties and corrosion resistance, as well as good adsorption, which can be decorated by dyeing treatment. However, the treatment of electroplating solution and electrolyte and the possible pollution problems also restrict its application development and popularization.

3D printing (additive manufacturing) is a rapid prototyping technology developed in the 1980s. Using its layer-by-layer printing characteristics, the SDMs can be prepared by changing the types of powder or wire. However, 3D printing has many barriers in mass production and has certain advantages in small-scale manufacturing, especially in high-end customized products. The SDMs can also be prepared by surface welding and laser cladding. Spraying technology is divided into cold spraying and thermal spraying. Thermal spraying is a technology that uses a certain heat source to heat the sprayed material to a molten or semi-molten state, atomize it with a high-pressure airflow, and spray it onto a pretreated workpiece surface with a certain speed to form a coating. The coating and the substrate constitute SDMs with different compositions and properties. Generally, the thickness of the coating prepared by spraying is thicker, but the interface bonding effect is poor. The interface bonding is mainly mechanical bonding with a small amount of metallurgical bonding.

3.5 Combined composite technology

As mentioned above, the composite technologies of SDMs are various, and the appropriate composite technology can be selected according to the characteristics and requirements of the matrix metal and the cladding metal. However, different types of composite technologies have their own advantages and disadvantages, and the application scenario is relatively single. In order to further improve the interface composite effect of SDMs and give full play to the advantages of various composite technologies, the combined composite technology has been developed. For example, the combination of various solid-solid composite technologies, and the combination of solid-liquid composite technology and solid-solid composite technology, the combination of liquid-liquid composite technology and solid-solid composite technology. The advantages and disadvantages of the combined composite technology for preparing SDMs are shown in **Table 1**. It can be seen from the table that the first-step composite technology in the combined composite technology has prepared the composites. The second-step composite technology is a further improvement of the first-step composite technology, aiming at improving the composite effect of the composites.

4. Classification of SDMs

4.1 Copper matrix cladding

4.1.1 Copper-aluminum cladding

Copper-aluminum cladding is an important SDM and has the characteristics of higher electric conductivity and thermal conductivity than aluminum, lower density than copper, and lower cost. The interface is metallurgical bonding, which can effectively reduce the electrochemical corrosion at the interface. The copper-aluminum cladding is mainly prepared by solid-solid composite method and solid-liquid composite method. The solid-solid composite method mainly includes the explosive composite method, the welding composite method, the rolling composite method, the extrusion composite method, etc. The solid-solid composite method has problems such as a complex process, high production cost, low interface bonding strength, and poor product quality stability. The solid-liquid composite method includes cast-rolling, core-filling continuous casting, hot-dip, etc. The solid-liquid composite method is limited by the types of aluminum alloys and has fewer production varieties. At present, copper-aluminum cladding products of 1xxx series, 3xxx series, and 8xxx series of aluminum alloys can be produced; copper-aluminum cladding products of 2xxx series, 4xxx series, 5xxx series, 6xxx series, and 7xxxx series cannot be produced. The thickness of

Combined composite technology	Advantages	Disadvantages
Powder metallurgy + rolling	Process simple, low cost, expanded types of SDMs.	Interface control difficulty.
Cast-rolling + rolling	Fewer defects, stable process, higher production efficiency, and low production cost.	Susceptible to air humidity.
Core-filling continuous casting + rolling	Higher interface bonding strength.	Thicker interface, process complex, single size, limited application.
Core-filling continuous casting + drawing	Higher product qualification rate, good electrical conductivity, higher interface bonding strength.	Process complex, single size, limited application.
Casting + cold extrusion	Low cost, fine grain.	Low production efficiency.
Explosive + rolling	Wide range of applications, excellent mechanical properties, fast forming.	Environmental pollution, security risk, control difficultly.

Table 1.
The advantages and disadvantages of the combined composite technology.

the copper layer is limited. Currently, copper-aluminum cladding has a copper layer volume fraction of 1.5–75%. Each preparation process has its unique advantages, and the copper-aluminum cladding prepared also has its own characteristics. However, with the development of technology, there is an increasing demand for copper-aluminum cladding. It is urgent to fully tap into the potential of various preparation processes and comprehensively utilize the advantages of various single preparation processes to prepare multifunctional and cost-effective copper-aluminum cladding.

Copper-aluminum cladding is gradually becoming an effective alternative to copper or copper alloys in more and more fields while ensuring their electric conductivity, thermal conductivity, and other properties. In the background of cost reduction and efficiency increase, copper-aluminum cladding is bound to be more widely promoted and applied in power batteries, energy storage power stations, photovoltaic power generation, power, electronics, communication devices, LED lighting, building decoration, and other fields.

4.1.2 Copper-steel cladding

Copper-steel cladding combines the performance characteristics of copper and steel. It not only has the advantages of good electrical conductivity, thermal conductivity, and beautiful decoration of copper but also has the characteristics of high strength of steel. The composites can reduce the amount of precious metals, and the cost is also effectively reduced. Copper-steel cladding is mainly prepared by explosive bonding, explosive + rolling bonding, and diffusion bonding. Copper-steel cladding has been widely used in various fields, and its performance has been validated. From both practical and economic perspectives, copper-steel cladding has great development prospects. In the civilian field, copper-steel cladding has good applications in the fields of cookware and building decoration materials.

Copper-steel cladding prepared using high-performance copper alloys instead of copper or copper alloys for roofs and door and window frames is not only more durable but also lower in cost. In the field of electrical and electronics, copper-steel cladding is widely used in electronic components, such as conductive springs made of traditional copper or copper alloys. By replacing them with copper-steel cladding wires, it ensures good conductivity and sufficient flexibility while also significantly reducing costs. In addition, copper-steel cladding can be used in the automotive industry to manufacture components such as gearbox worm gears and bushings. In the metallurgical industry, copper-steel cladding can be used as electrodes for electroplating technology, etc. In short, with the continuous development of science and technology, the requirements for material properties are becoming increasingly strict, and copper-steel cladding will also play an increasingly important role.

Copper and copper alloys are typical functional materials. It has excellent electrical and thermal conductivity. If it is to be used as a structural functional material. The lamination of copper alloy with steel, stainless steel, and other materials is an ideal solution. To reduce weight, it can be combined with aluminum alloy to form a layered metal.

4.2 Aluminum matrix cladding

4.2.1 Aluminum-aluminum cladding

Aluminum alloy is one of the lightest metal structural materials, which has the advantages of low density, high specific strength, good oxidation resistance, and corrosion resistance. However, for some special application scenarios, a single aluminum alloy material can no longer meet the requirements for alloy performance. Aluminum-aluminum cladding has been widely used in more and more fields due to its lightweight, designability, and strong impact resistance. They are mainly prepared by rolling composite, explosive composite, and casting composite.

The application of aluminum-aluminum cladding in the field of armored materials has strong advantages. It is usually prepared with aluminum alloys with higher strength and aluminum alloys with lower strength but better toughness and weldability. The United States has developed 7039/7005 and 7039/5083 aluminum-aluminum cladding for tanks; Italy has developed a composite armor plate suitable for light armored combat vehicles. The core of the plate is made of Al-Zn-Mg-Cu alloy, and the panel and back plate are made of Al-Zn-Mg alloy. Increasing the interlayer gap will reduce the bulletproof performance of the aluminum-aluminum cladding. The properties, thickness, stacking sequence, and number of layers of the matrix metal and cladding metal have significant effects on the bulletproof performance of the armor material.

7A52 aluminum alloy belongs to the Al-Zn-Mg heat treatment, which strengthens high-strength aluminum alloy. It has a high specific strength and good fracture toughness and has been widely used in welding structural parts of armored vehicles, gun racks, tanks, and other weapons and equipment. 7A62 aluminum alloy is developed by further adding Zn and Mg alloy elements into 7A52 aluminum alloy, which has higher strength, better hot workability, and corrosion resistance. 7A01 aluminum alloy has good ductility and formability, but its strength is low. The aluminum-aluminum cladding prepared by hot rolling of these three alloys, which retained the good physical, chemical, and mechanical properties of each layer, has excellent protective performance.

4.2.2 Aluminum-steel cladding

Aluminum-steel cladding not only has the characteristics of high strength and low cost of steel, but also has the characteristics of corrosion resistance, low density, and good electrical and thermal conductivity of aluminum. It is widely used in light industry, shipbuilding, metallurgy, aviation, transportation, energy, and other fields. There are many preparation methods of aluminum-steel cladding, including explosive composite, rolling composite, extrusion composite, casting composite, diffusion composite, hot-dipping composite, and so on. However, there are significant differences in physical and chemical properties such as melting point, coefficient of linear expansion, thermal conductivity, and mechanical properties between aluminum and steel. During the preparation process of aluminum-steel cladding, thermal stress is high and cracks are prone to occur. Considering the composite effect and the adaptability of industrial production, it is mainly industrially produced by explosive composite, casting composite, and rolling composite.

Aluminum-steel cladding forms a continuous diffusion zone through the interdiffusion of aluminum and steel atoms at the interface; that is, the metallurgical bonding interface is formed. Whether the metallurgical bonding interface is formed, the distribution and thickness of the intermetallic interface directly determine the composite effect of aluminum-steel cladding. The explosive composite method has the advantages of small equipment investment and a simple process, but the process parameters are difficult to control accurately, there are potential safety hazards, and there is high noise, which limits its application. For the casting composite method, there are also some difficulties in its preparation process. Firstly, the metal surface is prone to oxidation, which directly affects the composite effect. Secondly, the poor wettability between the aluminum melt and the steel plate can also affect the composite effect. It is necessary to prepare a transition layer through processes such as hot-dip, chemical deposition, and electroplating to improve the wettability between the aluminum melt and the steel plate. Again, due to the extremely low solid solubility of Fe in Al, it is easy to form intermetallic. Therefore, the interface of the casting composite will form a large number of hard and brittle phases such as Al_5Fe and $Al_{13}Fe_4$, which will directly reduce the interface bonding strength.

In order to prepare aluminum cladding with a good bonding interface, it is necessary to optimize the casting temperature of aluminum melt, the preheating temperature of the steel plate, the preheating temperature of the die, the holding temperature, and the holding time. Through these measures, the wettability of preparation transition is improved. After proper heat treatment, the interface microstructure can be optimized. In addition, ultrasonic vibration, electric field, and magnetic field can be applied to the melt during the preparation process to improve the flow field, temperature field, and pressure field of the melt.

Rolling composite has the advantages of high productivity, good stability, easy industrial automation production, and the ability to prepare large-sized products. It has become one of the main production technologies for aluminum-steel cladding. With the increasing demand for high-performance aluminum-steel cladding, new preparation technologies are emerging. These processes include: different temperature rolling, asymmetric rolling, corrugated-flat rolling, solid-liquid casting rolling, and so on. The continuous emergence of new technologies, new materials, new processes, and new equipment will push the application of aluminum-steel cladding to a new height.

4.2.3 Aluminum-titanium cladding

Aluminum-titanium cladding has the advantages of lightweight, excellent corrosion resistance, and high strength and stiffness, and has been widely used in various fields such as construction, chemical equipment, aerospace, and military equipment. Due to the high deformation resistance of titanium at room temperature and strong work hardening, aluminum-titanium cladding is often prepared by hot rolling composites or cold rolling composites + heat treatment. Stacking aluminum and titanium plates, followed by vacuum hot pressing diffusion and rolling composites, the nanoscale phases will be formed at the interface. When the thickness of the aluminum layer is about 10–67%, the aluminum-titanium cladding has good strength and plasticity. The intermetallic compounds at the Al-Ti cladding interface are usually $AlTi_3$, TiAl, Al_3Ti, Ti_2Al_5 and $TiAl_2$. Of the five compounds, Al_3Ti is the easiest to form, while Ti_2Al_5 and $TiAl_2$ need to be transformed from Al_3Ti.

Therefore, Al_3Ti compound is preferentially formed through diffusion at the interface. Al_3Ti compound has good hardness and compressive strength but poor plasticity and toughness. The solid solubility of aluminum in titanium is greater than that of titanium in aluminum. The Al and Ti atoms will interdiffuse first until the atomic concentration exceeds the solid solubility, and then the intermetallic phases will be formed.

4.3 Titanium matrix cladding

4.3.1 Titanium-copper cladding

Titanium has a series of advantages, such as low density, high melting point, high specific strength, and excellent corrosion resistance, and is widely used as an anode material in electrode materials. However, the electric conductivity of titanium is worse than that of copper. In order to improve the electric conductivity of titanium-based anodes, titanium-copper cladding is prepared, which will reduce the voltage and power loss of the electrolytic cell, improve work efficiency, and reduce production costs. This electrode has the advantages of high specific strength, good electric conductivity, and good corrosion resistance. It is one of the most promising composites in corrosive environments.

The lattice of titanium is hexagonal close-packed, the melting point is 1677°C, and the linear expansion coefficient is 8.2×10^{-6}/K, while the melting point of copper is 1084°C and the linear expansion coefficient is 16.6×10^{-6}/K. Thus, titanium and copper are difficult to composite. At present, the preparation of titanium-copper cladding is mainly through explosive composite, hot pressing diffusion composite, extrusion composite, rolling composite, and so on. Explosive composite has the problems of easy leakage welding, more microcracks, and an unfriendly environment.

For thinner titanium plates, the deformation of titanium is large during the bonding process, so the application of explosive composite is limited. The hot pressing diffusion composite is a method of bonding the metal plate at high temperatures by applying pressure to promote atomic diffusion. The titanium-copper cladding prepared by hot pressing diffusion composite will form a certain thickness of the intermetallic, which will affect the performance of the electrode material. Extrusion composite and rolling composite can also effectively prepare titanium-copper cladding. With the increasing demand for the high performance of titanium-copper cladding, the preparation of titanium-copper cladding by corrugated-flat rolling has been applied and has shown unique advantages.

4.3.2 Titanium-steel cladding

Steel has the advantages of high strength, stable performance, and low cost, but the corrosion resistance of steel is poor. Titanium is an excellent, lightweight, corrosion-resistant material with a low density, but its price is expensive. Titanium-steel cladding is a kind of composite with high strength, good corrosion resistance, lightweight, and low cost. At present, many kinds of titanium-steel cladding have been developed, such as TA2/Q235B, TA10/Q345, TA10/Q245, and so on. With the development of technology, simply improving corrosion resistance is no longer sufficient to meet the requirements. It is necessary to develop and prepare complex parts of titanium-steel cladding, which require high formability. Titanium-steel cladding with good formability, such as TA1/Q235 and TA1/304, has been gradually developed. Titanium-steel cladding is mainly used in petrochemical equipment, cross-sea bridges, and other fields.

With the continuous expansion of the application fields of titanium-steel cladding, the titanium-steel cladding prepared by explosive composite, rolling composite, explosive composite + rolling composite, diffusion composite, and extrusion composite cannot meet the higher requirements of the market for its performance and size specifications. In order to better leverage the advantages of titanium-steel cladding, new technologies and equipment such as vacuum rolling composite, corrugated-flat rolling composite, different temperature rolling composite, asynchronous rolling composite, and cold spraying + rolling composite have been continuously developed and used to prepare titanium-steel cladding. The formation of metallurgical bonding interfaces in titanium-steel cladding is the foundation for achieving a good composite effect, but various brittle intermetallics such as TiC, FeTi, and Fe_2Ti will be formed at the interface. If the thickness of the intermetallic layer at the interface is too large, it is not conducive to the improvement of interface bonding strength. When the thickness of the intermetallic layer is about 1–2 μm, the interface bonding strength can reach 260–320 MPa. According to the law of diffusion, temperature has the most significant impact on the intermetallic layer. Therefore, it is crucial to control the rolling temperature and heat treatment temperature reasonably to prepare titanium-steel cladding with high interface bonding strength. The rolling temperature of titanium-steel cladding is generally between 850°C and 950°C. In addition, in order to reasonably control the formation of intermetallic at the interface, introducing a thin metal intermediate layer at the bonding interface to prevent interdiffusion of elements can effectively improve the bonding quality and enhance the comprehensive performance of the titanium-steel cladding. At present, commonly used intermediate layer metals include copper, niobium, nickel, molybdenum, vanadium, pure iron, and IF steel.

4.4 Metalized polymer film

Metalized polymer film is prepared by depositing a layer of copper, aluminum, zinc, and other metals on the surface of a polymer substrate through techniques such as physical vapor deposition, chemical plating, and electroplating. It is a typical SDM, which has great application prospects in fields such as battery current collectors, capacitors, semiconductors, etc. Metallic plastic current collector (MPCC) can significantly improve the energy density and safety of batteries by reducing the thickness and weight of the current collector. Moreover, due to the insulation, thermal shrinkage, and melting properties of the polymer itself, it can enhance the safety of batteries, thus attracting a lot of attention from researchers in the industry. In MPCC, copper and aluminum are generally selected for the metal layer, and polyester (PET), polypropylene (PP), and polyamide (PI) are generally selected for the polymer matrix. Metalized film is widely used in the field of capacitors, especially in film capacitors, due to the high dielectric constant, high insulation resistance, and good heat resistance. In addition, metalized film can also be used for surface connection in semiconductor technology to connect various components.

Metalized polymer film prepared by magnetron sputtering deposition has the characteristics of small temperature rise, stable metal layer, strong interface force, good repeatability, and is suitable for continuous large-area film, but the deposition efficiency is relatively low and the target utilization rate is low (usually less than 30%). In order to improve the purity and uniformity of the deposited metal layer on the substrate surface, multi-target co-sputtering and low-pressure high-speed sputtering can be adopted. This technique improves the effective ionization rate of the gas, thereby improving the deposition speed. Evaporation coating can also prepare metalized film; the vacuum evaporation can evaporate more metal atoms than magnetron sputtering. Therefore, after magnetron sputtering, evaporation can be used to increase the thickness of the metal layer on the substrate and effectively improve the uniformity.

In order to prepare metalized polymer films by chemical plating, the surface of the polymer substrate should be degreased and cleaned first, and then the surface should be roughened with $K_2Cr_2O_7$ and concentrated H_2SO_4 or 10% $KMnO_4$ or 10% NaOH. Then, the sensitizer is uniformly adsorbed on the substrate surface to improve the catalytic activity of the surface and enhance the interface bonding between the metal layer and the polymer. The advantage of chemical plating is that a uniform film can be prepared on the insulator without DC power supply equipment. In addition, chemical plating has the advantages of being environmentally friendly, low cost, high efficiency, and so on.

The metalized polymer film prepared by the electroplating process is to connect the polymer substrate with the cathode in the electrolyte solution (such as copper sulfate), while the metal acts as an anode (such as a copper ball encapsulated in a titanium blue anode bag). Under the action of external direct current, copper atoms are oxidized to Cu^{2+}, which enters the electrolyte solution and supplements the metal ions. The Cu^{2+} in the solution migrates to the surface of the substrate, obtains electrons, and reduces them to copper atoms, and then forms a copper layer. The metalized film prepared by electroplating has good interface bonding strength.

Due to its weight reduction effect, metalized polymer films have been tried for automotive batteries. This increases the energy density of the battery. Due to the extremely high safety requirements of electric vehicles, so far, the technology and products are still in the process of industrialization.

But battery giants and technology companies in China and the United States have invested huge enthusiasm and money in research. Japanese, Korean, and European companies are also actively cooperating to carry out applied research.

5. Interface bonding mechanism of SDMs and regulation of interface microstructure and properties

5.1 Interface bonding mechanism of SDMs

SDMs are widely used in more and more fields due to their excellent performance. The interface bonding strength directly determines the performance of SDMs. In general, the intermetallic layer will be formed during the preparation process or the heat treatment process, which is the basis for the formation of the metallurgical bonding interface. Intermetallic compounds are usually brittle phases. A too-thick, brittle layer is not conducive to interface bonding strength. Therefore, it is necessary to strictly control the interface thickness to ensure that SDMs have a high interface strength.

Accurately revealing the interface bonding mechanism is very important for rationally regulating the interface microstructure and improving the interface bonding strength. At present, a lot of research works have been done on the interface bonding mechanism of SDMs. The mainstream interface bonding mechanisms include: mechanical meshing mechanism, film mechanism, dislocation mechanism, metal bond mechanism, energy mechanism, diffusion mechanism, three-stage mechanism, and so on.

1. Mechanical meshing mechanism

 When two kinds of metal with rough surfaces are composited under large pressure, the contacted metal surfaces will gradually mesh with each other, thus forming a tightly bonded composite. This kind of mechanical meshing is a simple physical bonding, and there is no metallurgical bonding. It generally exists in the SDMs prepared by cold rolling composites, cold extrusion composites, and other cold composite technologies. Due to the low interface bonding strength, the application field is limited. It is often necessary to carry out heat treatment and other processes to improve the interface bonding effect.

2. Film mechanism

The surface of the metal will produce work hardening during the grinding pretreatment process, and a thin surface layer with poor plasticity will be formed. During the rolling deformation process, the brittle film and the surface oxide film will be broken. The fresh metal surface with good plasticity is exposed under the action of rolling pressure, and the atomic diffusion occurs on the contact surface, so as to realize the composite between metal plates. The film mechanism is suitable for explaining the interface bonding process of cold rolling, and the interface bonding effect can be improved by adjusting the rolling parameters.

3. Dislocation mechanism

The dislocation mechanism believes that the bonding between the two metals is the result of plastic deformation of the metal in the bonding zone. When the two metals come into contact with each other and produce plastic deformation, dislocations will migrate to the metal surface with the rupture of the oxide film, resulting in a step with an atomic spacing height. Under the action of rolling force, the dislocations on the interface accumulate continuously, accelerate the rupture of the oxide film on the metal surface, reduce the plastic deformation resistance, and indirectly promote the bonding between metals. The dislocation mechanism is a useful supplement to the film mechanism, but it is not suitable for the bonding process without plastic deformation.

4. Metal bond mechanism

Burton M.S. first proposed the metal bond mechanism. When the surface spacing of two metals reaches the distance that the two metal atoms can attract each other, the bonding of two metals can be realized. In the metal bond mechanism, the spacing should reach the atomic level. When the spacing between the two atoms decreases, the attraction gradually increases. When the spacing between the two atoms reaches two times the normal atomic spacing, the attraction is the largest. At this time, further reducing the atomic spacing, the attraction between the two atoms will gradually decrease. When the spacing reaches the normal atomic spacing, the attraction is zero, and the two metal atoms form a metal bond. Although the metal bond mechanism is a relatively mature theory, it cannot reasonably explain the bonding in a low-temperature environment.

5. Energy mechanism

The energy mechanism was first proposed by Cline C.L. in 1966. This mechanism analyzes the interface bonding process from the perspective of energy based on the atomic activation energy. According to the energy mechanism, when the two metals contact each other, even if the two atoms meet other requirements, the two atoms will not bond when the two atoms do not have the lowest energy to achieve the bonding. Only when the atoms have sufficient energy and are activated can the bonding of the two metals be achieved within a certain atomic spacing. As the temperature increases, the energy of the atoms will gradually increase, which is conducive to the bonding of the two metals. The energy mechanism is a supplement and improvement to the metal bond mechanism.

6. Diffusion mechanism

The diffusion mechanism points out that when the two metals are contacted, under the action of external heat or deformation heat, the temperature of the metal at the interface will be increased, and the metal atoms at the interface will be activated, thus forming a diffusion zone through interdiffusion. The atoms in the diffusion zone can continue to diffuse, so that the thickness of the interface layer will gradually increase, and solid solutions or intermetallic will be formed in the diffusion zone. The diffusion mechanism explains the bonding of the two metal interfaces from the perspective of metallurgy, but it cannot explain the phenomenon that as the diffusion layer increases to a certain thickness, a large number of brittle intermetallic formed lead to a decrease in the interface bonding strength.

7. Three-stage mechanism

The three-stage mechanism is proposed on the basis of the above mechanism, which can well explain the interface bonding mechanism of SDMs prepared by the rolling composite technology. According to this mechanism, the interface formation process is divided into three stages. The first stage is the physical contact stage, which mainly forms physical contact bonding through plastic deformation. The two metal atoms are close to each other at the contact interface, and weak chemical bonds can be formed at this stage. The second stage is the physical and chemical stage. The contact interface produces mutual physical and chemical effects; the chemical bond will be formed, and the bonding between two metals will be realized. The third stage is the diffusion stage. With the continuous diffusion at the interface, the diffusion layer will be formed, the interface bonding strength will also be improved, and finally the effective bonding is realized.

With the advancement of science and technology, more and more interface bonding mechanisms have been proposed to reveal the interface bonding mechanism of SDMs. However, it is not difficult to find that various mechanisms have certain limitations. Therefore, in order to prepare high-performance SDMs, it is very important to study the interface bonding mechanism and continuously clarify and reveal the interface bonding mechanism, which is also one of the important research directions of SDMs.

5.2 Effects of annealing on the interface microstructure of SDMs

The interface of SDMs prepared by room temperature composite technology, such as cold rolling composite and cold extrusion composite, is mainly based on a mechanical bonding interface. Even the interface of SDMs prepared by thermal composite technology will also have some mechanical bonding interface [10]. At the same time, there is a certain stress at the interface of SDMs after severe plastic deformation. These factors will reduce the interface bonding strength of SDMs. In order to improve the interface microstructure, transform the mechanical bonding into metallurgical bonding, eliminate or reduce the internal stress, and effectively improve the comprehensive mechanical properties, it is necessary to carry out appropriate annealing heat treatment. During annealing, the growth of intermetallic compounds is induced due to the mutual diffusion of elements. The thickness of the compound layer increases gradually due to heat. New compound phases are formed where no intermetallic compounds exist.

The schematic diagram of growth behavior of intermetallic during annealing is shown in **Figure 10**. The thickness of intermetallic compounds increases with the increase of annealing temperature and time. The effect of annealing temperature on composite phase is more significant. However, too thick an intermetallic layer is not conducive to the improvement of interface bonding strength. Therefore, it is necessary to reasonably control the annealing temperature and time during the annealing process to obtain the best annealing effect.

5.3 Methods to improve the interface bonding strength of SDMs

The interface bonding strength of SDMs directly determines the cladding effect of SDMs. The interface bond strength can be measured by stripping, stretching, and shearing methods. Although the interface bonding strength of the SDMs can be improved by proper annealing treatment. However, the interface bonding strength has certain genetic characteristics. If the original bonding effect of the SDMs is poor, it is difficult to obtain better interface bonding strength even after annealing treatment. Therefore, it is very important to improve the interface bonding strength of the SDMs during the preparation process. With the continuous progress of composite technology, new technologies continue to appear. These technologies include: asynchronous rolling, corrugated plate rolling, different temperature rolling, anaerobic rolling, and other new composite technologies. These new processes can effectively improve the interface bonding strength of SDMs.

If there is a significant difference in the plastic deformation capacity of the two metals, the degree of deformation of the two metals is not consistent. At this time,

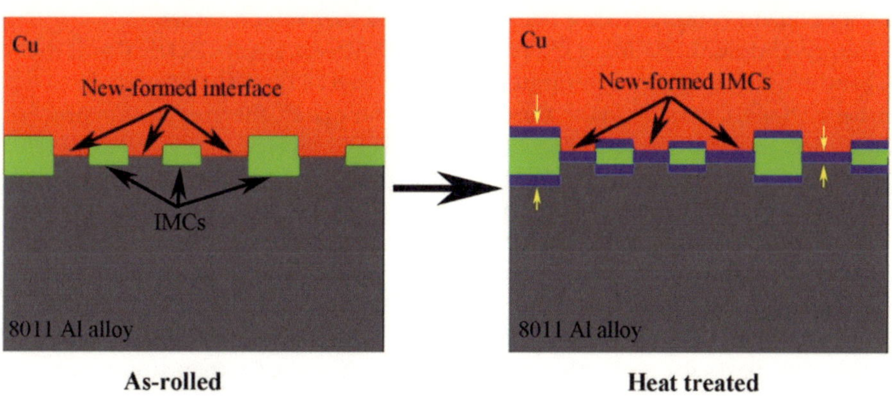

Figure 10.
Schematic diagram of growth behavior of intermetallic during annealing.

if the traditional flat rolling process is used to prepare SDMs, the compound effect may be poor. To solve this problem, the asynchronous rolling composite method was developed. The different speed of the two work rolls is the characteristic of asynchronous rolling, which is different from the traditional rolling process. Generally speaking, metals that are not easy-to-deform correspond to faster work roller speeds. The metal that is easy-to-deform corresponds to the work roller with a slower speed. Compared with synchronous rolling composite, asynchronous rolling composite can reduce the rolling force and the energy consumption, enhance the interface bonding effect, improve the shape quality, and then obtain the SDMs strip.

The interface bonding of SDMs can be effectively improved when the special-shaped roller is used to prepare SDMs. Among them, corrugated-flat rolling is an efficient special-shaped roller rolling method developed in recent years to improve the interface bonding strength. The corrugated-flat rolling adopts a new type of roller with a special corrugated structure on the surface, which corresponds to the difficult-to-deform metal, and the other working roller is a flat roller corresponding to the easy-to-deform metal. The periodic deformation will be formed on the difficult-to-deform metal surface during the preparation process, and then increasing the bonding area between the two metal plates and improving the interface bonding strength. Different temperature rolling is a new rolling composite method developed to solve the problem of significant differences in deformation resistance between two metals during the rolling preparation process. By applying different preheating temperatures to the two metals, the synergistic deformation of the two metals in the rolling preparation process can be better realized. In addition, oxygen-free rolling can effectively reduce the surface oxidation of the metal, which can effectively improve the interface bonding strength of the SDMs.

In summary, with the continuous development of composite technology, more and more new and efficient composite technologies will be gradually developed to prepare high-performance SDMs. High-performance and multifunctional SDMs are an important development trend of cladding.

6. Application of SDMs

SDMs have been widely used in transportation, mobile communication, energy, household appliances, building decoration, military equipment, and other fields due to their excellent comprehensive performance and cost-effectiveness.

Copper-aluminum cladding is a type of SDM with excellent electrical and thermal conductivity. Copper-aluminum cladding is used as filter shell cover plates for 5G base stations. Compared with pure copper cover plates, the preparation process is simplified, and the cost can be saved by 10–50%. At present, the metal shell of radio frequency devices for 6G products is also gradually beginning to use copper-aluminum cladding for preparation. In the field of photovoltaic power generation, the photovoltaic welding strips and other components have been partially replaced by copper-aluminum cladding, which not only meets the performance requirements but also reduces the cost.

In the field of energy storage, battery current collectors require good electric conductivity, plasticity, strength, and corrosion resistance. Copper-aluminum cladding is being tried as a current collector material for a new type of battery. There is a huge market demand for copper-aluminum cladding in the fields of vehicle secondary batteries, positive and negative conversion components of electric vehicles, and positive and negative connectors of energy storage power stations. In addition, copper-aluminum cladding also has a huge market demand in the fields of power system connection components, transition rows, LED lighting, building decoration panels, and so on.

As a new type of kitchenware material, stainless steel-aluminum cladding has attracted much attention in recent years. The stainless-steel aluminum cladding combines the bright, easy-to-clean features of the stainless-steel surface. The density of aluminum is small; it can reduce weight, and the thermal conductivity is high, so that the pot is heated evenly. Therefore, the comprehensive performance of kitchenware is excellent. It has become the preferred material for high-end kitchenware. It is used as car-body material, decorative parts, and heat exchange parts of vehicles, and high-speed trains can achieve significant weight reduction effects and have played a great role in promoting the development of the field of lightweight transportation. Compared with copper heat exchangers, those with aluminum-aluminum cladding have the advantages of light weight, corrosion resistance, high reliability, low cost, and easy recovery. Therefore, it has gradually replaced copper in the selection of vehicle heat exchanger materials.

Titanium-aluminum cladding is used as a mobile phone shell, which not only has the characteristics of light weight and high strength, but also has a good metal texture. Copper/stainless steel/copper cladding, stainless steel/aluminum/stainless steel cladding, and aluminum/copper/aluminum cladding are used as mobile phone medium plate materials, which have the characteristics of high strength, high thermal conductivity, easy welding, and light weight, and have broad market application prospects. Copper-steel cladding is also used in architectural decoration, power electronics, the military industry, and other fields.

In addition, SDMs such as titanium-steel cladding, nickel-steel cladding, and steel-steel cladding have good application prospects in fields such as light industry, metallurgy, environmental protection, aerospace, and so on. The application of partially SDMs is shown in **Table 2**.

SDMs	Applications
Al/Cu	Cookware, roof and wall plate, heat exchanger, special engraving plates, electrical components
Al/Fe	Reflectors in electric heaters, automobile silencers, brake disks.
Ag/Cu	Electrical components and appliances.
Cu/Fe	Bullet jacket.
Al/Cu/steel	Cookware: roast and bowl for induction heater.
Al/steel/Al	Automotive exhaust systems.
Cu/stainless steel/Cu	Communication cable shields buried in acidic soil.
Ni/stainless steel/Cu	Bottom battery cap.
Ti/stainless steel/Ni	Bipolar electrode in fuel cell.
Ag/Cu/Ag	Coin (warm rolled).
Ni/Cu/Ni	Coin (warm rolled).

Table 2.
Applications of SDMs.

At present, SDMs have been well applied in various industries and will play an increasingly important role. With the development of the economy and society, the demand for functional SDMs is increasing. The development of new SDMs with special functional characteristics is also an important research and development direction.

7. Conclusions and prospects

The SDMs have excellent comprehensive performance by combining the characteristics of the matrix material and cladding material and have broad market application prospects in many fields. SDMs prepared by traditional technology cannot meet the requirements of the market. We should strengthen research and development and basic theoretical research to develop SDM materials with excellent performance to meet the market demand for new materials.

1. The new and efficient composite technologies should be developed, the configuration design, stacking sequence, layer thickness ratio, and other process parameters should be optimized to prepare SDMs with better composite effects. In-depth analysis of the evolution mechanism of intermetallic interstitials during annealing is helpful to further improve the interfacial bonding strength of SDMs.

2. Develop new functional SDMs, fully utilize the thermal, electrical, optical, magnetic, and other functional characteristics, and broaden the application fields of SDMs. It is necessary to vigorously promote the application of SDMs in emerging and rapidly developing fields. At the same time, it is necessary to carry out the research of SDMs waste recycling technology.

3. Strengthen the basic theoretical research and carry out in-depth research works on the interface bonding mechanism, synergistic deformation mechanism, heat and mass transfer model, interface microstructure evolution and prediction, heterogeneous strengthening mechanism, and failure mechanism under multi-field service conditions of SDMs.

4. The development of emerging industries and industrial products has put forward higher requirements for materials. Layered metal is one of the effective solutions to cope with material upgrading. Fields such as electric vehicles, power batteries, energy storage devices, artificial intelligence servers, and low-altitude aircraft are all of great interest in layered metals. Research and application cases show that SDMs have strong adaptability and great development value. Cladding metals is facing great development opportunities.

Author details

Huisheng Cai[1], Siqi Yang[1] and Huaqing Li[2,3]*

1 Inner Mongolia Key Laboratory of New Materials and Surface Engineering, School of Material Science and Engineering, Inner Mongolia University of Technology, Hohhot, China

2 Materials Academy, JITRI, Suzhou, Jiangsu, China

3 Jiangsu CHINACLD Composite Materials Co., LTD, China

*Address all correspondence to: 13861311766@163.com

IntechOpen

© 2025 The Author(s). Licensee IntechOpen. This chapter is distributed under the terms of the Creative Commons Attribution License (http://creativecommons.org/licenses/by/4.0), which permits unrestricted use, distribution, and reproduction in any medium, provided the original work is properly cited.

References

[1] Wadsworth J, Lesuer DR. Ancient and modern laminated composites—From the great pyramid of Gizeh to Y2K. Materials Characterization. 2000;**45**(4-5):289-313

[2] Li L, Nagai K, Yin F. Progress in cold roll bonding of metals. Science and Technology of Advanced Materials. 2008;**9**(2):023001

[3] Saito Y, Utsunomiya H, Tsuji N, et al. Novel ultra-high straining process for bulk materials-development of the accumulative roll-bonding (ARB) process. Acta Materialia. 1999;**47**(2):579-583

[4] Mo TQ, Chen ZJ, Li BX, et al. Tailoring of interface structure and mechanical properties in ARBed 1100/7075 laminated composites by cold rolling. Materials Science and Engineering A. 2019;**755**:97-105

[5] Wang T, Wang Y, Bian L, et al. Microstructural evolution and mechanical behavior of Mg/Al laminated composite sheet by novel corrugated rolling and flat rolling. Materials Science and Engineering A. 2019;**765**:138318

[6] Li S, Luo C, Bashir M-U, et al. Interface structures and mechanical properties of corrugated + flat rolled and traditional rolled Mg/Al clad plates. Rare Metals. 2021;**40**:2947-2955

[7] Hoseini Athar MM, Tolaminejad B. Weldability window and the effect of interface morphology on the properties of Al/Cu/Al laminated composites fabricated by explosive welding. Materials and Design. 2015;**86**:516-525

[8] Zhang J p, Huang H g, Zhao R d, et al. Cast-rolling force model in solid–liquid cast-rolling bonding (SLCRB) process for fabricating bimetal clad strips. Transactions of Nonferrous Metals Society of China. 2021;**31**(3):626-635

[9] Xiao-ke T, Xin-hua L, Guo-liang X. Interfacial microstructure and properties of copper cladding aluminum alloy composites. Chinese Journal of Rare Metals. 2022;**46**(7):862-872

[10] Cai H, Wang Q, Zhang N, et al. Shear behavior of Cu/Al/Cu trilayered composites prepared by high-temperature oxygen-free rolling. Journal of Alloys and Compounds. 2024;**1004**:175857

Chapter 5

Iron: From Basic Chemistry to Modern C−H Bond Functionalization Catalysts

Hamad H. Al Mamari

Abstract

Iron (Fe) is a first-row transition metal that offers several advantages, including low cost, Earth abundance, and environmental safety. These benefits are particularly significant compared to other transition metals from the second-row and beyond. Unlike precious metals such as palladium and platinum, iron is readily available, accessible, affordable, low in toxicity, and recyclable. This chapter aims to provide an overview of the importance of iron in various aspects of life and highlight the significant impact of iron catalysis. It begins by examining the occurrence of iron in nature and its importance to the environment and human health. The chapter then discusses iron compounds, focusing on their uses and applications in chemistry in general and organic synthesis in particular. This includes the role of iron compounds as catalysts and reagents in various organic synthetic transformations, such as electrophilic aromatic substitution reactions, cross-coupling reactions, cycloaddition reactions, and oxidation and reduction chemistries. The chapter also discusses developments in iron-catalyzed C−H bond functionalization, including iron catalysts inspired by biological systems. It then continues by covering photocatalyzed C−H bond functionalization. The chapter then highlights the importance of artificial intelligence and machine learning in catalyst design, which could be applied in iron catalysis. Given the green features of iron and iron catalysis, represented by Earth abundance, low cost, low toxicity, and recyclability, the chapter concludes with the role of iron catalysis in green chemistry and the preservation of the environment.

Keywords: iron, iron catalysis, oxidation, reduction, substitution, C−H functionalization, photoredox catalysis, olefin metathesis, green chemistry

1. Introduction

Iron (Fe) is situated in the first-row of transition metals in the periodic table. It is a metal with an atomic number of 26 and an average molar mass of 55.85 g/mol. At atmospheric pressure, iron exists in three allotropes: α-Fe, γ-Fe, and δ-Fe. The element has four isotopes: ^{54}Fe, ^{56}Fe, ^{57}Fe, and ^{58}Fe, with respective abundances of 5.85%, 91.75%, 2.12%, and 0.29%. Iron is the most abundant element on Earth and the second most abundant transition metal in the Earth's crust. Its ground-state electronic configuration is [Ar] $4s^2\ 3d^6$. The most common oxidation states of iron are +2 and +3. In iron complexes, iron exhibits an electronic configuration of [Ar] $4s^0\ 3d^8$. Iron(II) typically has a coordination number of six and an octahedral geometry. Iron is versatile and capable of adopting various oxidation states ranging from −2 (−II) to +6 (+VI). This remarkable characteristic enables iron to catalyze a wide array of chemical transformations including oxidation and reduction reactions. The flexibility of iron to switch between different oxidation states plays a crucial role in various catalytic organic synthesis reactions. Additionally, iron is considered Lewis acidic due to its ability to accept electron pairs.

1.1 Iron in nature

Iron exists in its ores primarily as oxides. Notable iron ores include hematite (iron(III) oxide, Fe_2O_3), magnetite (Fe_3O_4), and siderite (iron(III) carbonate, $FeCO_3$).

Naturally, iron is the most abundant transition metal in biological systems. One of the most recognized forms of natural iron is in the blood of humans and other mammals. In the human body, iron is present in heme b within hemoproteins and cytochromes, which are molecular components of hemoglobin and myoglobin (**Figure 1**). In heme b (protoporphyrin IX), iron binds to four molecules of pyrrole units, which are five-membered nitrogen-containing heterocycles. Essentially, heme is a porphyrin bound to iron [1–3].

An essential biological function of iron-containing heme is to transport oxygen in the bloodstream of humans and other vertebrates. This capability is attributed to the iron's binding affinity for oxygen (O_2) within heme. The Lewis acidity of iron in hemoglobin and myoglobin facilitates the binding with Lewis acidic oxygen. This crucial biological interaction enhances the solubility of oxygen in the blood.

Iron is also found in cytochromes, such as cytochrome P450, which contain an iron-porphyrin unit or heme b, similar to hemoglobin and myoglobin.

1.2 Preface

Iron catalysis plays a significant role in both chemistry as a whole and in organic chemistry specifically. It has garnered substantial attention and has numerous applications in the field of synthetic organic chemistry. The purpose of this book chapter is to provide readers with an overview of the impact of iron catalysis and to highlight

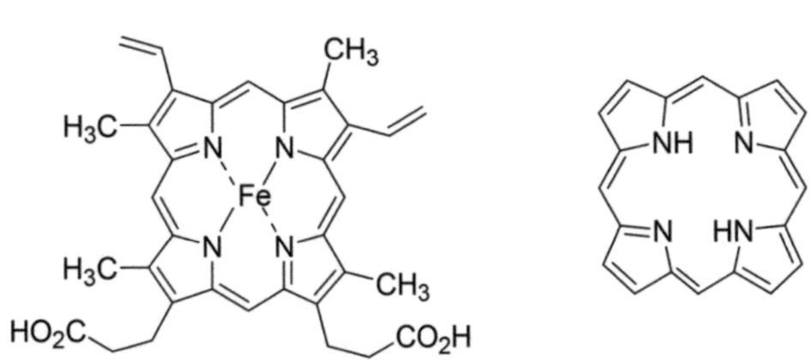

Protoporphyrin IX, heme B — porphyrin

Figure 1.
Iron in protoporphyrin IX, heme B.

its applications in synthetic organic chemistry. This chapter is not intended to serve as a comprehensive review of iron catalysis in organic chemistry. For those seeking a detailed and thorough analysis of methods and reactions in organic chemistry, it is recommended to refer to selected comprehensive and insightful reviews on the subject [4–8].

1.3 Prospective

Iron chemistry has seen significant advancements in recent years. Iron catalysis has evolved from early reactions, such as oxidation, reduction, and electrophilic aromatic substitution, to more complex processes like cross-coupling and cycloaddition reactions. The impressive capabilities of iron catalysis are now evident in areas such as C–H functionalization and carbonyl olefin metathesis. Recent developments have also introduced iron photoredox catalysis, which holds great potential, especially in the functionalization of inert C–H bonds. In the near future, science may witness exciting breakthroughs in this area of iron catalysis. Furthermore, iron's environmental advantages, including its abundance on Earth, low toxicity, cost-effectiveness, and recyclability, make it a valuable contributor to green chemistry and the reduction of greenhouse gases like CO_2.

The following sections discuss the evolution of iron catalysis from early reactions to the most recent developments.

2. Oxidation

Iron catalysis has been effectively utilized to epoxidize alkenes using hydrogen peroxide (H_2O_2), resulting in the formation of corresponding epoxides. Beller et al. reported an epoxidation method employing inexpensive $FeCl_3{\cdot}6H_2O$ as the iron catalyst, along with pyridine-2,6-dicarboxylate and an amine (**Figure 2**) [9].

Additionally, iron has played a role in the Baeyer-Villiger reaction of ketones, converting them into the corresponding esters or lactones [10]. This method utilizes Fe_2O_3 as the iron catalyst and molecular oxygen (**Figure 3**) [11].

Furthermore, iron is employed in a Wacker-type process to transform terminal alkenes into their corresponding aldehydes [12]. This reaction uses PhIO as the oxidant and [Fe(2,5-Cl_2TPP)OTf] as the iron catalyst (**Figure 4**). The Fe catalyst is generated *in situ* by reacting [Fe(2,5-Cl_2TPP)Cl] with $AgSO_3CF_3$, allowing a wide

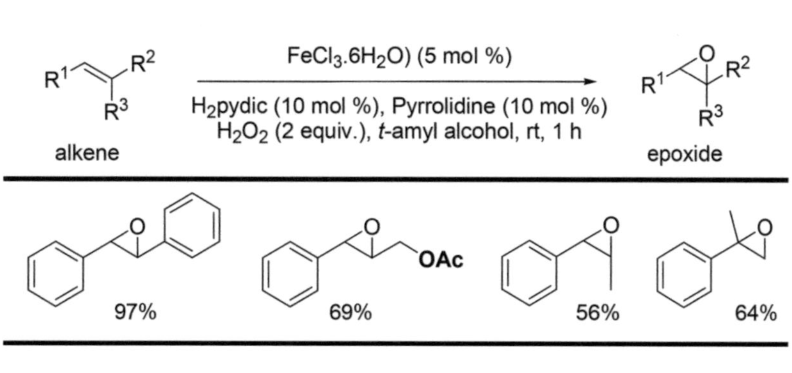

Figure 2.
Epoxidation of alkenes using $FeCl_3$ catalyst.

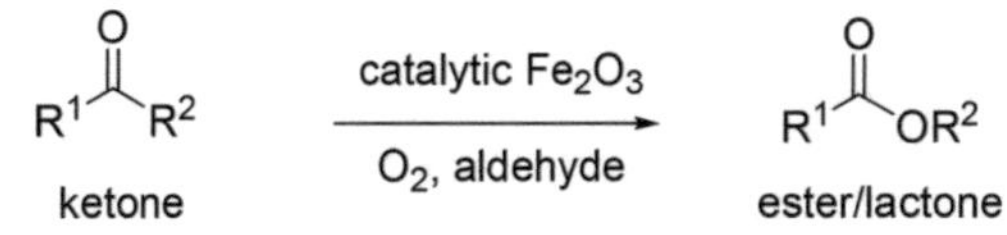

Figure 3.
Epoxidation of alkenes using $FeCl_3$ catalyst.

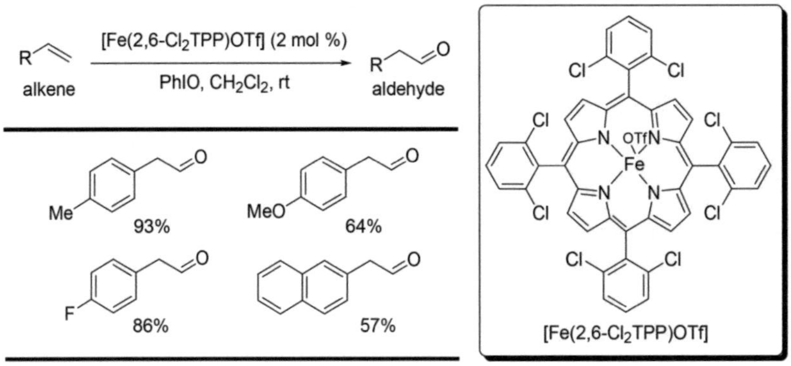

Figure 4.
Fe-catalyzed Wacker-type oxidation of alkenes to aldehydes.

range of alkenes to undergo this iron-catalyzed E1 reaction, producing the aldehydes in good yields [12].

In another application, commercially available photocatalytic α-Fe_2O_3 has been used to oxidize aryl boronic acids to phenols [13]. This protocol involves the use of the photocatalytic iron oxide catalyst, oxygen, and solar visible light (see **Figure 5**). A substantial variety of aryl boronic acids have been tested, yielding corresponding phenols in high yields [13].

Moreover, iron(III) chloride serves as a Lewis acid in the oxidation of sulfides to sulfones using $KMnO_4$. The oxidation reaction can be enhanced through the Lewis acidic properties of $FeCl_3$ (**Figure 6**) [14]. This process begins with the formation of a Lewis complex, resulting from the interaction between the Lewis basic permanganate (MnO_4^-) and the Lewis acidic $FeCl_3$.

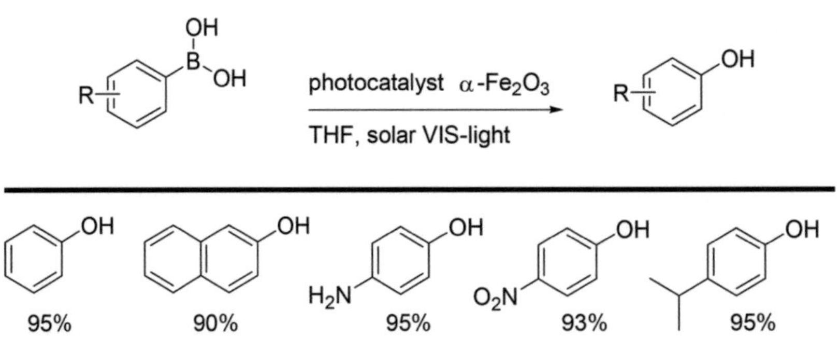

Figure 5.
Iron oxide-catalyzed oxidation of arylboronic acids to phenols.

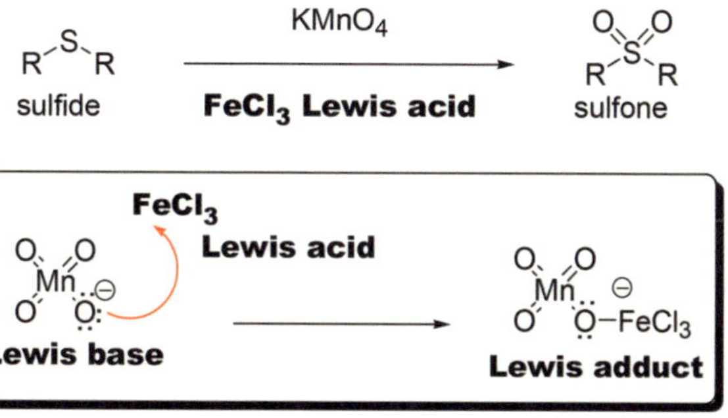

Figure 6.
Oxidation of sulfides to sulfones using $KMnO_4$ accelerated by Lewis acidic $FeCl_3$.

3. Reduction

Iron catalysis has proven to be important for the reduction of alkenes to their corresponding alkanes. Bench-stable $FeCl_2$ was utilized as a precatalyst in the hydrogenation of alkenes using hydrogen (H_2) in the presence of a catalytic amount of a tetradentate ligand and a catalytic amount of a Grignard reagent at low temperature (**Figure 7**) [15].

The iron-catalyzed hydrogenation method has been applied to a wide range of alkenes, yielding excellent results and high yields of the corresponding alkanes. Additionally, this method has been extended to a reductive cross-coupling of vinyl halides with stoichiometric amounts of Grignard reagents, which act as both activators and coupling partners (**Figure 7**) [15].

An asymmetric iron-catalyzed hydrogenation/reduction of aryl ketones has also been reported [16]. This method employs H_2, the inexpensive $Fe_3(CO)_{12}$, and a chiral ligand (**Figure 8**).

A wide range of aryl ketones has been hydrogenated to generate the corresponding chiral alcohols with high yields and significant enantiomeric excess (ee) [16].

Iron and its catalysts are also employed to reduce nitroarenes to yield corresponding anilines. This transformation is valuable due to the versatile synthetic utility of

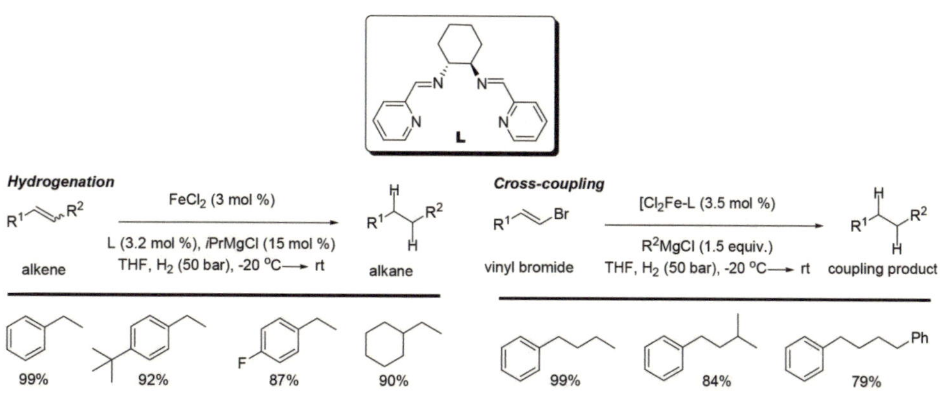

Figure 7.
Iron-catalyzed hydrogenation of alkenes and reductive cross-coupling of vinyl halides with Grignard reagents.

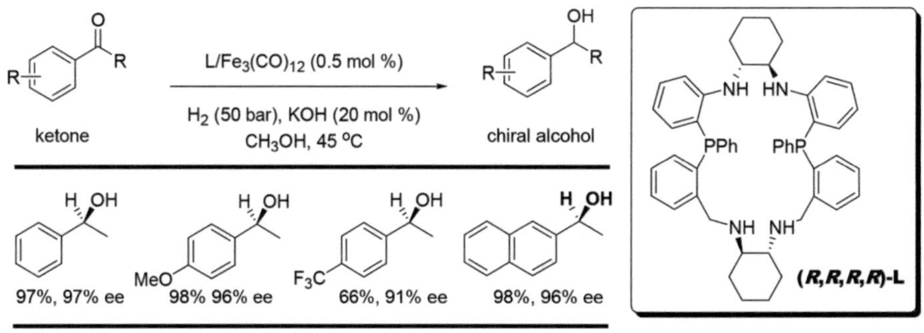

Figure 8.
Iron-catalyzed asymmetric hydrogenation of ketones to give chiral alcohols.

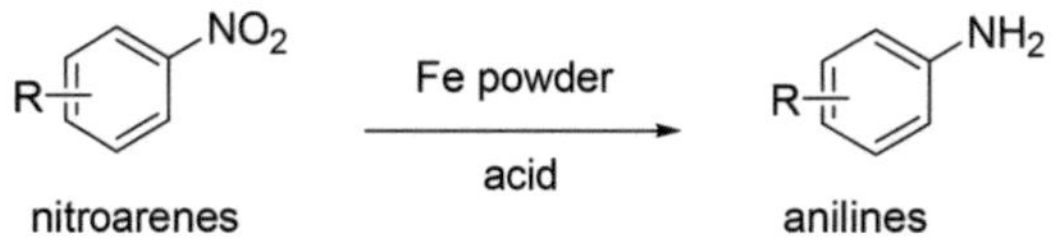

Figure 9.
Iron-promoted reduction of nitroarenes to anilines.

anilines in organic synthesis. Typically, iron powder is used in the presence of concentrated acids such as hydrochloric acid or acetic acid (see **Figure 9**).

Recent developments have emerged in the iron-promoted or iron-catalyzed reduction of nitroarenes to anilines [4–8].

4. Substitution

4.1 Nucleophilic substitution

The iron-catalyzed benzylation of 1,3-dicarbonyl compounds has been reported using inexpensive iron(III) hexahydrate, $FeCl_3{\cdot}6H_2O$ [17]. This nucleophilic substitution reaction utilizes a β-ketoester and nitromethane (CH_3NO_2), which acts as both the base and solvent, along with a benzyl alcohol or benzyl acetate, and a catalytic amount of $FeCl_3{\cdot}6H_2O$ (5 mol%) (**Figure 10**) [17].

The iron-catalyzed nucleophilic substitution relies on acidic 1,3-dicarbonyl compounds, such as the reported β-ketoesters and 1,3-diketones, which are distinctly acidic at the methylene position flanked by the two carbonyl groups. A suitable base is used to deprotonate the compounds at this methylene position, forming an α-carbanion that resonates to the corresponding enolate. This nucleophilic intermediate subsequently participates in the iron-catalyzed nucleophilic substitution reactions. A variety of 1,3-dicarbonyl compounds were successfully benzylated, yielding corresponding products in good yields [17].

Additionally, a straightforward iron-catalyzed amidation of benzyl alcohols with unsubstituted amides has been reported [18]. This reaction involves an amide and either benzyl or allyl alcohol, employing a catalytic amount of the inexpensive iron(III) chloride, $FeCl_3$, as the catalyst (**Figure 11**).

The iron-catalyzed amidation reaction produced the corresponding products in high yields [18].

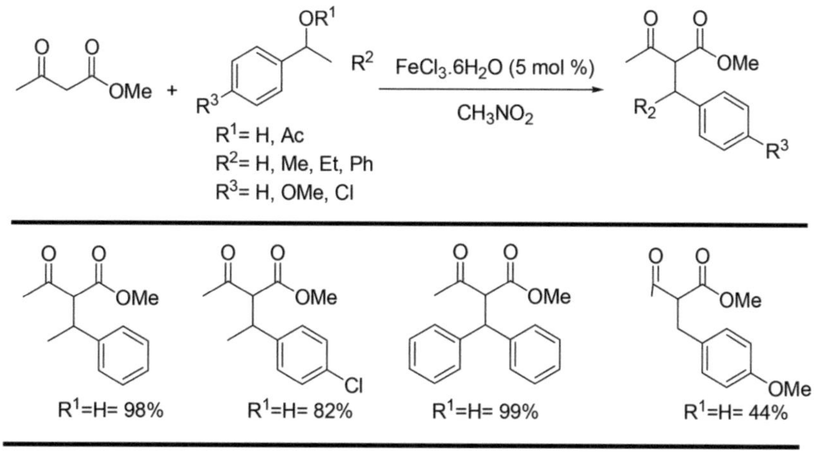

Figure 10.
Iron-catalyzed benzylation of 1,3-dicarbonyl compounds.

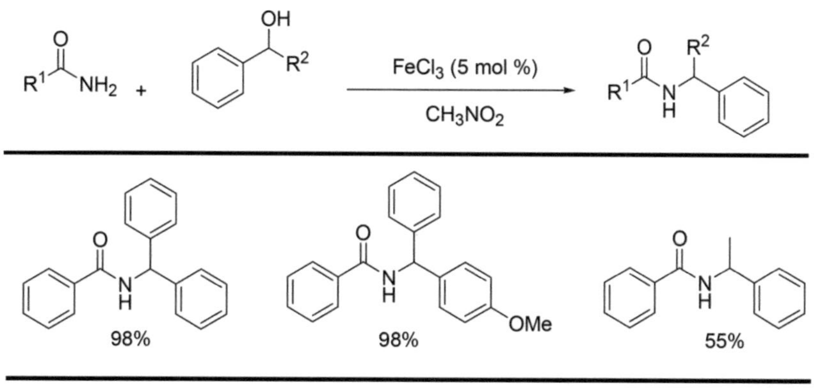

Figure 11.
Iron-catalyzed amidation.

4.2 Electrophilic aromatic substitution

Iron(III) halides are used as Lewis acid catalysts in the electrophilic aromatic substitution of benzene. Typically, iron(III) bromide ($FeBr_3$) is employed with bromine

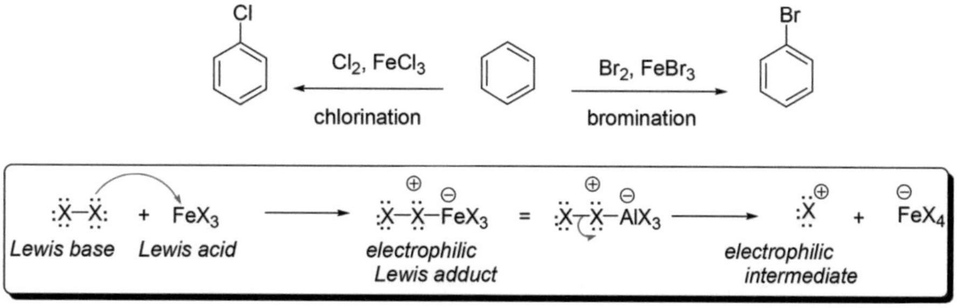

Figure 12.
Halogenation of benzene catalyzed by FeX_3.

(Br_2) for the bromination of benzene, while iron(III) chloride ($FeCl_3$) is used for the chlorination of benzene with chlorine (Cl_2).

The iron Lewis acid catalyst reacts with the Lewis basic halogen to form an electrophilic Lewis adduct intermediate. In this Lewis adduct, the terminal halogen atom carries a partial positive charge, making it more susceptible to nucleophilic attack by benzene. Alternatively, the Lewis adduct can produce an electrophilic halogen intermediate with a positive charge, which facilitates an enhanced nucleophilic attack by benzene (**Figure 12**).

5. Cycloaddition

The utility of iron catalysis in organic synthesis has been expanded to include cycloaddition reactions. Fu et al. reported an iron-catalyzed catalytic asymmetric synthesis of β-lactams [19]. This method involves a [2 + 2] cycloaddition between ketenes and *N*-triflylimines, resulting in β-lactams with good diastereoselectivity and enantioselectivity (**Figure 13**).

Iron catalysts have also been successfully utilized in [4 + 2] cycloaddition reactions. Cramer et al. demonstrated an enantioselective iron-catalyzed [4 + 2] cycloaddition involving inactivated dienes [20]. A range of dienes with both electron-donating and electron-withdrawing groups participated in this [4 + 2] Fe-catalyzed process, yielding cyclohexenes with excellent yields and high enantioselectivity (refer to **Figure 14**) [20].

Additionally, Kurahashi, Matsubara, and coworkers reported an iron-catalyzed [4 + 2] cycloaddition of aldehydes and dienes [21]. This method showcases a hetero Diels-Alder [4 + 2] cycloaddition of inactivated aldehydes and simple 1,3-dienes, utilizing a cationic iron(III) porphyrin catalyst, [Fe(TPP)]BF4 (see **Figure 15**). The corresponding cycloadducts were obtained in good yields [21].

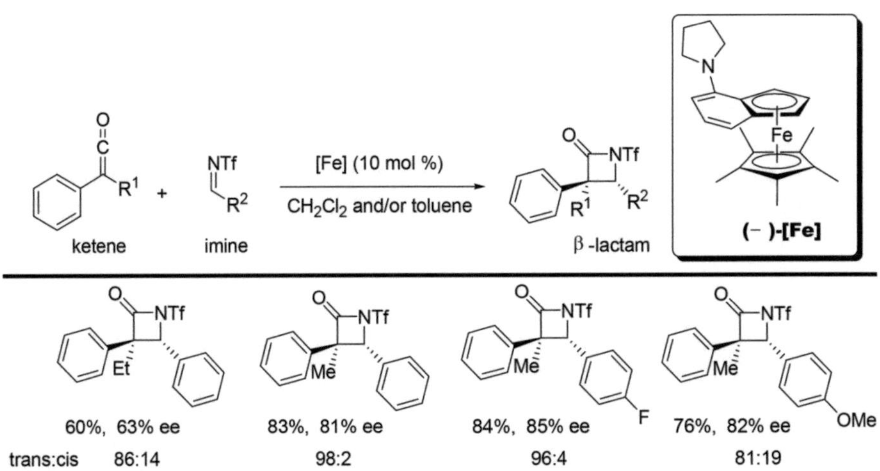

Figure 13.
Iron-catalyzed [2 + 2] cycloaddition between ketenes and N*-triflyl imines to give β-lactams.*

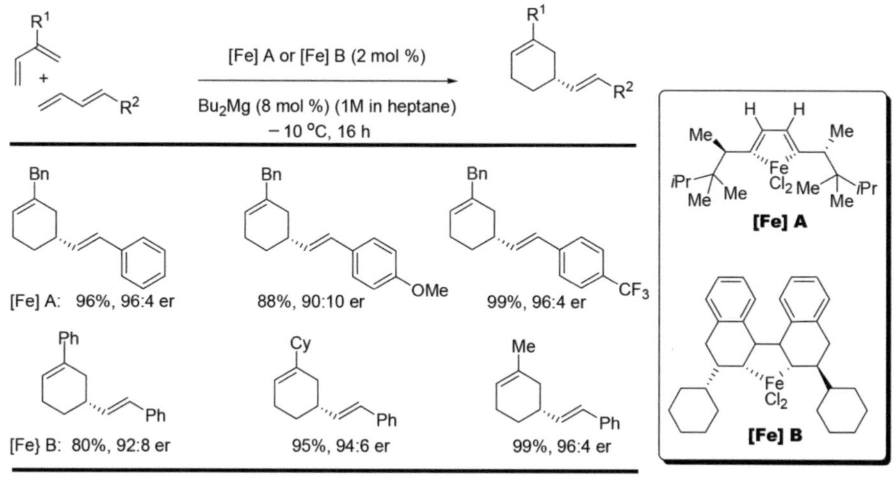

Figure 14.
Enantioselective iron-catalyzed [4 + 2] cycloaddition reactions of inactivated dienes.

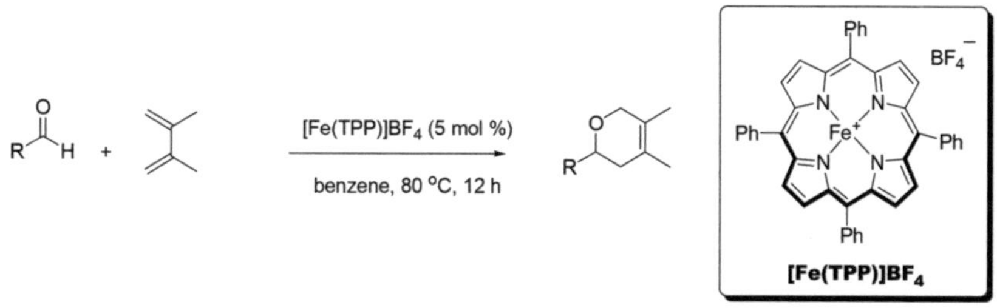

Figure 15.
Iron-catalyzed [4 + 2] cycloaddition reactions of inactivated dienes.

6. Cross coupling

Cross-coupling reactions are a significant strategy for forming carbon-carbon (C–C) bonds. This strategy involves the reaction between a nucleophilic organometallic reagent and an electrophilic aryl or alkenyl halide (or pseudohalide) in the presence of a transition metal catalyst [22–24].

Iron catalysts have played a crucial role in facilitating cross-coupling reactions. For instance, Tamura and Kochi demonstrated an iron-catalyzed cross-coupling of methyl magnesium bromide with alkenyl bromides [25]. This method involves an electrophilic alkenyl bromide with a specific double bond geometry and utilizes $FeCl_3$ as the iron catalyst (**Figure 16**). The reaction leads to the formation of sp^3-sp^2 C–C bonds in a stereospecific manner, preserving the double bond geometry of the alkenyl bromide; Z-alkenyl bromides yield Z-alkenes, while E-alkenyl bromides yield E-alkenes [25].

Another notable contribution in iron-catalyzed cross-coupling reactions was reported by Cahiez et al. [26]. The reported method employs $Fe(acac)_3$ to catalyze reactions involving various alkenyl halides as electrophiles and Grignard reagents.

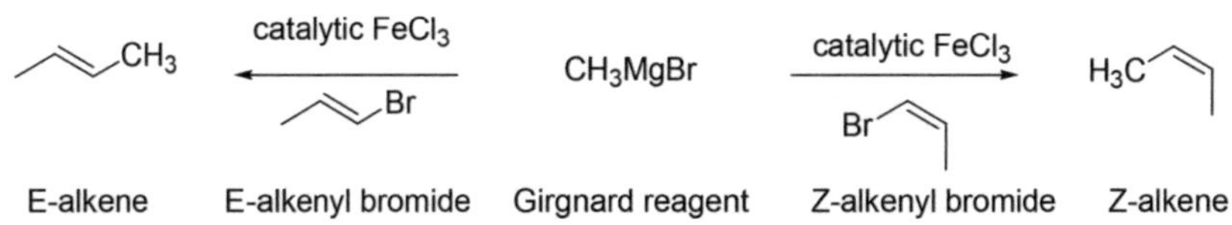

Figure 16.
Iron-catalyzed cross-coupling of alkenyl bromides with Grignard reagents.

Notably, the presence of stoichiometric amounts of *N*-methylpyrrolidone (NMP) as a solvent additive was essential for enhancing the yields of the cross-coupled products. When THF was used as a solvent without NMP, the resulting yields of cross-coupled products were very low (**Figure 17**) [26].

Additionally, Fürstner et al. made significant contributions to iron-catalyzed cross-coupling reactions by demonstrating the successful cross-coupling of various aryl chlorides, triflates, and tosylates with different Grignard reagents [27]. The method employed $Fe(acac)_3$ as the catalyst, resulting in the formation of corresponding cross-coupled products with good yields (**Figure 18**).

Nakamura et al. provided further advancements in iron-catalyzed cross-coupling reactions, successfully coupling secondary and primary alkyl halides with a variety of electronically different aryl magnesium bromides using $FeCl_3$ as the catalyst [28]. The use of TMEDA, (N,N,N′,N′-tetramethylethylenediamine), was critical to suppress undesired side reactions, such as olefin formation (**Figure 19**) [28].

Overall, the developed iron-catalyzed cross-coupling reactions have proven compatible with various functional groups, yielding excellent results [28].

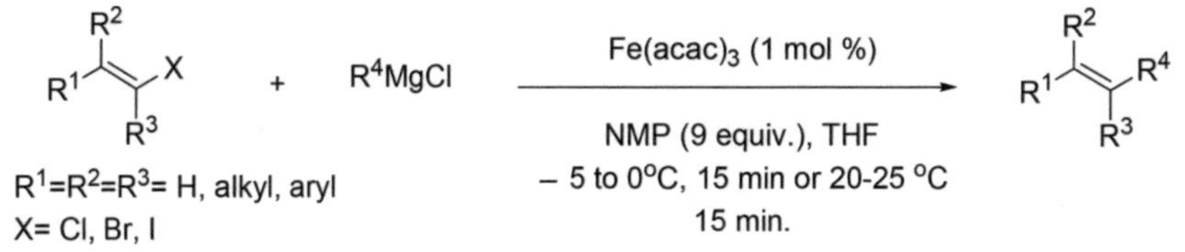

Figure 17.
Iron-catalyzed cross-coupling of alkenyl halides with Grignard reagents.

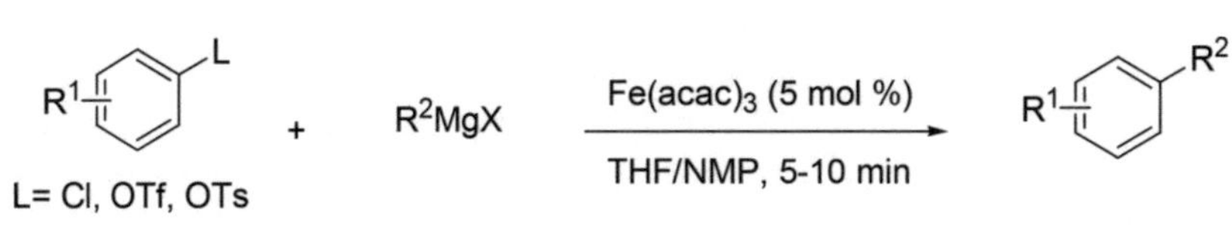

Figure 18.
Iron-catalyzed cross-coupling of aryl halides/sulfonates with Grignard reagents.

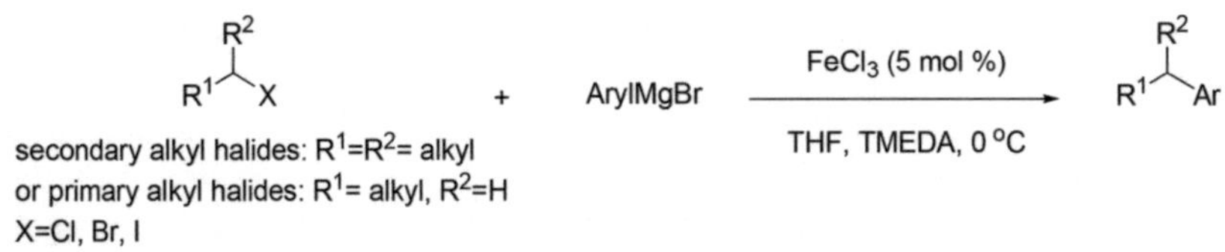

Figure 19.
Iron-catalyzed cross-coupling of aryl halides with Grignard reagents.

7. Protection of alkenes

Simple iron complexes, such as $Fe(CO)_5$ and $Fe_2(CO)_9$, (**Figure 20**) can coordinate with alkenes to form alkene-iron complexes [1].

Carbon-carbon double bonds are considered electron-dense and thus nucleophilic. This chemical property allows alkenes to react with electrophiles in electrophilic addition reactions (**Figure 21**). Alkene-iron complexes can be prepared by heating or irradiation. By complexing with an iron carbonyl compound, the nucleophilicity of the C═C bond is significantly reduced, rendering it electrophilic (**Figure 21**). This change occurs due to the electron-withdrawing ability of the CO ligands attached to iron. Consequently, the electronic nature of alkenes shifts from nucleophilic to electrophilic. This reversal in electronic character can have significant implications in synthetic organic chemistry [1].

The coordination between alkenes and iron complexes, along with the accompanying change in the electronic nature of alkenes, can be utilized in two key ways. The complexation of a C═C bond with an iron carbonyl serves as both activation toward nucleophiles and protection against electrophilic attack. Unlike other functional groups, such as alcohols, diols, ketones, and aldehydes, protecting alkenes by alternative means is challenging. Thus, using iron complexes for their protection is a synthetically useful strategy. This approach is particularly beneficial for compounds containing multiple double bonds with varying steric hindrance or accessibility when a selective electrophilic addition is desired at a more hindered double bond. The less hindered double bond can be selectively protected by complexation with an iron complex, facilitating chemoselective electrophilic addition to a more hindered or less accessible double bond.

Similarly, this protection strategy can be applied to enynes. By complexing the C═C bond with an iron complex, the C−C triple bond in the alkyne can be primed for chemoselective electrophilic addition.

Iron complexes such as $[CpFe(CO)_2]^+$, where Cp is cyclopentadienyl and abbreviated as Fp, can coordinate with alkenes to form more stable alkene-iron complexes. These Fp alkene-iron complexes carry a positive charge, making them cationic, which in turn increases their reactivity toward nucleophiles. Fp can be prepared through the oxidative or reductive cleavage of the $Cp_2Fe_2(CO)_2$ dimer.

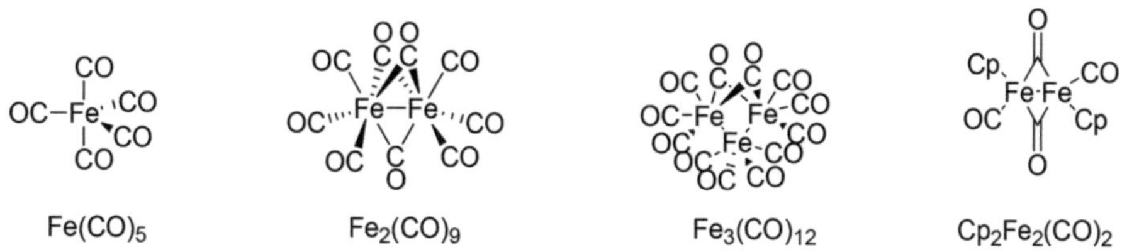

Figure 20.
Iron complexes.

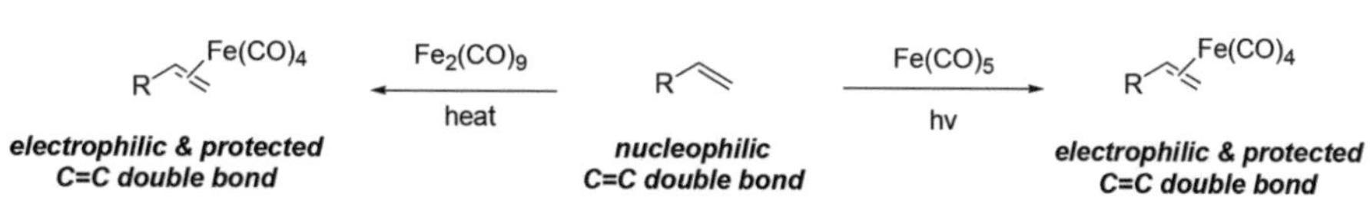

Figure 21.
Reversal of alkene electronic nature by complexation with iron complexes.

8. Protection of dienes

Iron complexes such as $Fe(CO)_5$, $Fe_2(CO)_9$, and $Fe_3(CO)_{12}$ (**Figure 1**) can coordinate with conjugated 1,3-dienes to form 1,3-diene-Fe η^4 complexes. These diene-Fe complexes can be very stable and are prepared by treating a conjugated 1,3-diene with $Fe(CO)_5$, $Fe_2(CO)_9$, or $Fe_3(CO)_{12}$ through heating or irradiation (**Figure 22**).

The formation of a stable diene-Fe complex serves as protection for the nucleophilic diene, rendering the protected diene inactive toward typical electrophilic addition reactions. Additionally, protected s-cis dienes as η4 complexes with Fe should not undergo the typical Diels-Alder reactions that they would normally participate in under standard conditions.

9. Carbonyl olefin metathesis

The versatility of iron catalysis has been successfully applied to carbonyl olefin metathesis. Catalytic iron(III) chloride ($FeCl_3$) has been used to facilitate intramolecular ring-closing carbonyl metathesis with alkenes, resulting in the formation of cyclopentenes [29]. This method utilizes readily available and inexpensive $FeCl_3$ as the iron catalyst (**Figure 23**).

This remarkable advancement showcases the Lewis acidity of iron and its implications for catalysis in organic synthesis. The carbonyl olefin ring-closing metathesis is noted for being operationally simple and effective under mild reaction conditions.

10. C—H bond functionalization

10.1 Recent developments

The functionalization of inert C—H bonds has become a powerful strategy for constructing new chemical bonds [30–36]. This transformation of inert C—H bonds into reactive ones has been catalyzed by various transition metals [30–36]. Notably, iron has been employed to catalyze C—H bond functionalization reactions [37].

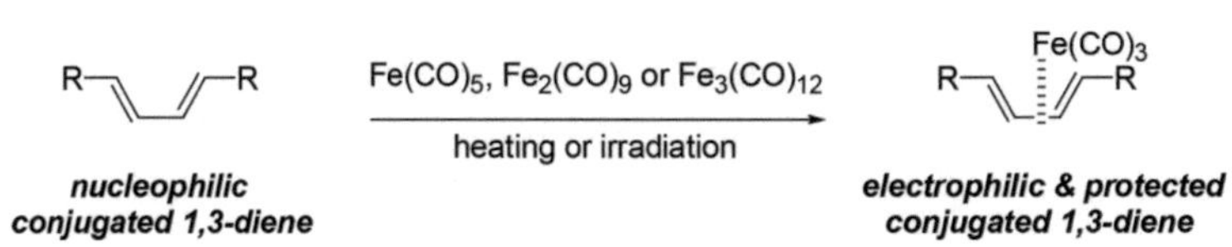

Figure 22.
Protection of conjugated 1,3-dienes as η^4 complexes with iron.

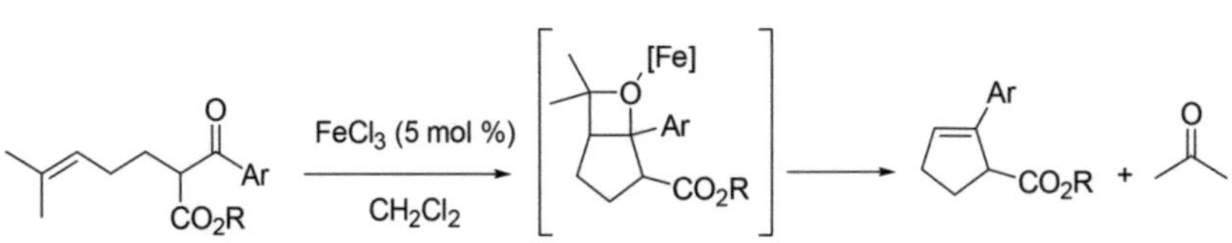

Figure 23.
Iron-catalyzed carbonyl olefin metathesis.

Nakamura et al. reported on the iron-catalyzed methylation of C—H bonds in carboxylic acids and ketones using trimethylaluminum as a methyl source, in the presence of a tridentate phosphine ligand [38]. The C—H bond methylation utilizes $Fe(acac)_3$ as the iron catalyst alongside a dimethylamino triphosphine ligand (**Figure 24**).

A broad range of methylated carboxylic acids and ketones have been synthesized using the iron-catalyzed methylation methods developed by Nakamura et al. [38].

Recently, Ackermann et al. disclosed a novel modular triazole-based directing group. They introduced the triazolyl(dimethyl)methyl (TAM) group, which was shown to undergo efficient ruthenium-catalyzed *ortho*-C—H bond arylation of benzamides using aryl bromides [39]. In a further expansion of the synthetic utility of the triazole directing group, Ackermann et al. reported an iron-catalyzed *ortho*-C—H bond arylation of triazole benzamides using Grignard reagents [40]. This method features the use of the TAM group to direct iron-catalyzed C—H bond arylation with Grignard reagents (**Figure 25**).

Consequently, various benzamides featuring the novel triazolyl(dimethyl)methyl (TAM) group underwent efficient iron-catalyzed C—H bond arylation with Grignard reagents. A range of C—H bond arylation products was obtained, exhibiting good yield and a wide tolerance of functional groups.

Additionally, iron catalysis has been extended to C—H borylation [41]. A well-defined iron catalyst, $[FeMe_2(dmpe)_2]$, has been shown to catalyze the dehydrogenative C—H borylation of arenes (**Figure 26**). This method employs pinacolborane as the boron source and requires UV irradiation; no borylation reaction occurs without it [41].

10.2 Mechanistic aspects of iron-catalyzed C—H bond functionalization

Directed C—H bond functionalization relies on the Lewis basicity of the directing group and the Lewis acidity of the transition metal. Bidentate directing groups have

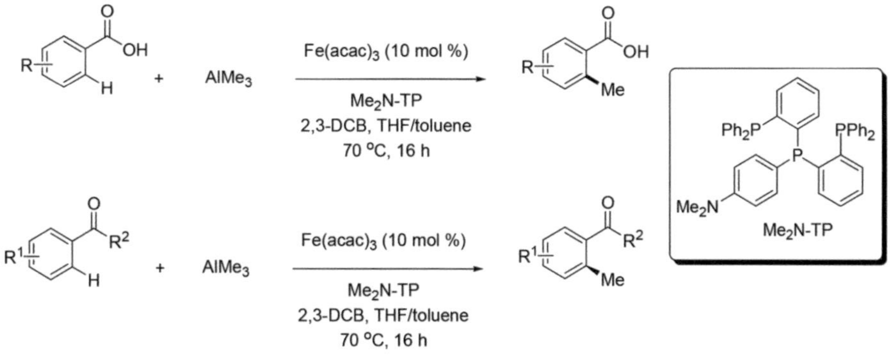

Figure 24.
Iron-catalyzed C—H bond methylation of carboxylic acids and ketones using a tridentate phosphine ligand.

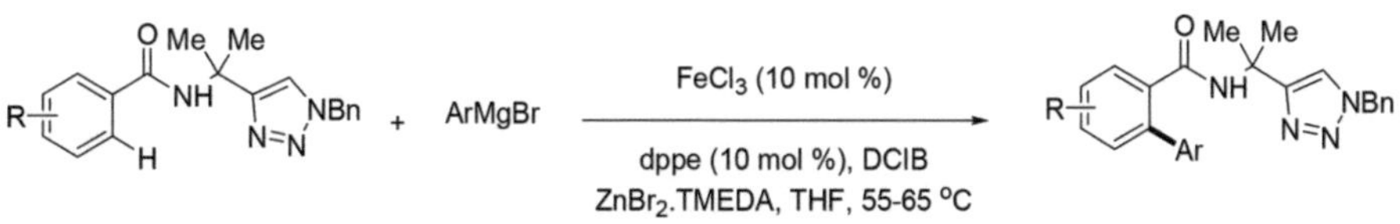

Figure 25.
Iron-catalyzed C—H bond arylation of benzamides bearing triazole-based directing group with Grignard reagents.

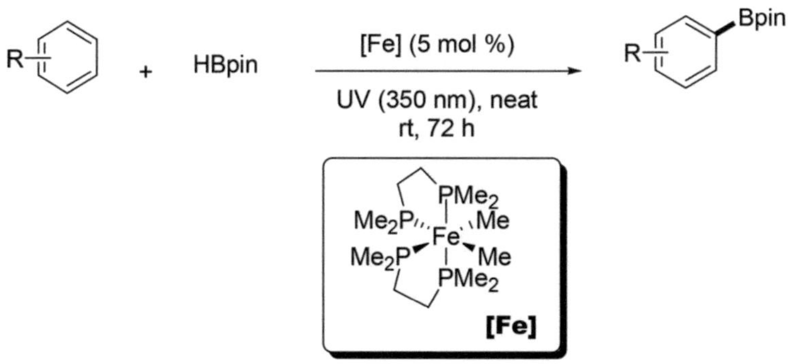

Figure 26.
Fe-catalyzed C—H bond borylation of arenes.

two Lewis basic atoms that are appropriately positioned to facilitate the formation of double five-membered chelates. A proposed mechanism for iron-catalyzed C−H bond functionalization, assisted by a bidentate directing group such as 8-aminoquinoline, is illustrated in **Figure 27**.

In this mechanism, an amide containing the *N*,*N*-bidentate directing group 8-aminoquinoline coordinates with Lewis acidic iron, which has a bisphosphine ligand, to form complex I (**Figure 27**). Agostic interactions between the iron center and the γ-C−H bond lead to C−H bond cleavage, resulting in the formation of bis-five-membered chelate II [42]. Subsequently, oxidative addition yields intermediate III, which, after reductive elimination, produces the functionalized product (**Figure 27**).

A similar mechanism can be applied to the iron-catalyzed C−H methylation of carboxylic acids and ketones, as reported by Nakamura et al. (**Figure 28**) [38]. The Lewis basic atom of the carbonyl group coordinates with the Lewis acidic iron complex, which is stabilized by a triphosphine ligand, resulting in the formation of Lewis complex I. Agostic interactions between the iron metal center and the γ-C−H bond facilitate the cleavage of the C−H bond, leading to the generation of intermediate II. Subsequent oxidative addition produces intermediate III, which then undergoes reductive elimination to yield the methylated product (**Figure 28**).

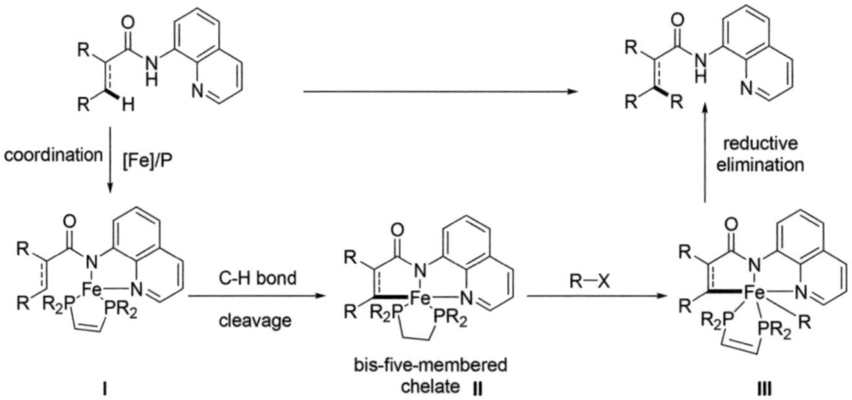

Figure 27.
A possible mechanism for Fe-catalyzed C—H bond functionalization assisted by a N,N-*bidentate directing group.*

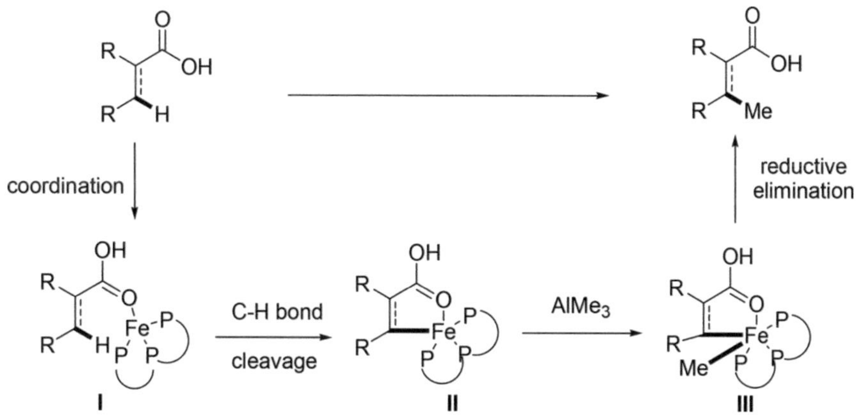

Figure 28.
Possible mechanism for Fe-catalyzed C—H bond methylation of carboxylic acids and ketones.

10.3 Bioinspired iron catalysts for C−H activation

Nonheme iron-dependent oxygenase Rieski enzymes [43] have been shown to functionalize C−H bonds in cholesterol towards its metabolism (**Figure 29**) [44].

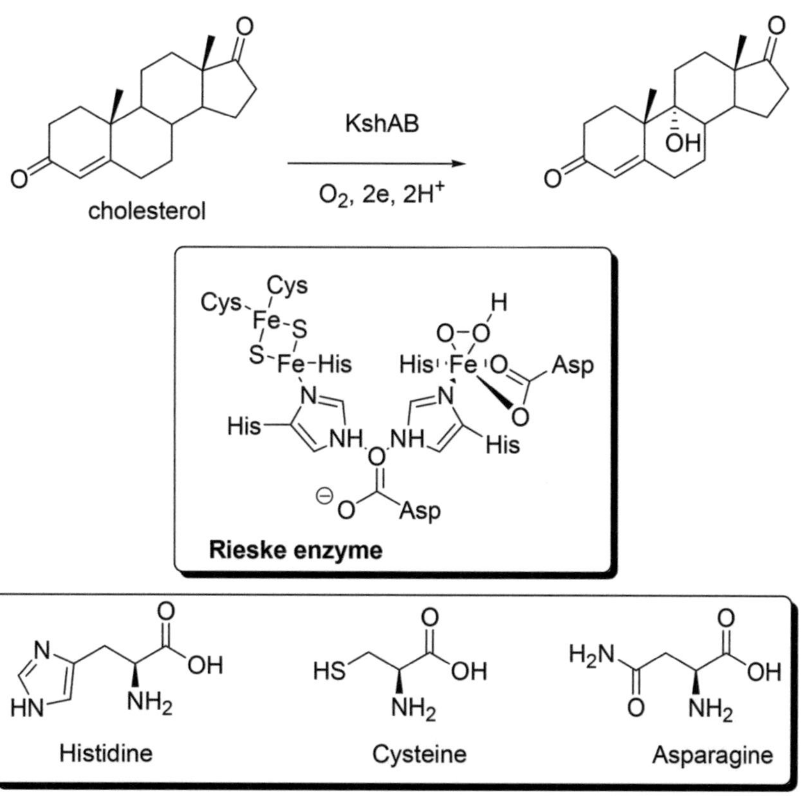

Figure 29.
Cholesterol oxidation by Rieske enzyme.

Recent advancements have been made in the design and synthesis of nonheme iron-based oxidants that imitate enzyme activity [45]. These nonheme iron oxidants have been successful in oxidizing C−H and C=C bonds with good yields and enantioselectivities. Additionally, a bioinspired iron(II) catalyst has demonstrated the ability to achieve stereoselective oxidation of inactive aliphatic C−H bonds using H_2O_2 [46]. This oxidation process employs an iron catalyst coordinated with a bispyrrolidine-bispyridine-based ligand (**Figure 28**). The developed bioinspired iron catalyst has effectively oxidized tertiary inactivated sp^3-H bonds across various substrates, resulting in the formation of the corresponding tertiary alcohols (**Figure 30**) [46].

The efficiency of the developed bioinspired iron catalyst outperformed bacteria in the selective oxidation reaction in the oxidation of (+)-artimicinin (**Figure 31**) [46].

The developed bioinspired catalyst demonstrated superior performance compared to bacteria in the selective oxidation of (+)-artemisinins (**Figure 31**). Additionally, the nonheme iron catalyst was utilized to selectively oxidize remote $C(sp^3)-H$ bonds in methylene groups [47]. This selective oxidation was efficient and produced the corresponding ketones in good yields (**Figure 32**).

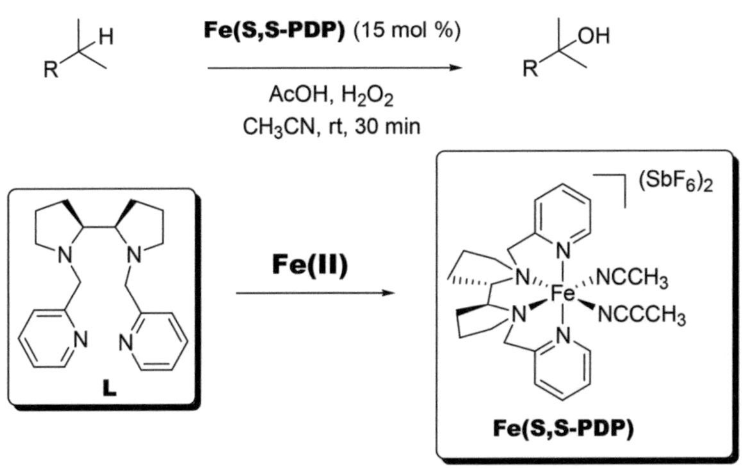

Figure 30.
Stereoselective C(sp³)-H oxidation by a nonheme iron catalyst.

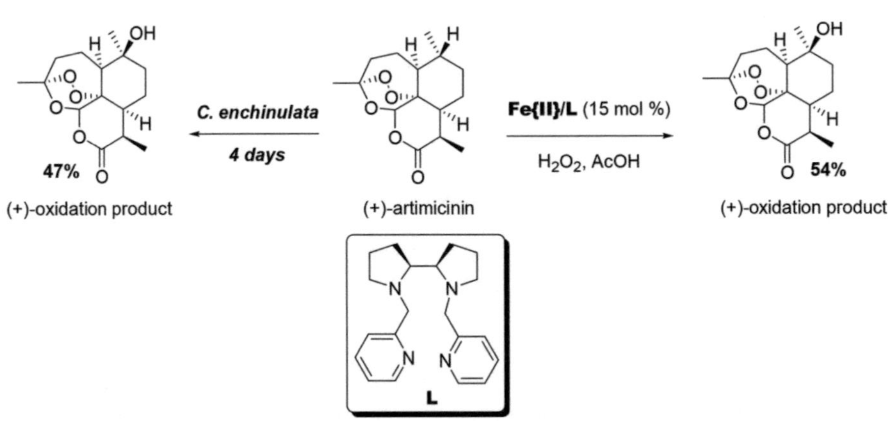

Figure 31.
Stereoselective C—H oxidation by a nonheme iron catalyst.

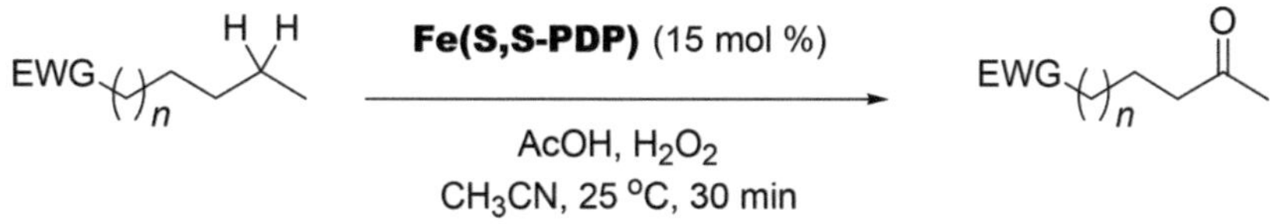

Figure 32.
Selective C(sp³)—H methylene oxidation by a nonheme iron catalyst.

10.4 The prospect of photoredox iron-catalyzed C−H activation

Recently, progress has been made in the field of visible light-mediated photoredox catalysis [48, 49]. Photocatalysts that rely on second-row transition metals face challenges due to their scarcity and high cost, as well as their requirement for high-energy light for irradiation. This has led to a search for less expensive and more environmentally friendly alternatives for photoredox catalysts [50]. Transition metal-based photocatalysts, such as those made from ruthenium [51] and iridium, [52] are well-known. However, both of these metals are costly and toxic. As a result, researchers have turned to photoredox catalysis that utilizes earth-abundant transition metals, such as iron. Several iron-based catalysts have been identified as efficient photoredox catalysts, including those featuring bipyridine [53], phenanthroline [54], terpyridine [55], and acetylacetone (**Figure 33**) [56].

Iron photoredox catalysis has been applied in intramolecular C−H oxygenation of 2-biphneylcarboxylic acids [57]. The method employs $Fe(NO_3)_3{\cdot}9H_2O$ as an iron precatalyst and a bipyridine bis carboxylic acid as a ligand under irradiation with a 40 W LED light (**Figure 34**).

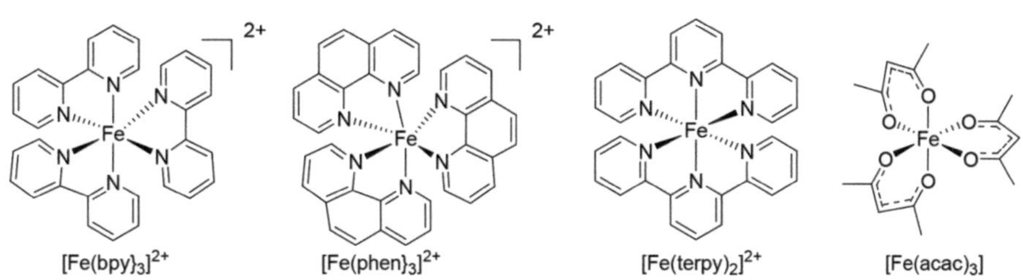

Figure 33.
Iron-based photoredox catalysts.

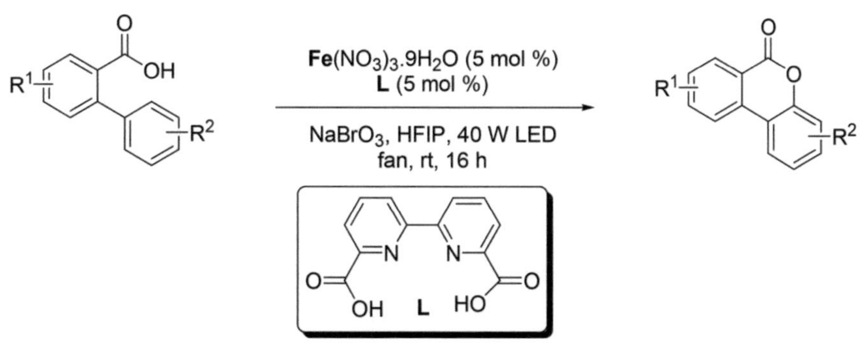

Figure 34.
Intramolecular C—H oxygenation of 2-biphenylcarboxylic acids by iron photoredox catalyst.

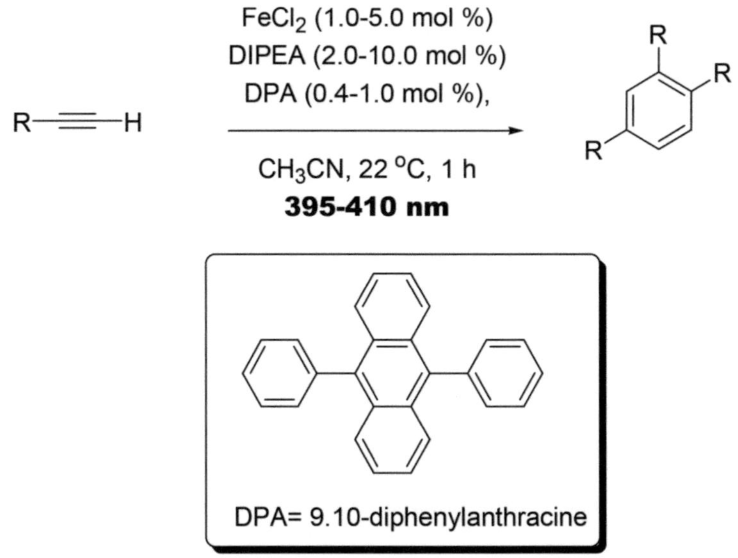

Figure 35.
Combination of iron catalysis and photocatalysis in cyclotrimerization of terminal alkynes.

Various 2-biphenylcarboxylic acids have undergone intramolecular iron-photocatalyzed C−H oxygenation, resulting in the corresponding products being produced in good yields [57]. Additionally, the combination of iron catalysis and photocatalysis has facilitated the cyclotrimerization of terminal alkynes [58]. This method employs $FeCl_2$ as the iron precatalyst for cyclotrimerization, 9,10-diphenylanthracene (DPA) as the photocatalyst, and diisopropylamine (DIPEA) as the catalytic reductant. The reaction is carried out under visible light irradiation using blue LEDs with wavelengths ranging from 395 to 410 nm (**Figure 35**) [58].

The combination of photocatalysis with C−H functionalization has garnered significant interest from the scientific community. Transition metals commonly used in photocatalytic C−H bond functionalization include ruthenium (Ru) and iridium (Ir) [59]. With the introduction of iron-based photoredox catalysts, such as those illustrated in **Figure 33**, there is potential for applying these iron-based catalysts in C−H bond functionalization. However, the use of iron in photocatalytic C−H bond functionalization remains unexplored.

11. Artificial intelligence (AI) and machine learning (ML) toward catalyst design

Artificial intelligence (AI) and machine learning (ML) have ushered in a new era in science and technology through the use of data-driven algorithms, pattern recognition, and accurate predictions. Recent advancements in AI and ML are set to transform the approach to scientific research [60]. The array of existing catalysts and ligands required for catalysis is vast, and the number of catalytic reactions in chemistry and organic synthesis is extensive. Designing and synthesizing catalysts, as well as developing sustainable catalytic systems and methods, are crucial for advancing new catalytic reactions.

AI and ML have been increasingly applied in the field of catalysis [61]. For instance, AI techniques have been utilized in heterogeneous catalysis [62–66], as well as in the design of catalysts employing earth-abundant elements [67]. AI has played a significant role in discovering new catalyst systems and in applying computational techniques alongside ML to identify transition metal catalysis [68–71]. An example of this application is a study by Foppa et al., who employed Subgroup Discovery (SGD) [72], an AI approach, to identify physicochemical parameters. This methodology combined AI techniques with high-throughput experimentation to synthesize catalysts supported by SiO_2, containing Ru, W, and P, which were then tested for the oxidation of propylene [72].

Machine learning, particularly Bayesian optimization [73], has been employed by Norskov et al. in the design and optimization of catalysts for fuel cells [74]. Recognizing the vast amount of available information about catalysis and catalyst design, Wang and others developed an AI-based method aimed at enhancing catalyst optimization through the integration of existing literature knowledge and practical experimentation [75]. This AI workflow was specifically applied to improve the production of ammonia.

12. Heterogeneous iron catalysis

The advantages of using iron in catalysis stem from its ready availability, accessibility, low cost, abundance, low toxicity, selectivity, and recyclability, making iron a highly attractive option in organic synthesis [4, 76]. Iron (Fe) is the most abundant and least expensive transition metal, especially when compared to other first-row transition metals like Ni, Mn, Co, and Cu [4, 76]. It is also more abundant and affordable than second-row transition metals such as Pd, Ru, Rh, Ir, and Pt [4, 76].

These beneficial characteristics have led to the widespread application of iron catalysis in industrial organic synthesis. Consequently, heterogeneous iron catalysis has become commonly used in various organic synthesis processes [77]. Specific applications include C−H bond functionalization, C−C cross-coupling reactions, and asymmetric synthesis [5, 78].

13. Iron catalysis and its role in green chemistry

The use of catalysis in chemistry is one of the 12 principles of green chemistry [79]. Unlike second-row transition metals, which are expensive and toxic, first-row transition metals are generally less toxic, more abundant in the Earth's crust, less expensive, and environmentally friendly. Among these, iron is the most abundant first-row transition metal. Therefore, utilizing iron as a catalyst enhances the valuable application of catalysis in line with green chemistry principles. Given the earth's abundance and low toxicity of iron, its use aligns with another green chemistry principle: "Less Hazardous Chemical Synthesis." This principle focuses on avoiding the use of toxic chemicals and minimizing waste or byproducts that can harm the environment. Additionally, iron catalysis relates to the "Design of Benign Chemicals" principle, contributing to the overarching goal of "Prevention". By employing iron as a catalyst, the reduction of waste at the source can be achieved [79]. As a result, iron catalysis is well-aligned with the goals of green chemistry and sustainability.

It has been successfully used to convert CO_2 into synthesis gas, specifically hydrogen (H_2) and carbon monoxide (CO), under photocatalytic conditions [80]. This method represents a valuable conversion of harmful greenhouse gases into useful feedstock—synthesis gas. In this process, $[Fe_3(CO)_{12}]$ serves as the iron precatalyst, a Ru catalyst acts as the photocatalyst, 2,2′-bipyridine functions as an electron reservoir, triethanolamine serves as an electron and proton donor, and N-methylpyrrolidone (NMP) is used as a solvent, all under visible light irradiation (see **Figure 36**) [80].

Iron catalysis [81], when combined with copper, has been shown to effectively photoelectrochemically reduce CO_2 to methane (CH_4) [82]. A binary iron-copper catalyst has demonstrated its efficiency as an electrocatalyst for the selective reduction of CO_2 to CH_4. Density functional theory (DFT) calculations indicate that copper and iron work in synergy to create a distorted angle of 126.05° for the O−C−O bond. This distortion facilitates a strong interaction with CO_2, enhancing its conversion into CH_4 [82]. Additionally, iron catalysis has been utilized for the conversion of CO_2 to carbon monoxide (CO) under electrochemical conditions [83]. Specifically, the iron-porphyrin catalyst $[Fe(TPP)]^{2-}$ has been employed in the electrocatalytic conversion of CO_2 into synthesis gas (**Figure 37**).

Iron has been utilized to reduce carbon dioxide (CO_2), a greenhouse gas that contributes to global warming. Iron sulfide (FeS) nanoparticles have been employed to convert gaseous CO_2 into methanol (CH_3OH) [84]. Additionally, FeS has been used in its pure form as well as doped with various transition metals such as nickel (Ni), manganese (Mn), cobalt (Co), and titanium (Ti). FeS has also been effective in catalyzing carbon fixation in terrestrial hot spring environments [84].

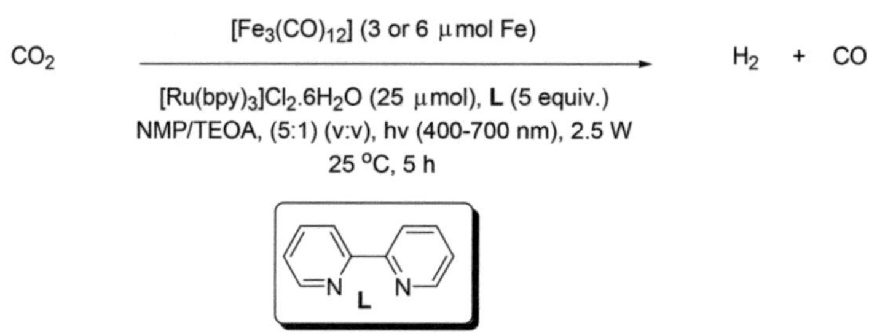

Figure 36.
Iron-catalyzed photocatalytic conversion of CO_2 into synthesis gas.

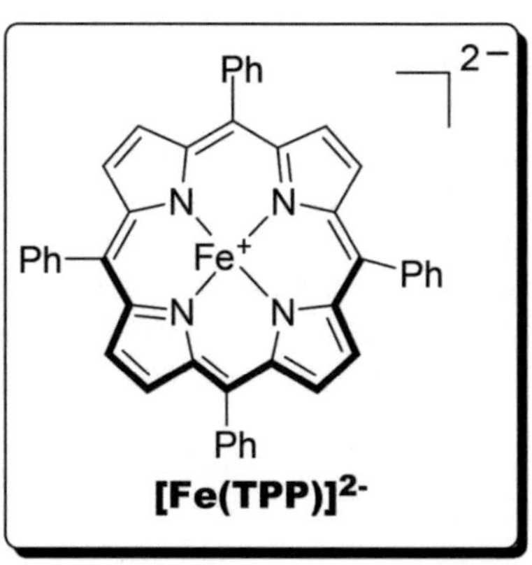

Figure 37.
$[Fe[TPP)]^{2-}$.

DOI: http://dx.doi.org/10.5772/intechopen.1009449

14. Limitations of iron catalysis

Iron has numerous advantages that make it a popular choice in catalysis, including its ready availability, accessibility, low cost, abundance, low toxicity, selectivity, and recyclability (vide supra) [4, 76]. These qualities have garnered significant interest from the scientific community. However, there are also limitations associated with using iron as a catalyst [8]. For instance, iron is not highly electron-dense, which can prevent it from sustaining certain chemical reactions or their specific conditions. This low electron density may lead to a rapid loss of ligands bound to the iron center, resulting in changes to the electronic state of iron. Consequently, this can decrease selectivity and complicate reaction pathways. Iron can exist in various oxidation states, which can further complicate catalytic cycles and lead to unintended synthetic routes. Additionally, certain chemical reactions may require preactivation of iron. Another drawback of iron catalysts is their low stereoselectivity, such as enantioselectivity. In contrast, the intermediates formed in organic reactions catalyzed by noble transition metals like palladium (Pd), rhodium (Rh), and ruthenium (Ru) are generally well-characterized. However, the intermediates in iron-catalyzed organic reactions are not as well understood. This uncertainty stems from the high lability of organoiron compounds, which can hinder their isolation and characterization. As a result, there is limited understanding of the mechanisms involved in iron-catalyzed organic transformations [8]. Furthermore, iron and its complexes face challenges regarding selectivity and control in redox reactions. Iron complexes can participate in both two-electron transfer and single-electron transfer processes, with the latter potentially interfering with the former. This lack of control over electron transfer presents a significant limitation in iron catalysis. Current research is focused on addressing these limitations of iron. Some strategies involve combining iron precatalysts with ligands typically associated with noble metals, while others explore the use of bimetallic complexes (Fe-M) that incorporate two first-row transition metals as catalysts in organic synthesis.

15. Conclusion

Iron is a highly attractive transition metal due to its low cost, abundant availability, and environmentally friendly nature. It plays a critical role in biological systems, particularly in its presence in heme b, which facilitates the transport of oxygen throughout the bloodstream in humans and other mammals. Unlike precious metals such as palladium and platinum, iron is readily accessible, affordable, low in toxicity, selective, and recyclable. As a transition metal, iron can exist in various oxidation states, allowing it to participate in numerous catalytic cycles across a wide range of synthetic organic reactions. Iron catalysis is evident in many organic reactions, where it facilitates the formation of new carbon-carbon (C—C) and carbon-X (C—X) bonds. Its significance is particularly notable in classical halogenation reactions of benzene, aiding in the synthesis of versatile halogenated benzenes through electrophilic aromatic substitution. Moreover, iron catalysis extends to oxidation processes, Baeyer-Villiger reactions, and cycloaddition reactions, such as Diels-Alder [4 + 2] cycloaddition. The importance of iron catalysis is also highlighted in cross-coupling reactions, where it is widely used for robust C—C bond formation. The elegance of iron complexes is showcased by iron carbonyls, which can reverse the nucleophilic nature of alkenes and conjugated 1,3-dienes, making them electrophilic and susceptible to nucleophilic attack. Additionally, iron catalysis has made substantial contributions to catalytic C−H bond functionalization reactions, including methylation, arylation, and the versatile and powerful C−H borylation. Inspired by nature, bioinspired iron catalysis has emerged in the realm of C−H bond functionalization. The impact of photoredox C−H bond functionalization is significant, although the catalysis is typically executed by ruthenium (Ru) and iridium (Ir). The potential for iron-photocatalyzed C−H bond functionalization remains to be explored in the near future. Iron possesses many advantages, such as its ready availability, accessibility, low cost, abundance, low toxicity, selectivity, and recyclability, which play a vital role in green chemistry. The impact of iron catalysis is evident in three main principles of green chemistry: prevention, catalysis, and the design of benign chemicals. Its role in green chemistry is demonstrated by the reduction of CO_2 to synthesis gas, CO, and H_2. While iron catalysis has many advantages and benefits, it does have some limitations, including the electronic nature of Fe and issues related to stereoselectivity. Current trends in iron catalysis aim to minimize these limitations. Given the green features of iron (*vide supra*), iron catalysis catalysis will contribute to various green processes such as green C−H bond functionalization and sustainability-driven processes.

Author details

Hamad H. Al Mamari
Department of Chemistry, College of Science, Sultan Qaboos University, Muscat, Sultanate of Oman

*Address all correspondence to: halmamari@squ.edu.om

IntechOpen

© 2025 The Author(s). Licensee IntechOpen. This chapter is distributed under the terms of the Creative Commons Attribution License (http://creativecommons.org/licenses/by/4.0), which permits unrestricted use, distribution, and reproduction in any medium, provided the original work is properly cited.

References

[1] Pearson AJ. Iron Compounds in Organic Synthesis. London: Academic Press; 1994

[2] Roat-Malone RM. Bioinorganic Chemistry, A Short Course. New Jersey: Wiley; 2007

[3] Sono M, Roach MP, Coulter ED, Dawson JH. Heme-containing Oxygenases. Chemical Reviews. 1996;**96**:2841-2887. DOI: 10.1021/cr9500500

[4] Plietker B, editor. Iron Catalysis in Organic Chemistry, Reaction and Applications. Wiley VCH: Weinheim; 2008

[5] Bauer I, Knölker HJ. Iron catalysis in organic synthesis. Chemical Reviews. 2015;**115**:3170-3387. DOI: 10.1021/cr500425u

[6] Bolm C, Legros J, Le Paih J, Lorenzo ZL. Iron-catalyzed reactions in organic synthesis. Chemical Reviews. 2004;**104**:6217-6254. DOI: 10.1021/cr040664h

[7] Guðmundsson A, Bäckvall JE. On the use of iron in organic chemistry. Molecules. 2020;**25**:1349-1369. DOI: 10.3390/molecules25061349

[8] Fürstner A. Iron catalysis in organic synthesis: A critical assessment of what it takes to make this base metal a multitasking champion. ACS Central Science. 2016;**2**:778-789. DOI: 10.1021/acscentsci.6b00272

[9] Anilkumar G, Bitterlich B, Gelalcha FG, Tseab MK, Beller M. An efficient biomimetic Fe-catalyzed epoxidation of olefins using hydrogen peroxide. Chemical Communications. 2007:289-291. DOI: 10.1039/b612048b

[10] Ten Brink GJ, Arends IWCE, Sheldon RA. The Baeyer–Villiger reaction: New developments toward greener procedures. Chemical Reviews. 2004;**104**:4105-4123. DOI: 10.1021/cr030011l

[11] Murahashi I, Oda Y, Naota T. Fe_2O_3-catalyzed baeyer-villiger oxidation of ketones with molecular oxygen in the presence of aldehydes. Tetrahedron Letters. 1992;**33**:7557-7560. DOI: 10.1016/S0040-4039(00)60823-7

[12] Chen GQ, Xu ZJ, Zhou CY, Che CM. Selective oxidation of terminal aryl and aliphatic alkenes to aldehydes catalyzed by iron(III) porphyrins with triflate as a counter anion. Chemical Communications. 2011;**47**:10963-10965. DOI: 10.1039/C1CC13574K

[13] Sawant SD, Hudwekar AD, Aravinda Kumar KAA, Venkateswarlu V, Singh PP, Vishwakarma RA. Ligand- and base-free synthesis of phenols by rapid oxidation of arylboronic acids using iron(III) oxide. Tetrahedron Letters. 2014;**55**:811-814. DOI: 10.1016/j.tetlet.2013.12.003

[14] Lai S, Lee DG. Lewis acid assisted permanganate oxidations. Tetrahedron. 2002;**58**:9879-9887. DOI: 10.1016/S0040-4020(02)01227-9

[15] Frank DJ, Guiet L, Käslin A, Murphy E, Thomas SP. Iron-catalysed alkene hydrogenation and reductive cross-coupling using a bench-stable iron(II) precatalyst. RSC Advances. 2013;**3**:25698-25701. DOI: 10.1039/C3RA44519D

[16] Li Y, Yu S, Wu X, Xiao J, Shen W, Dong Z, et al. Iron catalyzed asymmetric hydrogenation of ketones. Journal of the American Chemical Society.

2014;**136**:4031-4039. DOI: 10.1021/ja5003636

[17] Kischel J, Mertins K, Michalik D, Zapf A, Beller M. A general and efficient iron-catalyzed benzylation of 1,3-dicarbonyl compounds. Advanced Synthesis and Catalysis. 2007;**349**:865-870. DOI: 10.1002/adsc.200600497

[18] Jana U, Maiti S, Biswas S. An efficient $FeCl_3$-catalyzed amidation reaction of secondary benzylic and allylic alcohols with carboxamides or p-toluenesulfonamide. Tetrahedron Letters. 2008;**49**:858-862. DOI: 10.1016/j.tetlet.2007.11.176

[19] Lee EC, Hodous BL, Bergin E, Shih C, Fu GC. Catalytic asymmetric Staudinger reactions to form â-lactams: An unanticipated dependence of Diastereoselectivity on the choice of the nitrogen substituent. Journal of the American Chemical Society. 2005;**127**:11586-11587. DOI: 10.1021/ja052058p

[20] Braconi E, Cramer N. Crossed regio- and enantioselective iron-catalyzed [4+2]- cycloadditions of unactivated dienes. Angewandte Chemie International Edition. 2022;**61**:e202112148 (1 of 5). DOI: 10.1002/anie.202112148

[21] Fujiwara K, Kurahashi T, Matsubara S. Cationic iron(III) porphyrin-catalyzed [4 + 2] cycloaddition of unactivated aldehydes with simple dienes. Journal of the American Chemical Society. 2012;**134**:5512-5515. DOI: 10.1002/anie.202112148

[22] Dietrich F, Stang PJ. Metal-Catalyzed Cross-Coupling Reactions. Weinheim: Wiley-VCH; 1998

[23] de Meijere A, Bräse S, Oestreich M, editors. Metal Catalyzed Cross-Coupling Reactions and More. Vol. 3. Hoboken, NJ, USA: Wiley; 2013. DOI: 10.1002/9783527655588

[24] Campeau LC, Hazari N. Cross-coupling and related reactions: Connecting past success to the development of new reactions for the future. Organometallics. 2019;**38**:3-35. DOI: 10.1021/acs.organomet.8b00720

[25] Tamura M, Kochi JK. Vinylation of Grignard reagents. Catalysis by iron. Journal of the American Chemical Society. 1971;**93**:1487-1489. DOI: 10.1021/ja00735a030

[26] Cahiez G, Avedissian H. Highly stereo- and chemoselective iron-catalyzed alkenylation of organomagnesium compounds. Synthesis. 1998;**8**:1199-1205. DOI: 10.1055/s-1998-2135

[27] Alois Fürstner A, Leitner A. Iron-catalyzed cross-coupling reactions of alkyl-grignard reagents with aryl chlorides, tosylates, and triflates. Angewandte Chemie, International Edition. 2002;**41**:609-612. DOI: 10.1002/1521-3773(20020215)41:4%3C609::AID-ANIE609% 3E3.0.CO;2-M

[28] Nakamura M, Matsuo K, Ito S, Nakamura E. Iron-catalyzed cross-coupling of primary and secondary alkyl halides with aryl Grignard reagents. Journal of the American Chemical Society. 2004;**126**:3686-3687. DOI: 10.1021/ja049744t

[29] Ludwig JR, Zimmerman PM, Gianino JB, Schindler CS. Iron(iii)-catalysed carbonyl–olefin metathesis. Nature. 2016;**533**:374-379. DOI: 10.1038/nature17432

[30] Shilov AE, Shul'pin GB. Activation of C—H bonds by metal complexes.

Chemical Reviews. 1997;**97**:2879-2932. DOI: 10.1021/cr9411886

[31] Kakiuchi F, Chatani N. Catalytic methods for C–H bond functionalization: Application in organic synthesis. Advanced Synthesis and Catalysis. 2003;**345**:1077-1101. DOI: 10.1002/adsc.200303094

[32] Rouquet GN, Chatani N. Catalytic functionalization of C(sp^2)–H and C(sp^3)–H bonds by using Bidentate directing groups. Angewandte Chemie, International Edition. 2013;**52**:11726-11743. DOI: 10.1002/anie.201301451

[33] Sambiagio C, Schönbauer D, Blieck R, Dao-Huy T, Pototschnio G, Schaaf P, et al. A comprehensive overview of directing groups applied in metal-catalyzed C–H functionalization chemistry. Chemical Society Reviews. 2018;**47**:6603-6743. DOI: 10.1039/C8CS00201K

[34] Al Mamari HH, Stefane B, Zugelj HB. Metal-catalyzed C–H bond functionalization of phenol derivatives. Tetrahedron. 2020;**76**:130925-130936. DOI: 10.1016/j.tet.2020.130925

[35] Požgan F, Groselj U, Svete J, Štefane B, Al Mamari HH. Recent advances in the nickel-catalyzed alkylation of C–H bonds. Molecules. 2024;**29**:1917. DOI: 10.3390/molecules29091917

[36] Al Mamari HH. Ir-catalyzed ortho-C–H Borylation of aromatic C(sp^2)-H bonds of carbocyclic compounds assisted by N-bearing directing groups. Reactions. 2024;**5**(2):318-337. DOI: 10.3390/reactions5020016

[37] Shang R, Ilies L, Nakamura E. Iron-catalyzed C–H bond activation. Chemical Reviews. 2017;**117**:9086-9139. DOI: 10.1021/acs.chemrev.6b00772

[38] Shang R, Ilies L, Nakamura E. Iron-catalyzed ortho C–H methylation of aromatics bearing a simple carbonyl group with methylaluminum and tridentate phosphine ligand. Journal of the American Chemical Society. 2016;**138**:10132-10135. DOI: 10.1021/jacs.5b04818

[39] Al Mamari HH, Diers E, Ackermann L. Triazole-assisted ruthenium-catalyzed C–H Arylation of aromatic amides. Chemistry - A European Journal. 2014;**20**:9739-9743. DOI: 10.1002/chem.201403019

[40] Gu Q, Al Mamari HH, Graczyk K, Diers E, Ackermann L. Iron-catalyzed C(sp^2)-H and C(sp^3)-H Arylation by Triazole assistance. Angewandte Chemie, International Edition. **2014**;*53*:3868-3871. DOI: 10.1002/anie.201311024

[41] Dombray T, Werncke CG, Jiang S, Grellier M, Vendier L, Bontemps S, et al. Iron-catalyzed C–H borylation of arenes. Journal of the American Chemical Society. 2015;**137**:4062-4065. DOI: 10.1021/jacs.5b00895

[42] Omae I. Cyclometalation Reactions, Five-Membered Ring Products as Universal Reagents. Tokyo: Springer; 2014

[43] Barry SM, Challis GL. Mechanism and catalytic diversity of Rieske non-heme iron-dependent oxygenases. ACS Catalysis. 2013;**3**:2362-2370. DOI: 10.1021/cs400087p

[44] Oloo WN, Que L Jr. Bioinspired nonheme iron catalysts for C–H and C=C bond. Accounts of Chemical Research. 2015;**48**:2612-2621. DOI: 10.1021/acs.accounts.5b00053

[45] Kal S, Xu S, Que L Jr. Bio-inspired Nonheme iron oxidation catalysis: Involvement of Oxoiron(V) oxidants in

cleaving strong C−H bonds. Angewandte Chemie, International Edition. 2020;**59**:7332-7349. DOI: 10.1002/anie.201906551

[46] Chen MS, White MC. A predictably selective aliphatic C–H oxidation reaction for complex molecule synthesis. Science. 2007;**318**:783-787. DOI: 10.1126/science.1148597

[47] Chen MS, White MC. Combined effects on selectivity Fe-catalyzed methylene oxidation. Science. 2010;**327**:566-571. DOI: 10.1126/science.1183602

[48] Narayanam JMR, Stephenson CRJ. Visible light. Photoredox catalysis: Applications in organic synthesis. Chemical Society Reviews. 2011;**40**:102-113. DOI: 10.1039/B913880N

[49] Marzo L, Pagire SK, Reiser O, Konig B. Visible-light photocatalysis: Does it make a difference in organic synthesis? Angewandte Chemie International Edition. 2018;**57**:10034-10072. DOI: 10.1002/anie.201709766

[50] de Groot LHM, Aleksandra Ilic A, Schwarz J, Warnmark K. Iron photoredox catalysis—past, present, and future. Journal of the American Chemical Society. 2023;**145**:9369-9388. DOI: 10.1021/jacs.3c01000

[51] Lytle FE, Hercules DM. Luminescence of Tris(2,2'-Bipyridine) ruthenium(II) dichloride. Journal of the American Chemical Society. 1969;**91**:253-257. DOI: 10.1021/ja01030a006

[52] Dedeian K, Djurovich PI, Garces FO, Carlson G, Watts RJ. A new synthetic route to the preparation of a series of strong photoreducing agents: Fac-Tris-Ortho-Metalated complexes of iridium(III) with substituted 2-phenylpyridines. Inorganic Chemistry. 1991;**30**:1685-1687. DOI: 10.1021/ic00008a003

[53] Gualandi A, Marchini M, Mengozzi L, Natali M, Lucarini M, Ceroni P, et al. Organocatalytic enantioselective alkylation of aldehydes with [Fe(bpy)3]Br2 catalyst and visible light. ACS Catalysis. 2015;**5**:5927-5931. DOI: 10.1021/acscatal.5b01573

[54] Parisien-Collette S, Hernandez-Perez AC, Collins SK. Photochemical synthesis of Carbazoles using an $[Fe(phen)_3](NTf_2)_2/O_2$ catalyst system: Catalysis toward sustainability. Organic Letters. 2016;**18**:4994-4997. DOI: 10.1021/acs.orglett.6b02456

[55] Hernandez-Perez AC, Collins SK. A visible-light-mediated synthesis of carbazoles. Angewandte Chemie International Edition. 2013;**52**:12696-12700. DOI: 10.1002/anie.201306920

[56] Maçôas EMS, Kananavicius R, Myllyperkio P, Pettersson M, Kunttu H. Ultrafast electronic and vibrational energy relaxation of $Fe(acetylacetonate)_3$ in solution. The Journal of Physical Chemistry. A. 2007;**111**:2054-2061. DOI: 10.1021/jp066271z

[57] Xia S, Hu K, Lei C, Jin J. Intramolecular aromatic C−H acyloxylation enabled by iron photocatalysis. Organic Letters. 2020;**22**:13851389. DOI: 10.1021/acs.orglett.0c00002

[58] Neumeier M, Chakraborty U, Schaarschmidt D, de la Pena V, O'Shea VP, Perez-Ruiz R, et al. Combined photoredox and iron catalysis for the cyclotrimerization of alkynes. Angewandte Chemie, International Edition. 2020;**59**:13473-13478. DOI: 10.1002/anie.202000907

[59] Holmberg-Douglas N, Nicewicz DA. Photoredox-catalyzed

C–H functionalization reactions. Chemical Reviews. 2022;**122**:1925-2016. DOI: 10.1021/acs.chemrev.1c00311

[60] Emmert-Streib F. From the digital data revolution toward a digital society: Pervasiveness of artificial intelligence. Machine Learning and Knowledge Extraction. 2021;**3**:284-298. DOI: 10.3390/make3010014

[61] Yang W, Fidelis TT, Sun WH. Machine learning in catalysis, from proposal to practicing. ACS Omega. 2020;**5**:83-88. DOI: 10.1021/acsomega.9b03673

[62] Maier WF. Early years of high-throughput experimentation and combinatorial approaches in catalysis and materials science. ACS Combinatorial Science. 2019;**21**:437-444. DOI: 10.1021/acscombsci.8b00189

[63] Goldsmith BR, Esterhuize JA, Liu J, Bartel CJ, Sutton C. Machine learning for heterogeneous catalyst design and discovery. AICHE Journal. 2018;**64**:2311-2323. DOI: 10.1002/aic.16198

[64] Eyke NS, Koscher BA, Jensen KF. Toward machine learning-enhanced high-throughput experimentation. Trends in Chemistry. 2021;**3**:120-132. DOI: 10.1016/j.trechm.2020.12.001

[65] Chen L, Chen Z, Xue Y, Su B, Chen W, Pang X, et al. High-entropy alloy catalysts: High throughput and machine learning-driven design. Journal of Materials Informatics. 2022;**2**:19. DOI: 10.20517/jmi.2022.23

[66] Jorge Benavides-Hernandez J, Dumeignil F. From characterization to discovery: Artificial intelligence, machine learning and high-throughput experiments for heterogeneous catalyst design. ACS Catalysis. 2024;**14**:11749-11779. DOI: 10.1021/acscatal.3c06293

[67] Xu G, Cai C, Zhao W, Yonghua Liu Y, Tao WT. Rational design of catalysts with earth-abundant elements. WIREs Computational Molecular Science. 2023;**13**:e1654. DOI: 10.1002/wcms.1654

[68] Suzuki K, Toyao T, Maeno Z, Takakusagi S, Shimizu KI, Takigawa I. Statistical analysis and discovery of heterogeneous catalysts based on machine learning from diverse published data. ChemCatChem. 2019;**11**:4537-4547. DOI: 10.1002/cctc.201900971

[69] Li Z, Ma X, Xin H. Feature engineering of machine-learning chemisorption models for catalyst design. Catalysis Today. 2017;**280**:232-238. DOI: 10.1016/j.cattod.2016.04.013

[70] Li Z, Achenie LEK, Xin H. An adaptive machine learning strategy for accelerating discovery of perovskite electrocatalysts. ACS Catalysis. 2020;**10**:4377-4384. DOI: 10.1021/acscatal.9b05248

[71] Samuel Mace S, Xu Y, Nguyen BN. Automated transition metal catalysts discovery and optimisation with AI and machine learning. ChemCatChem. 2024;**16**:e202301475 (2 of 12). DOI: 10.1002/cctc.202301475

[72] Foppa L, Sutton C, Ghiringhelli LM, De S, Löser P, Schunk SA, et al. Learning design rules for selective oxidation catalysts from high-throughput experimentation and artificial intelligence. ACS Catalysis. 2022;**12**:2223-2232. DOI: 10.1021/acscatal.1c04793

[73] Snoek J, Larochelle H, Adams RP. Practical Bayesian optimization of machine learning algorithms. In: Advances in Neural Information Processing Systems 25: 26th Annual Conference on Neural Information Processing Systems 2012; 3-6 December 2012. Vol. 25. Lake Tahoe, Nevada, United States; 2012

[74] Nørskov J, Bligaard T, Rossmeisl J. Towards the computational design of solid catalysts. Nature Chemistry. 2009;**1**:37-46. DOI: 10.1038/nchem.121

[75] Lai NS, Tew YS, Zhong X, Yin J, Li J, Yan B, et al. Artificial intelligence (AI) workflow for catalyst design and optimization. Industrial and Engineering Chemistry Research. 2023;**62**:17835-17848. DOI: 10.1021/acs.iecr.3c02520

[76] Plietker B. Topics in Organometallic Chemistry-Iron Catalysis: Fundamentals and Applications. Berlin: Springer Verlag; 2011. p. 33

[77] Baruah MJ, Dutta R, Zaki MEA, Bania KK. Heterogeneous iron-based catalysts for organic transformation reactions: A brief overview. Molecules. 2024;**29**:3177-3206. DOI: 10.3390/molecules29133177

[78] Rana S, Biswas JP, Paul S, Paik A, Maiti D. Organic synthesis with the most abundant transition metal–iron: From rust to multitasking catalysts. Chemical Society Reviews. 2021;**50**:243-472. DOI: 10.1039/D0CS00688B

[79] Manahan SE. Green Chemistry and the Ten Comandments of Sustainability. 2nd ed. Columbia, Missouri, USA: ChemChar Research, Inc. Publishers; 2005

[80] Alsabeh PG, Rosas-Hernández A, Barsch E, Junge H, Ludwig R, Beller M. Iron-catalyzed photoreduction of carbon dioxide to synthesis gas. Catalysis Science & Technology. 2016;**6**:3623-3630. DOI: 10.1039/C5CY01129A

[81] Zhoua B, Ou P, Pant N, Cheng S, Vanka S, Chu S, et al. Highly efficient binary copper–iron catalyst for photoelectrochemical carbon dioxide reduction toward methane. PNAS. 2020;**117**:1330-1338. Available from: https://www.jstor.org/stable/26897547

[82] Bonetto R, Crisanti F, Sartorel A. Carbon dioxide reduction mediated by iron catalysts: Mechanism and intermediates that guide selectivity. ACS Omega. 2020;**5**:21309-21319. DOI: 10.1021/acsomega.0c02786

[83] Tarrago M, Ye S, Neese F. Electronic structure analysis of electrochemical CO_2 reduction by iron-porphyrins reveals basic requirements for design of catalysts bearing noninnocent ligands. Chemical Science. 2022;**13**:10029-10047. DOI: 10.1039/D2SC01863B

[84] Nan J, Luo S, Tran QP, Fahrenbach AC, Lu WM, Hu Y, et al. Iron sulfide-catalyzed gaseous CO_2 reduction and prebiotic carbon fixation in terrestrial hot springs. Nature Communications. 2024;**15**:10280-10290. DOI: 10.1038/s41467-024-54062-y

Chapter 6

Exploring Cobalt, Nickel, and Palladium Complexes for the Hydrogenation Reactions: A Review

Oluwasegun Emmanuel Olaoye and Fortunate Phenyo Sejie

Abstract

Catalyst design and synthesis are among the most important aspects of catalysis, especially for hydrogenation reactions. This work, through a systematic review of relevant literature, seeks to identify key parameters for the application of homogeneous cobalt, nickel, and palladium catalysts in hydrogenation reactions, particularly in the partial hydrogenation of biodiesel. The hydrogenation of biodiesel using the Co, Ni, and Pd catalysts stabilized by different ligands is dependent on the activity of the catalyst and its interaction with a ligand that can stabilize it at low oxidation states during the catalytic cycle. A Pd catalyst coordinated to a monodentate nitrogen ligand shows the highest activity, giving 100% conversion of methyl linoleate (C18:2) to methyl stearate (C18:0). The Pd (II) catalyst with a bidentate ligand, however, shows high activity but low stability as it reduces to Pd(0) species. The product selectivity depends on the time; hence, the desired product or composition mixture of the biodiesel is achieved by kinetic control of the reaction process. The optimized composition of the partially hydrogenated biodiesel sample oil sample was found to be 18.2% C18:0, 49.2% C18:1, 8.1% C18:2, and 23.4% C16:0, which showed improvement of the oxidative stability by 3 h. The use of homogeneous catalysts in biodiesel studies remains a challenge due to the difficulty in separating the catalysts from the hydrogenated material. Highly active catalysts are unsuitable for partial biodiesel hydrogenation as they produce excessive saturated substrates, which are undesirable.

Keywords: hydrogenation reactions, cobalt complexes, nickel complexes, palladium complexes, biodiesel hydrogenation

1. Introduction

Hydrogenation is the addition of hydrogen atoms across the unsaturated bonds. The strong covalent bonds of the hydrogen atoms make the splitting of the molecule have large energy barriers [1]. To overcome this, a catalyst is necessary for the addition of hydrogen atoms to the olefinic bonds. This reaction plays a crucial role in various industries, including petrochemicals, pharmaceuticals, and biodiesel production [1, 2]. In the context of renewable energy, catalytic hydrogenation is particularly significant for improving the properties of biodiesel, a leading alternative to conventional diesel. Biodiesel, derived from vegetable oils or animal fats, contains polyunsaturated fatty acids that are prone to oxidative degradation. This instability affects its storage, performance, and overall suitability as a transportation fuel. Catalytic hydrogenation addresses these issues by selectively converting polyunsaturated fatty acids into monounsaturated or partially saturated forms, reducing the biodiesel's oxidative stability [2].

The use of a catalyst, however, subdivides the addition of the two hydrogen atoms to the unsaturated moiety into a few catalytic steps that are not restricted by any symmetry [1]. These include the *oxidative addition* of hydrogen to the metal center, *insertion* of the coordinated unsaturated system into a metal-hydrogen bond, and *reductive elimination* of the hydrogenation product. Metal hydrides are well-known in catalytic hydrogenation reactions. Examples of the classical hydride complexes include $HM_n(CO)_5$, $H_2Fe(CO)_4$, and $HCoCO_4$, and they are prepared by the protonation of the metal complex or reaction of the metal complex with hydrogen gas. Non-classical dihydride complexes have also been reported, inclusive of $H_2M(Pr_3)_2CO_3$ [3]. Metal hydrides can be either weak bases or acidic, examples include $H(CO)_5Co$, which is acidic, and the η^5-$(C_5H_5)_2ReH$ which is a weak base. The source of hydrogen atoms during a hydrogenation reaction can be from molecular hydrogen gas or hydrogen donors (transfer hydrogenation) that produce the gas *in situ* [2]. The hydrogenation reactions carried out using hydrogen donors instead of molecular hydrogen gas have gained popularity due to their affordability and simple laboratory setup [4].

Transfer dehydrogenation reactions, often referred to today as transfer hydrogenation (TH) processes, involve donor and acceptor molecules that are typically different from one another. The donor compound needs to have a low oxidation potential to effectively participate in the reaction [1, 5]. Several compounds that can give up hydrogen include alcohols, glycols, aldehydes, amides, ethers, amines, aromatic hydrocarbons, and formate salts (**Figure 1**). An excellent donor dehydrogenates to give environmentally friendly by-products and should enhance selectivity toward desired products [1]. Transfer hydrogenation has become the center of research because it uses molecules that are safe to handle compared to pressurized H_2 molecule and it is easy to handle. Also, the use of hydrogen donors influences interesting selectivity in the reduction of substrates. The major milestone in the development of transfer hydrogenation (TH) reactions was the discovery of their involvement with late transition metals from groups 8, 9, 10, and 11, spanning the first to third rows of the periodic table. Additionally, metals like ruthenium and rhodium emerged as highly active catalysts for these reactions [1, 5]. However, their toxicity and high costs posed significant environmental and economic concerns. This challenge has sparked renewed interest in organometallic chemistry, particularly in the pursuit of more affordable and abundant first-row metal catalysts for transfer hydrogenation processes [6, 7].

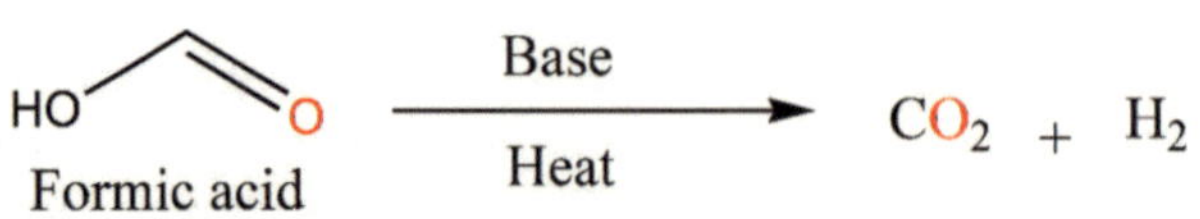

Figure 1.
Decomposition of formic acid to give hydrogen gas and carbon dioxide [5].

2. Hydrogenation of alkene/olefins

Hydrogenation reactions play a crucial role in the production of fine chemicals used in industries such as pharmaceuticals, agrochemicals, fragrances, cosmetics, and petrochemicals [6, 7]. These reactions enable the formation of single bonds from alkenes or alkynes, carbon-oxygen bonds from aldehydes, ketones, or esters, and carbon-nitrogen bonds (amines) from nitriles or imines [8]. Transition metals or organometallic compounds have been used as catalysts for the hydrogenation of substrates to the desired products. For instance, nickel, palladium, or platinum catalysts are required to transform alkenes and alkynes into alkanes. Chlorotris(triphenylphosphine)rhodium(I) $[RhCl(PPh_3)_3]$ Wilkinson's catalyst is among the most widely used catalysts for the selective hydrogenation of a variety of unsaturated substrates, especially alkene/olefins or alkynes, without affecting functional groups such as C=O, C–N, and NO_2, and its applications have been extensively reported [9]. The hydrogenation mechanism with Wilkinson's catalyst involves steps such as oxidative addition, alkene insertion, migratory insertion of hydrogen (the rate-determining step), and reductive elimination, making it a highly effective tool in catalytic applications [9].

Following the discovery of Wilkinson's catalyst, which is pivotal for alkene hydrogenation, researchers focused on rhodium-based complexes to advance homogeneous olefin hydrogenation [10, 11]. Osborn et al. [9] developed cationic rhodium complexes, specifically $[Rh(diene)(PR_3)_2]^+$ or $[Rh(cod)(PR_3)_2]^+$ (where R represents methyl or phenyl groups and cod stands for 1,5-cyclooctadiene). These complexes facilitate olefin hydrogenation, yielding a solvated species, $[RhH_2(PR_3)_2(S)_2]^+$ (where S denotes the solvent), upon exposure to hydrogen in solvents such as tetrahydrofuran (THF), acetone, or ethanol [10]. Under mild experimental conditions, these complexes produce an active catalyst that is essential for homogeneous hydrogenation. For this catalytic action to occur, the solvent ligand must dissociate before olefin insertion, following Wilkinson's catalytic mechanism [11]. Additionally, the cationic iridium complex $[PCy_3(cod)Ir(py)]^+PF_6^-$ (where Cy is cyclohexyl and py is pyridine), known as Crabtree's catalyst, exhibits exceptional efficacy in olefin hydrogenation, surpassing previous systems by a factor of 100 [12]. Unlike Wilkinson's system, which requires ligand dissociation to expose the active site to the substrate, Crabtree's catalyst allows the formation of $[Ir(olefin)_2(py)(PCy_3)]^+$, which subsequently reacts with H_2 to form a dihydride complex [13].

Apart from the early emergence of iridium, rhodium, and/or ruthenium complexes for the catalytic hydrogenation of unsaturated compounds, nickel and palladium complexes have also been employed to hydrogenate unsaturated/olefinic compounds. Chatterjee et al. [14] investigated the catalytic performance of [Ni(saloph)] (**1**) [15] (**Figure 2**) in the hydrogenation of cyclohexene and cyclooctene, converting them into cyclohexane and cyclooctane, respectively [14]. The reactions were carried out at a hydrogen pressure (H_2) of 60 bar and a temperature of 50°C. The catalyst (using ethanol as solvent, t = 5 h, 1% catalytic loading) displayed greater reactivity with cyclohexene, achieving a turnover frequency (TOF) of 13 h^{-1} compared to 7 h^{-1} for cyclooctene. The proposed reaction mechanism involves the formation of a high-valent $[Ni^{IV}(saloph)(H)_2]$ species, generated through the oxidative addition of hydrogen to the [Ni(saloph)] complex [14].

Shelvin et al. [16] reported active and enantio-selective phosphine nickel catalysts (**2–7**) for the asymmetric hydrogenation of olefins in α,β-unsaturated ester (β-methyl cinnamate). Various diphosphino ligands (**Figure 3**) combined with $[Ni(OAc)]_2$ were

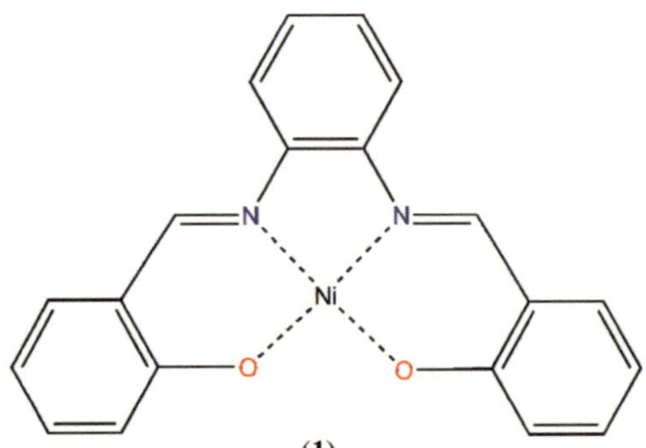

(1)

Figure 2.
Ni(saloph) complex explored for the hydrogenation of cyclohexene and cyclooctene [14].

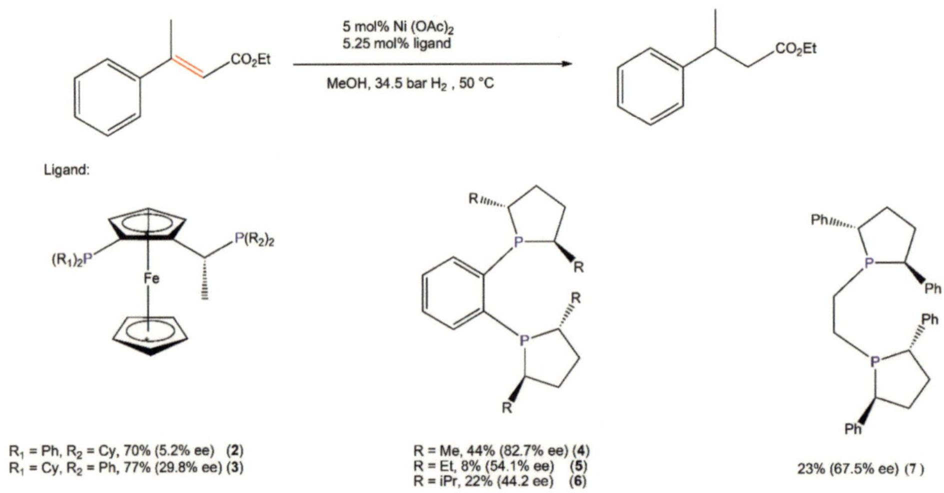

Figure 3.
Hydrogenation of β-methyl cinnamate catalyzed by bis(phosphino) nickel(II) complexes [16].

employed for the hydrogenation of olefins in β-methyl cinnamate. From their study, over 77% conversion and 89% enantioselectivity were obtained for the catalytic hydrogenation of β-methyl cinnamate with bis(phosphino)-nickel(II) complexes [16].

Similarly, palladium(II) complex of the form [Pd(PNP)Cl]Cl, containing N,N′-bis(diphenylphosphine)-2,6-diamopyridine, was reported by Schirme and his co-worker [17]. The palladium complexes were employed as a catalyst for the hydrogenation of styrene. The pincer-palladium catalysts, using ethanol at solvent at 60°C and 10 atm H_2 pressure, were efficient in the hydrogenation of styrene to the corresponding product [17]. Furthermore, Reddy investigated the kinetics of the hydrogenation of cyclohexene catalyzed by [Pd(P^N^P)Cl]Cl complexes (where PNP = N,N'-bis(diphenylphosphine) ethylbenzylamine) under mild experimental conditions (ethanol, 10–40°C and 1 atm H_2 pressure) [18]. Palladium(II) complexes (**8**) derived from protic HPNO acyl hydrazone, [Pd(PNO)X] (X=OAc, Cl, I) (**Figure 4**) were efficient in the hydrogenation of styrene under mild experimental conditions (MeOH, 40°C, 1 atm H_2 pressure, 1% catalyst loading) [19]. The catalyst bearing acetate group showed better catalytic activities, as compared to the one bearing chlorine and iodine with 100% conversion observed after 48 h.

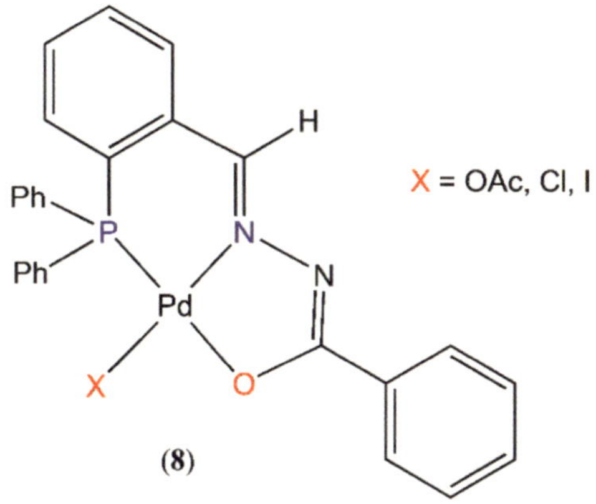

(8)

Figure 4.
Palladium(II) complexes bearing acyl hydrazone for hydrogenation reactions [19].

Gawron et al. [20] explored the effect of counterions on the cobalt complex $[Co(\eta^4\text{-cod})_2]^-$ (cod = 1,5-cyclooctadiene) (**9**) (**Figure 5**) for alkene hydrogenation. Using 3 mol% catalysts, they achieved 99% conversion of 1-octene under mild conditions (2 bar H_2, 30°C, 3 h). The study demonstrated that counterions significantly enhance the catalytic activity for 1-octene hydrogenation [20].

The efficiency of hydrogenation catalysts is determined by their activity, selectivity, and operational conditions. Wilkinson's catalyst ($RhCl(PPh_3)_3$) has long been a standard in olefin hydrogenation due to its exceptional activity and selectivity [9, 10]. However, the high cost and scarcity of rhodium have spurred interest in alternative catalysts, particularly cobalt (Co), nickel (Ni), and palladium (Pd) complexes. Cobalt complexes have shown promise for hydrogen activation through two-electron oxidative addition, closely mimicking the performance of noble metals like rhodium. These complexes are cost-effective and capable of operating under mild conditions, but their moderate or low turnover frequencies (TOFs) and occasional need for ligand modifications limit their overall efficiency [20]. Nickel complexes, valued for their abundance and affordability, are effective in hydrogenation when combined with pincer ligands that enhance electron donation. Despite these advantages, nickel catalysts typically exhibit lower TOFs and slower reaction rates compared to

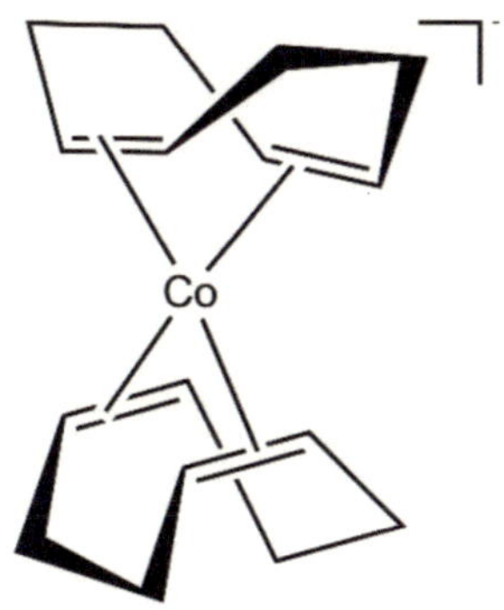

(9)

Figure 5.
A cobalt complex bearing 1,5 -cyclooctadiene ligand explored for the hydrogenation of 1-octene [20].

Wilkinson's catalyst [16]. Their requirement for higher temperatures and pressures further constrains their scalability. Palladium complexes provide a compelling alternative, balancing cost and performance. These catalysts achieve TOFs and selectivity levels comparable to Wilkinson's catalyst while operating under milder conditions. Palladium's resistance to catalyst poisoning and its ability to selectively hydrogenate alkenes without over-reduction make it particularly suitable for industrial applications [8]. **Table 1** shows a summary of the comparative analysis of the efficiency of cobalt, nickel, and palladium complexes versus Wilkinson's catalyst in olefin hydrogenation.

2.1 Homogeneous divalent nickel, palladium, and cobalt with nitrogen ligands and their applications in hydrogenation reactions

From the literature findings, nitrogen ligands have been proven to be valuable in stabilizing several homogeneous metal complexes that are widely used in hydrogenation reactions [21]. Nickel, an abundant and eco-friendly metal, is commonly used in hydrogenation reactions [22]. In a study comparing nickel-amide and nickel-imine catalysts for the transfer hydrogenation of 4-bromoacetophenone, the nickel-imine catalyst outperformed, achieving a 72% conversion under identical conditions, while the nickel-amide showed lower activities. Chiral diamine could also be used as an efficient ligand for palladium-catalyzed asymmetric hydrogenation of (E)-α-phenylcinnamic acid [23]. Palladium(II) complexes (**10** and **11**) with bidentate amine ligands, for instance, (S)-2-aminomethyl-1-ethyl pyrrolidine and (R,R)-1,2-diphenylethylenediamine, are reported to be active and efficient in the hydrogenation of olefin moiety in the cinnamic acid. Interestingly, 100% of the substrate is converted, yielding 76% *ee* conformer in the asymmetric hydrogenation of (E)-α-phenylcinnamic acid using (S)-2-aminomethyl-1-ethyl pyrrolidine as ligands. However, the use of (R,R)-1,2-diphenylethylenediamine led to a slight increase in enantioselectivity (**Figure 6**) [23].

The palladium(II) complex, bearing the N^O group, dates back to 1770 and has shown versatility as both homogeneous and heterogeneous catalysts, depending on the solvent [15, 24]. NHC ligand systems have demonstrated strong catalytic efficiency in the semi-hydrogenation of alkynes to produce Z-alkenes [15]. Additionally,

Catalyst	Substrate	TOF (h^{-1})	Reference(s)
Wilkinson	Hexene	650	[9–11]
	Cyclohexene	700	
	1-methylcyclohex-1-ene	13	
1	Cyclooctene	7	[14]
	Cyclohexene	13	
2–7	*β*-methyl cinnamate	n.d.	[16]
8	Styrene	209	[19]
9	1-octene	11	[20]

n.d = not determined.

Table 1.
Comparative analysis of the efficiency of Co, Ni-, and Pd- complexes versus Wilkinson's catalyst in olefin hydrogenation [9, 14–16].

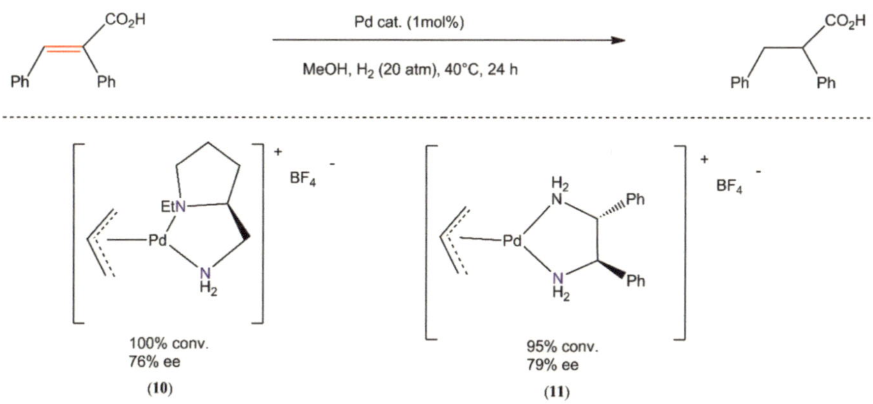

Figure 6.
Catalytic hydrogenation of olefin in cinnamic acid with complexes **10** *and* **11** *[23].*

the palladium complex with an imidazole moiety was among the first zero-valent catalysts known for its chemoselectivity and stereoselectivity in the semi-hydrogenation of both aliphatic internal alkynes and aromatic C=C bonds using formic acid as the hydrogen source [22]. Notably, product selectivity varies considerably with the choice of solvent, with MeCN and THF yielding different products under identical conditions [15]. Schiff base palladium complexes remain promising catalysts for biodiesel transfer hydrogenation. Similarly, pyridyl imine complexes of palladium, synthesized for various catalytic and biological applications, have been recognized for their selectivity in homogeneous catalysis, despite concerns over Pd(II) complex instability under molecular hydrogen pressure in the recent years. For instance, palladium complex bearing triaryl phosphine ligand effectively catalyzed styrene hydrogenation under mild conditions (40°C, 1 atm, and 1% catalyst load) [15]. Cobalt(II) complexes stabilized with nitrogen ligands present numerous catalytic opportunities, as nitrogen ligands help maintain lower oxidation states throughout catalytic cycles [25]. Unsaturated nitrogen ligands, such as imines, also serve as efficient π-acceptors and facilitate a deeper examination of cobalt chemistry compared to their saturated counterparts [25]. The catalytic use of cobalt complexes has been documented since the 1980s. A cobalt complex, $[Co(py)_6][(BPh_4)]$, for example, demonstrated high selectivity and activity for carbonyl hydrogenation at room temperature and atmospheric pressure [15]. Cobalt(II) complexes with nitrogen-based pincer ligands have also emerged as effective catalysts for the hydrogenation of CO_2, olefins, alkynes, ketones, aldehydes, and esters [15]. Although first-row transition metals such as cobalt are increasingly favored, the stability of their metal hydride or dihydride forms remains challenging. Notably, Lapointe et al. [26] investigated the catalytic activities of cobalt complexes (**12–14**) with bulky P^N^P ligands for alkene and alkyne hydrogenation (**Figure 7**). Their findings highlight the efficacy of these ligands in stabilizing cobalt centers while inhibiting undesirable side reactions. Furthermore, investigation of ligand effects revealed a two-electron process involving oxidative addition and reductive elimination, similar to noble metal reactivity. The best results were obtained at 100°C with 5 bar H_2 and 5 mol% catalyst loading [26].

Additionally, PNP-stabilized cobalt hydrides have also been reported in the literature, though the cobalt hydrides are expected to be highly unstable and not detectable using spectroscopic techniques [27]. Cobalt hydride complex **15** with P^N^P group (**Figure 8**) has been extensively studied as an ethylene hydrogenation catalyst, though

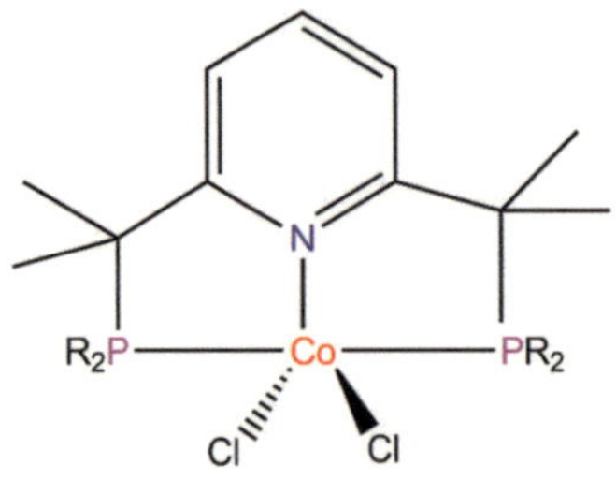

R = ^{i}Pr (**12**)
R = ^{t}Bu (**13**)

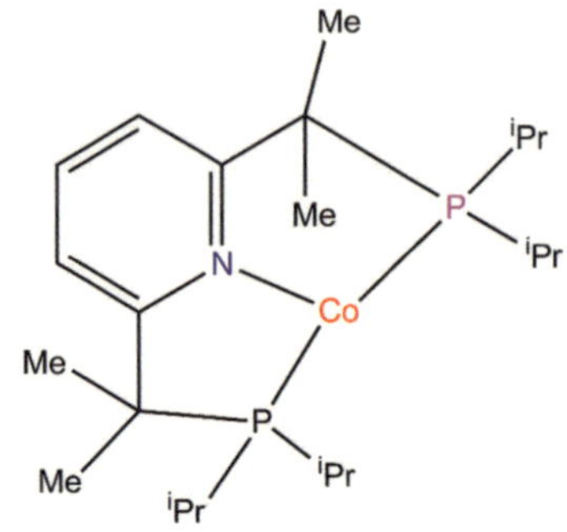

(**14**)

Figure 7.
*Cobalt catalysts (**12–14**) bearing P^N^P group for the hydrogenation of olefins [26].*

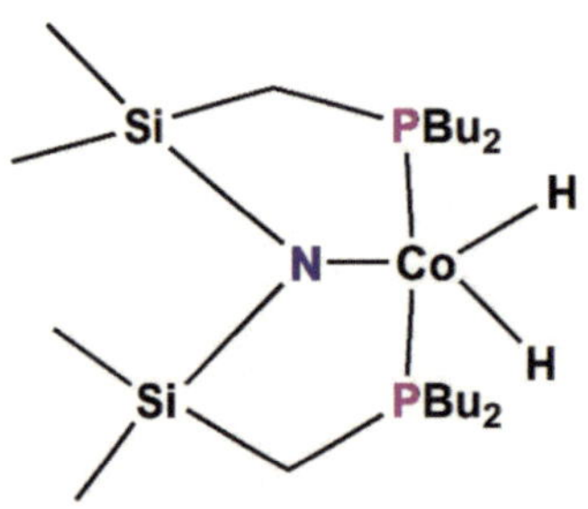

(**15**)

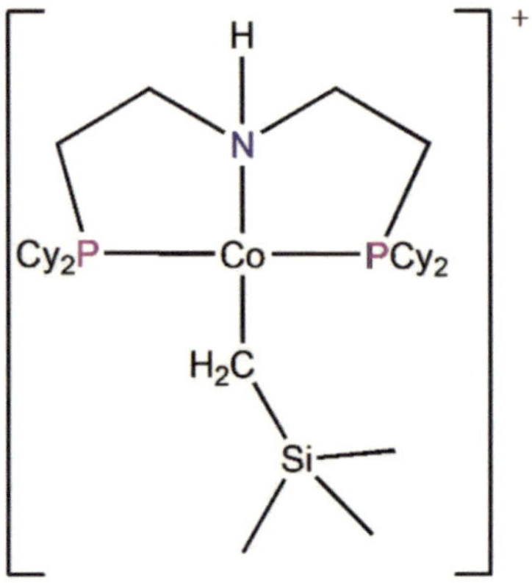

(**16**)

Figure 8.
Cobalt catalysts bearing P^N^P group for the hydrogenation of olefinic moiety in organic substrates [27, 28].

it showed limited activity with other alkenes [27]. Complex **16** was isolated as an ionic species stabilized with BAr_4F, which was used to hydrogenate methyl cyclohexanecarboxylate (MCHC) and ethyl cyclohexanecarboxylate (ECHC) [28]. The observed selectivities toward hydrogenation of the C=C were 66% and 2%, respectively (reaction conditions: substrate (0.5 mmol), catalyst loading (2 mol%), THF (1.0 mL), H_2 (55 bar), 120°C, 20 h).

3. Biodiesel production from seed oils

Biodiesel can be produced from plant sources (such as Jatropha and sunflower) through base or acid-catalyzed transesterification of fatty acid methyl esters (**Figure 9**) [29]. In our lab, we recently reported the biodiesel produced from indigenous seed oils, such as Jatropha [29]. In our study, we detailed a standard transesterification process for *Jatropha curcas* oil where 50 g of the seed oil is reacted with methanol at 60°C using 0.5 g of potassium hydroxide potassium hydroxide as a

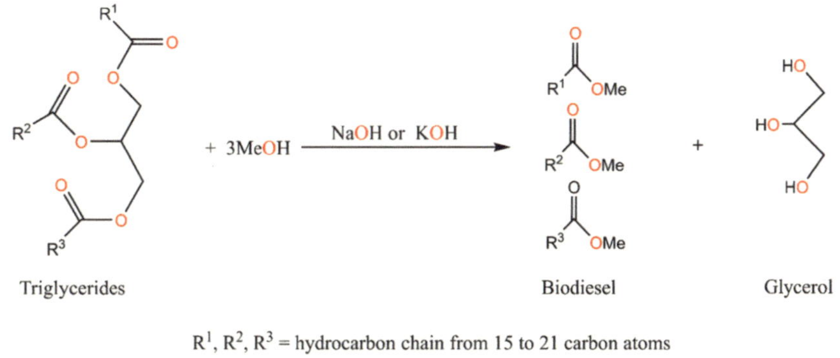

Figure 9.
The general equation for a transesterification reaction [29].

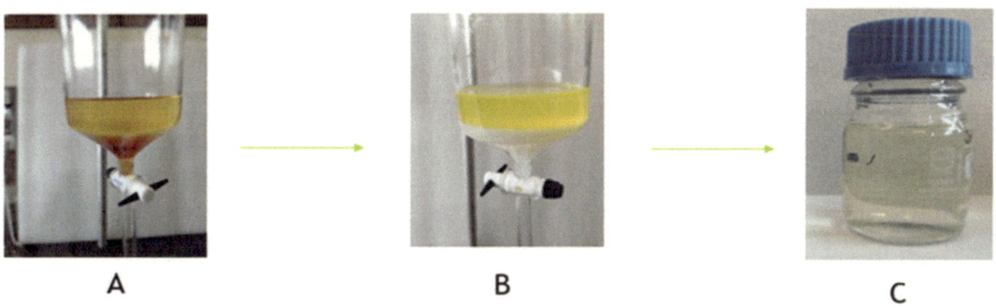

Figure 10.
A typical representation of biodiesel production.

catalyst [29]. The synthesized biodiesel was further purified by washing the product twice with warm distilled water to remove residual glycerol (**Figure 10**). A typical representation of a transesterification reaction is shown in **Figure 9**.

3.1 Hydrogenation of methyl linoleate using the homogeneous Ni, Co, and Pd divalent metal complexes stabilized with nitrogen ligands.

Many researchers have investigated the partial hydrogenation of fatty acid methyl esters (FAMEs) to improve the oxidative stability of biodiesel [30–32]. Biodiesel is composed of different amounts of FAMEs, of which some are unsaturated, while others are saturated. When a biodiesel sample contains high amounts of polyunsaturated components, the sample becomes prone to autooxidation reactions [29, 30]. Liu et al. [30] recently established 4 nm palladium nanoparticles dispersed in polyethylene glycol (PEG) to facilitate the selective/partial hydrogenation of methyl linoleate (**Figure 11**) [30]. Their study highlights the use of *in situ*-generated 4 nm palladium nanoparticles dispersed in PEG. The catalyst gave high selectivity toward methyl oleate (MO) without producing the saturated methyl stearate (MS) in the hydrogenation of methyl linoleate and sunflower oil [30], and the reported selectivities were attributed to the preferred coordination complex that is formed between the substrate and the palladium nanoparticles. This can either be coordinated through both C=C in the FAMEs or only one C=C. Furthermore, Carvalho and co-workers developed a novel catalytic system to address the low oxidative stability of biodiesel derived from soybean oil [32]. They utilized *in situ* generated palladium nanoparticles that are

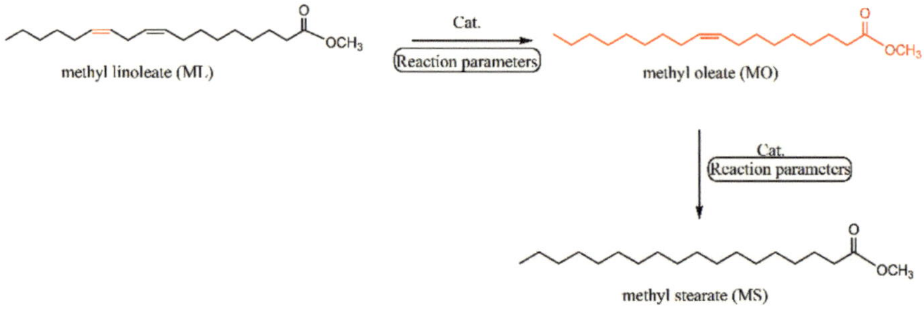

Figure 11.
Selective/partial hydrogenation of methyl linoleate (modeled biodiesel) [29].

dispersed in imidazolium-based ionic liquid (BMI·BF4) to enhance catalytic activity and reduce agglomeration. These catalysts also gave selectivity toward MO, and no MS was detected [32]. These two catalysts outperformed traditional heterogeneous catalysts like Pd/C in the partly hydrogenation of methyl linoleate with selectivity toward MO. In addition, the catalysts are recyclable up to three times without significant activity loss. Carvalho et al. also investigated reaction parameters such as temperature, hydrogen pressure, and catalyst concentration, highlighting the influence of the ionic liquid in stabilizing active species in different intermediates that are formed during the catalytic cycle, as well as enhancing selectivity [31]. In 2013, Cheng et al. [32] investigated the effects of different metal types (Ni and Pd) supported on SiO_2 or carbon as catalysts for the partial and/or full hydrogenation of cottonseed oil-derived fatty acid methyl esters (FAME) to improve oxidative stability. Their goal was to enhance the oxidative stability of biodiesel while maintaining favorable cold flow properties. Among the tested catalysts, Pd showed the highest catalytic activity and selectivity for producing mono-unsaturated FAME, which improves oxidative stability [32].

In our group, we investigated the homogeneous catalysts (**17–28**) shown in **Figures 12** and **13** for the hydrogenation of methyl linoleate and biodiesel, using hydrogen gas and hydrogen donor (formic acid), respectively. It is noted that the catalyst's efficiency depends on the active metal used to synthesize the catalyst and the ligand used to stabilize these metal centers [29, 33]. The catalysts in **Figure 8** are pyrazole complexes where sp^2-hybridized nitrogen from a monodentate ligand coordinates with the metal center, forming a square planar complex. Modifications to the ligand's R group, from hydrogen to bulky alkyl or ferrocene groups, introduce steric hindrance, potentially restricting bulky substrate access to the catalytic site. In **Figure 9**, a bidentate ligand with sp^2 and/or sp^3 hybridized nitrogen atoms coordinate to the metal center forming a five-membered ring between the ligand and the metal center.

As described in **Table 2**, complexes **17–28** were thoroughly investigated as catalysts for the catalytic performance of the methyl linoleate as modeled biodiesel [29, 33]. All six complexes (**17–22**) proved capable of catalyzing the molecular hydrogenation of methyl linoleate, resulting in the products depicted in **Figure 11**. However, complex **22** bearing the ferrocenyl ligand demonstrated superior catalytic activity in terms of selectivity toward methyl oleate (MO) under the same reaction conditions (40°C, 30 min), as shown in **Table 2**, entry 6 [29].

The structures of the nickel catalysts **17**, **24**, and **27** studied using DFT- MO6,6-31G** are shown in **Figure 14**. Due to the chelate effect, complexes **24** and **27** are expected to have enhanced stability compared to complex **17**. The optimized structures

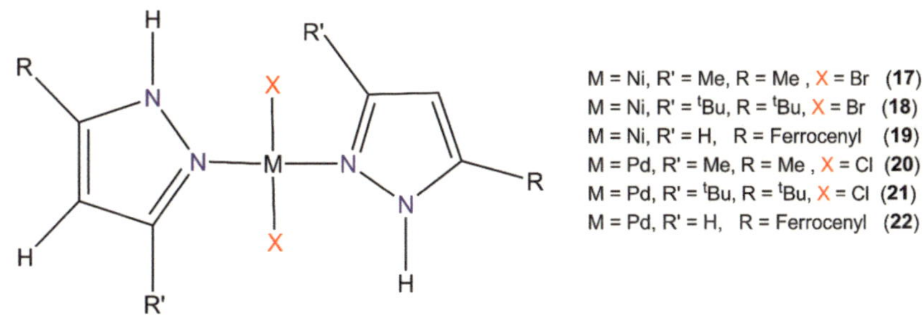

Figure 12.
Monomeric pyrazolyl nickel(II) and palladium(II) complexes (17–22) employed for the molecular hydrogenation reaction [29].

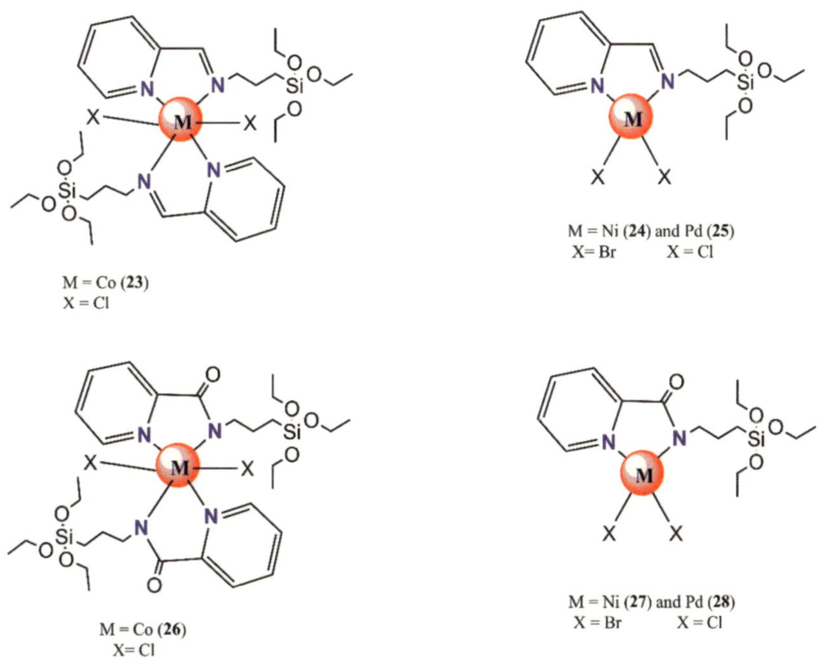

Figure 13.
The structure of imine (23–25) and amide (26–28) Co-, Ni-, and Pd-complexes explored for the transfer hydrogenation of biodiesel [33].

of the three metal complexes show expected behavior, with negative minimization energies indicating favored structures. Complexes with nickel coordinated to a chelation ligand exhibit lower energy, while those with a monodentate ligand show higher energy values. However, based on the experimental studies by Olaoye et al. [34] and Sejie et al. [35], the catalysts show to have enhanced catalytic activity with monodendate pyrazolyl and bidentate nitrogen ligands. This behavior could be attributed to the stability of the metal complexes before the activation of the nickel complexes [34, 35].

Additionally, Sejie et al. [33] evaluated complexes **23–28** for transfer hydrogenation of methyl linoleate using formic acid as a source of hydrogen [33]. The hydrogenation results reveal that the **23–28** complexes/catalysts produced varying compositions of saturated (MS) and unsaturated (MO) fatty acids. Complex 23 achieved complete hydrogenation (100%) to MS, while complexes **23** and **24** yielded lower percentages of MS (86% and

Entry	Complex	Time (h)	Conversion (%)	TON	TOF (h^{-1})	Amount of MO detected (%)[b]	Amount of MS detected (%)[b]
1[a]	**17**	0.5	35	42	84	65	4
2[a]	**18**	0.5	28	34	64	71	2
3[a]	**19**	0.5	29	35	70	71	3
4[a]	**20**	0.5	29	35	70	71	3
5[a]	**21**	0.5	100	120	240	0	100
6[a]	**22**	0.5	20	24	48	80	9
7[c]	**23**	12	100	650	54	14	86
8[c]	**24**	12	100	690	58	4	96
9[c]	**25**	12	100	730	61	0	100
10[c]	**26**	12	100	587	49	0	100
11[c]	**27**	12	100	587	49	0	100
12[c]	**28**	12	100	857	71	0	100
13[d]	**28**	6	62	624	104	28	3

[a]*Reaction conditions: 0.3 mmol of methyl linoleate; 3 μmol (1 mol%) of complex; 5 mL of methanol; 40°C; 5 bar.*
[b]*Conversions were estimated by ^1H NMR spectroscopy. Using dioxane as an internal standard. Each run was performed in duplicates.*
[c]*Reaction conditions (transfer hydrogenation): (ML 0.44 g, catalyst: 0.1 mol%), 12 h, 120°C, formic acid (40 mmol), KOH (10 mmol).*
[d]*= 6 h. Reaction mixture components were analyzed using GC-MS and/or ^1H NMR spectroscopy.*
TON in $mol_{substrate} mol^{-1}_{catalyst}$

Table 2.
Catalytic activities of complexes **17–28** *for the hydrogenation reaction of methyl linoleate using molecular hydrogen and formic acid [29, 33].*

96%, respectively) (**Table 2**, entries 7 and 8). Complexes **24–26**, based on amide ligands, efficiently converted 100% of ML to MS without side products, outperforming their imine-based counterparts (**23–24**) [33]. The amide catalysts' superior activity is attributed to the electron-withdrawing nature of the C=O group, enhancing the metal center's electrophilicity. Pd-based complexes exhibit the highest activity overall compared to the Ni- and Co-based complexes. Reducing reaction time to 6 h results in partial hydrogenation, allowing control over product composition, as evidenced by a mix of MS, MO, and ML (**Table 2**, entry 13).

3.2 Hydrogenation of biodiesel with nitrogen-based Co-, Ni-, and Pd-complexes

Our group investigated the partial/selective transfer hydrogenation of biodiesels produced from Jatropha seed oil using complexes **23–28** (**Figure 15**). The catalysts (**23–28**) were used to hydrogenate Jatropha biodiesel, which consists of different fatty acids and methyl esters of different chain lengths (**Figure 9**). The hydrogenation reactions were carried out under conditions designed for methyl linoleate (ML) hydrogenation. As demonstrated by earlier observations, all C18:1 and C18:2 were efficiently converted to C18:0 by catalysts **23–25**. On the other hand, less than 90% conversion of C18:1 and C18:2 to C18:0 was observed when using catalysts **26–28** [33].

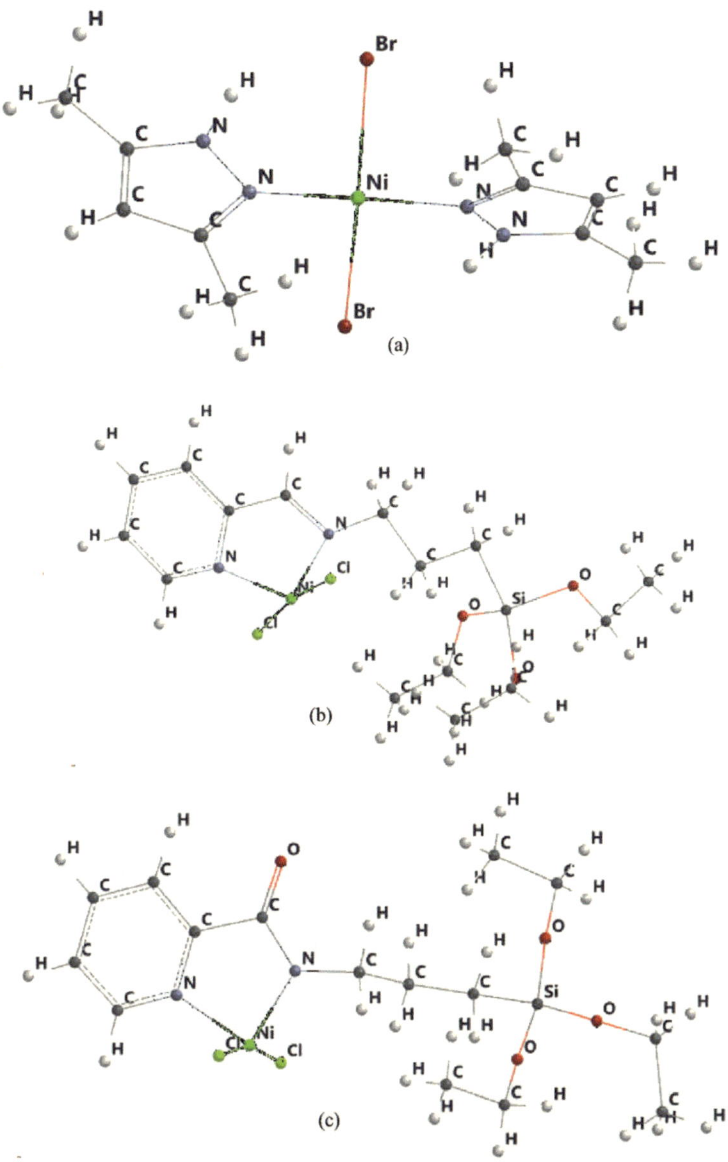

Figure 14.
The optimized geometry of nickel-pyrazole complex (17), imine complex (24), and amide complex (27) with minimization energies of −7265.03 kJ/mol, −2564.95 kJ/mol, and −1240.49 kJ/mol, respectively [34, 35].

The composition of the partially hydrogenated oil sample (H-BD1) was 18.2% C18:0, 49.2% C18:1, 8.1% C18:2, and 23.4% C16:0. This sample was tested for volatile water-soluble combustion products by passing oxygen gas over a heated sample of the oil sample (induced oxidation reaction), and the conductivity was determined over time [33]. The infrared (IR) spectrometer was used to study changes in the functional groups of the diesel before and after the oxidation reaction, after long-term storage. The findings showed that biodiesel samples with the higher levels of polyunsaturated fatty acids (C18:2) had shorter induction times, which indicate that they undergo oxidative reactions faster than those with lower levels of C18:2 components. The amount

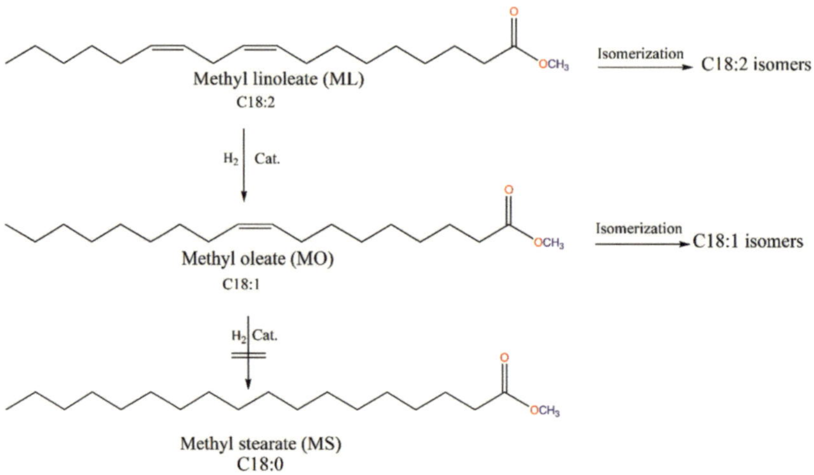

Figure 15.
Transfer hydrogenation of biodiesel with formic acid using complexes **21–26** *[33].*

of the monounsaturated fatty acid (C18:1) content was less significant in affecting the oxidation reactions as the hydrogenated sample showed a longer induction period compared to the hydrogenated sample despite the highest C18:1 content (41.54%). When compared to its un-hydrogenated counterpart, the partially hydrogenated sample has a higher induction time of 5 h compared to the un-hydrogenated samples, which had a 3-h induction period. However, the hydrogenation using homogeneous catalysts was shown to be less effective in improving the oxidative stability of the diesel as compared to blending the un-hydrogenated sample with 20% conventional diesel (LP5) with biodiesel. Catalytic activity was higher for the amide-based (**26–28**) compared to imine complexes (**23–25**). This was evident in the observed conversion of C18:2 to C18:0 [33].

Biodiesel hydrogenation enhances oxidative stability (in terms of selectivity) by converting unsaturated fatty acid methyl esters (FAME) into more stable mono or partially saturated esters. This process significantly influences biodiesel's storage and performance properties. A detailed comparison of catalysts, including cobalt (Co), nickel (Ni), and palladium (Pd), reveals advancements in catalytic efficiency, turnover metrics, and selectivity [29–33]. Cobalt-based complexes (**23** and **26**) exhibit high selectivity for methyl stearate (MS) formation, achieving complete conversion within extended reaction times (e.g., 12 h) as seen in **Table 3**. While cobalt is cost-effective and abundant, its lower activity compared to palladium stems from its higher activation energy, necessitating elevated temperatures and pressures [36]. Furthermore, nickel catalysts provide cost-effective alternatives for biodiesel hydrogenation but exhibit slower reaction rates compared to palladium-based systems. Chen et al. [32] observed that nickel catalysts had the lowest activity for the hydrogenation of biodiesel, indicating slower hydrogenation kinetics. Lastly, for palladium complexes, Liu et al. [30] reported the effectiveness of magnetic mesoporous palladium catalysts in selective hydrogenation of sunflower oil. Palladium systems achieved superior catalytic activity with a 92% yield of methyl oleate (MO) with full conversion of methyl linoleate. Pd also exhibited excellent reusability with no significant deactivation after multiple cycles [30]. The advantage of this is the high selectivity for mono-unsaturated products and low undesired products. Pd catalysts

Entry	Complex/ catalyst	Time (h)	Conversion (%)	Major product	Cost-effectiveness	Reference
1	**17**	0.5	35	MO	Very high	[29]
2	**18**	0.5	28	MO	Very high	[29]
3	**19**	0.5	29	MO	Very high	[29]
4	**20**	0.5	29	MO	Moderate	[29]
5	**21**	0.5	100	MS	Moderate	[29]
6	**22**	0.5	20	MO	Moderate	[29]
7	**23**	12	100	MS	High	[33]
8	**24**	12	100	MS	Very high	[33]
9	**25**	12	100	MS	Moderate	[33]
10	**26**	12	100	MS	High	[33]
11	**27**	12	100	MS	Very high	[33]
13	**28**	6	62	MO + MS	Moderate	[33]
14	Pd-PEG4000	3	100	MO	Moderate	[30]
15	Ni-SiO_2/Pd-C	1.5	100	MO + MS	Moderate	[32]

Table 3.
Summary of performance metrics of the cobalt, nickel, and palladium catalysts in the methyl linoleate and biodiesel hydrogenation [29–32].

operate efficiently under mild conditions, making them ideal for biodiesel optimization. The summary of the performance metrics of the oxidative stability (in terms of selectivity) improvement achieved by cobalt, nickel, and palladium catalysts is given in **Table 3**.

The Fourier molecular orbitals were calculated for the optimized geometries of the catalysts using the theoretical methods (**Figures 16–19**). This study is important as it helps to understand the catalytic activity and selectivity of the catalysts [37]. Methyl linoleate (HOMO of −9.0 eV and a LUMO of +0.4 eV) (**Figure 15**) interacts with the catalysts either through their HOMO or LUMO. Catalyst **17** is the most favorable due to its smaller energy gaps. The HOMO of catalyst **17** (−6.0 eV) interacts with the substrate's LUMO with a gap of 6.4, while its LUMO (−1.5 eV) interacts with the substrate's HOMO with a gap of 7.5 eV (**Figure 13**). These are significantly smaller compared to catalyst **24**, which have gaps of 8.7 eV and 10.5 eV, and catalyst **27**, with gaps of 9.5 eV and 10.8 eV. The smaller energy gaps of catalyst **17** enhance electron transfer in both donor-acceptor interactions, making it the most efficient catalyst for the hydrogenation reaction, as observed in the experimental data, where it takes short reaction times to hydrogenate substrates.

3.3 Conclusion and future perspectives

In this work, we reviewed the design, synthesis, characterization, and application of cobalt, nickel, and palladium complexes as catalysts for the hydrogenation of fatty acid methyl esters (FAMEs). These catalysts convert C18:2 in methyl linoleate and jatropha biodiesel into biodegradable C18:0 waxes Through kinetic control of the reaction time, it was possible to achieve partial hydrogenation after 6 h using

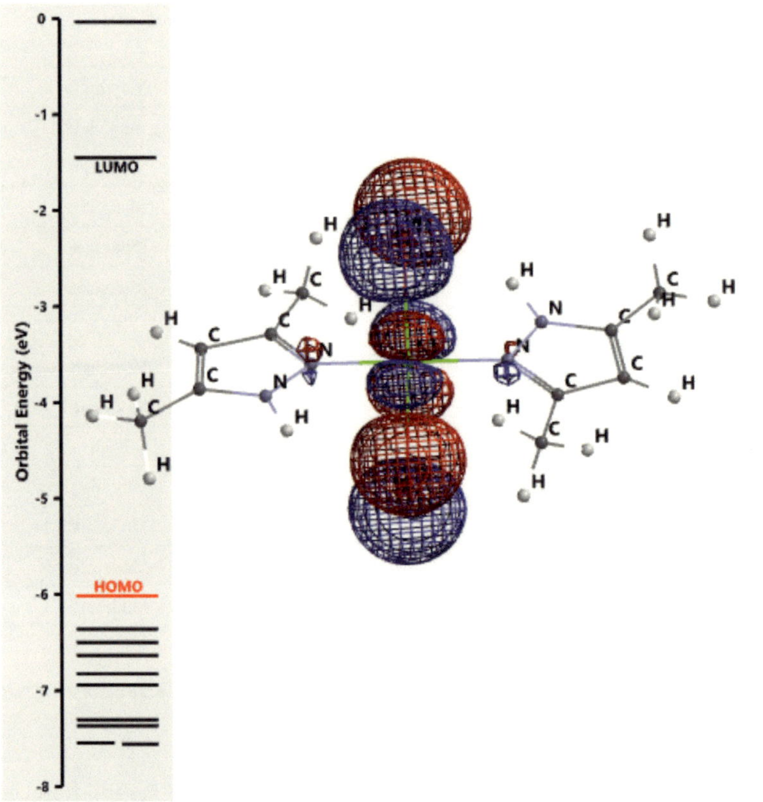

Figure 16.
Fourier molecular orbitals calculated for the optimized geometries catalyst **17** *[37].*

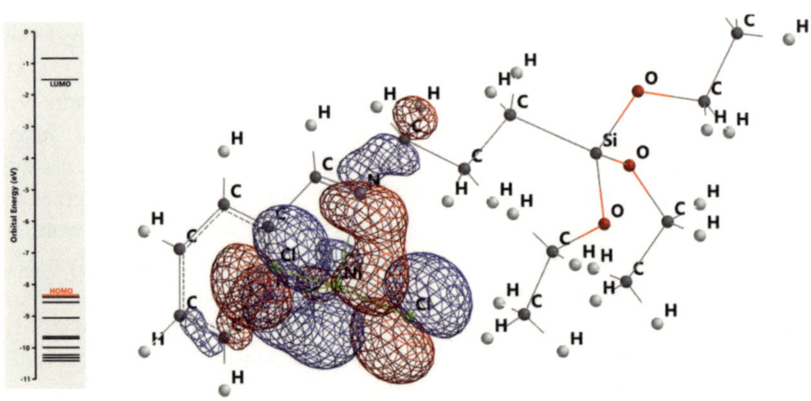

Figure 17.
Fourier molecular orbitals calculated for the optimized geometries catalyst **24** *[37].*

formic acid as the hydrogen donor. The catalyst stabilized with monodentate ligands is more stable and catalytically active compared to those with a bidentate nitrogen ligand which is supported by the DFT analysis. The exploration of cobalt, nickel, and palladium complexes for biodiesel hydrogenation highlights their versatility and potential to address the challenges associated with renewable energy and sustainable

DOI: http://dx.doi.org/10.5772/intechopen.1009451

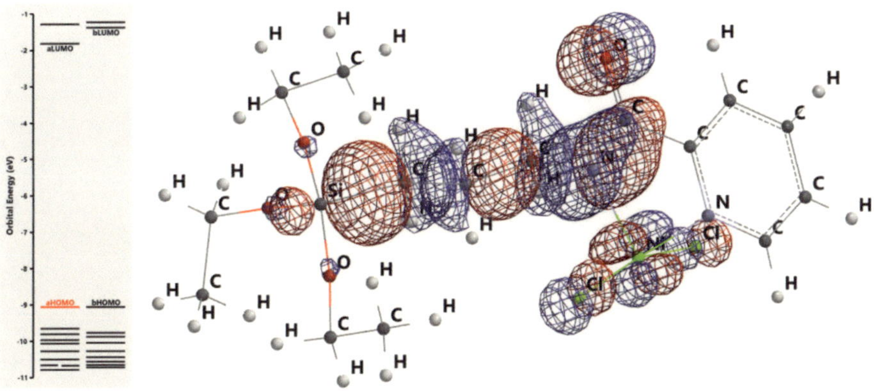

Figure 18.
Fourier molecular orbitals calculated for the optimized geometries of catalyst **27** *[37].*

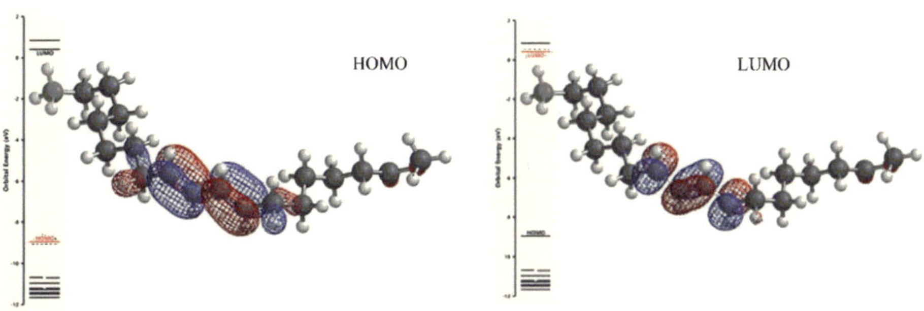

Figure 19.
Fourier molecular orbitals calculated for the methyl linoleate (modeled biodiesel) [37].

fuel production. These metal complexes, with their unique catalytic properties, have demonstrated efficacy in the hydrogenation of unsaturated fatty acids to improve the oxidative stability of biodiesel, in a time-dependent process. Though the catalyst converts over 50% of the substrate in a short period, controlled reaction kinetics can lead to the desired partially hydrogenated product. It is, however, difficult to control the isomerization of the C=C bonds in the C18:2 chains in the presence of the homogeneous Co, Ni, and Pd metal complexes. The catalyst structure proves to be a major contributor to the activity of the catalysts, while the product evolution from the four studies seems to be highly dependent on the reaction time. Palladium-based catalysts stand out for their superior selectivity and activity, offering a promising pathway for optimizing biodiesel's performance while minimizing the formation of fully saturated by-products. The integration of computational methods like density functional theory (DFT) can aid in unraveling the electronic and structural dynamics of these catalysts, paving the way for the rational design of next-generation systems. Additionally, leveraging renewable hydrogen sources and exploring catalytic systems that combine the strengths of homogeneous and heterogeneous catalysis could revolutionize biodiesel production. As the demand for clean and sustainable energy solutions grows, advancements in catalyst development will play a pivotal role in reducing reliance on fossil fuels and mitigating greenhouse gas emissions. By fostering innovation in catalytic chemistry, the energy sector can move closer to achieving a sustainable future.

Conflict of interest

The author(s) declare no potential conflicts of interest.

Author details

Oluwasegun Emmanuel Olaoye[1,2]* and Fortunate Phenyo Sejie[1]

1 Department of Chemistry, University of Botswana, Gaborone, Botswana

2 Department of Biochemistry and Industrial Chemistry, Caleb University, Lagos, Nigeria

*Address all correspondence to: segun.olaoye@yahoo.com

IntechOpen

© 2025 The Author(s). Licensee IntechOpen. This chapter is distributed under the terms of the Creative Commons Attribution License (http://creativecommons.org/licenses/by/4.0), which permits unrestricted use, distribution, and reproduction in any medium, provided the original work is properly cited.

References

[1] van Leeuwen PWNM. Homogeneous Catalysis: Understanding the Art. Dordrecht, Netherlands: Kluwer Academic Publishers; 2004. ISBN 1402031769

[2] Ashraful AM, Masjuki HH, Kalam MA, Rizwanul Fattah IM, Imtenan S, Shahir SA, et al. Production and comparison of fuel properties, engine performance, and emission characteristics of biodiesel from various non-edible vegetable oils. Energy Conversion and Management. 2014;**80**:202-228

[3] Bhaduri S, Mukesh D. Homogeneous Catalysis: Mechanisms and Industrial Applications. Hoboken, NJ, USA: John Wiley & Sons; 2014

[4] Werkmeister S, Neumann J, Junge K, Beller M. Pincer-type complexes for catalytic (De) hydrogenation and transfer (De) hydrogenation reactions: Recent progress. Chemistry–A European Journal. 2015;**21**(35):12226-12250

[5] Baidilov D, Hayrapetyan D, Khalimon AY. Recent advances in homogeneous base-metal-catalyzed transfer hydrogenation reactions. Tetrahedron. 2021;**98**:132435

[6] Saudan LA. Hydrogenation processes in the synthesis of perfumery ingredients. Accounts of Chemical Research. 2007;**40**(12):1309-1319

[7] Hansen KB, Hsiao Y, Xu F, Rivera N, Clausen A, Kubryk M, et al. Highly efficient asymmetric synthesis of sitagliptin. Journal of the American Chemical Society. 2009;**131**(25):8798-8804

[8] van Leeuwen PWNM. Homogeneous Catalysis: Understanding the Art. Dordrecht, Netherlands: Springer Science & Business Media; 2006

[9] Osborn JA, Jardine FH, Young JF, Wilkinson G. The preparation and properties of tris (triphenylphosphine) halogenorhodium (I) and some reactions thereof including catalytic homogeneous hydrogenation of olefins and acetylenes and their derivatives. Journal of the Chemical Society A: Inorganic, Physical, Theoretical. 1966:1711-1732

[10] Shapley J, Schrock R, Osborn J. Preparation and catalytic properties of some cationic iridium (III) and rhodium (III) dihydrido complexes. Journal of the American Chemical Society. 1969;**91**(10):2816-2817

[11] Schrock RR, Osborn JA. Catalytic hydrogenation using cationic rhodium complexes. I. Evolution of the catalytic system and the hydrogenation of olefins. Journal of the American Chemical Society. 1976;**98**(8):2134-2143

[12] Crabtree R. Iridium compounds in catalysis. Accounts of Chemical Research. 1979;**12**(9):331-337

[13] Crabtree RH, Demou PC, Eden D, Mihelcic JM, Parnell CA, Quirk JM, et al. Dihydrido olefin and solvento complexes of iridium and the mechanisms of olefin hydrogenation and alkane dehydrogenation. Journal of the American Chemical Society. 1982;**104**(25):6994-7001

[14] Chatterjee D, Bajaj H, Halligudi S, Bhatt K. Catalysis of alkene hydrogenation and oxidation by nickel-saloph complex; a novel bifunctional catalyst. Journal of Molecular Catalysis. 1993;**84**(1):L1-L5

[15] de Vries JG, Elsevier CJ. Handbook of Homogeneous Hydrogenation. Weinheim, Germany: Elsevier - Wiley-VCH; 2007

[16] Shevlin M, Friedfeld MR, Sheng H, Pierson NA, Hoyt JM, Campeau L-C, et al. Nickel-catalyzed asymmetric alkene hydrogenation of α, β-unsaturated esters: High-throughput experimentation-enabled reaction discovery, optimization, and mechanistic elucidation. Journal of the American Chemical Society. 2016;**138**(10):3562-3569

[17] Schirmer W, Floerke U, Haupt HJ. Preparation, properties, and molecular structures of a rigid tridentate chelate ligand N, N′-bis (diphenylphosphino)-2, 6-diaminopyridine with M (II) and M (0) transition metals (M (II): Ni, Pd, Pt; M (0): Cr, Mo, W). ChemInform. 1987;**18**(33)

[18] Reddy VVS. Synthesis, characterization and catalytic activity of a Pd (II) complex of bis (2-(diphenylphosphino) ethyl)-benzylamine in the homogeneous hydrogenation of cyclohexene. Journal of Molecular Catalysis. 1988;**45**(1):73-84

[19] Bacchi A, Carcelli M, Costa M, Leporati A, Leporati E, Pelagatti P, et al. Palladium (II) complexes containing a P, N chelating ligand part II. Synthesis and characterisation of complexes with different hydrazinic ligands. Catalytic activity in the hydrogenation of double and triple CC bonds. Journal of Organometallic Chemistry. 1997;**535**(1-2):107-120

[20] Gawron M, Gilch F, Schmidhuber D, Kelly JA, Horsley Downie TM, Jacobi von Wangelin A, et al. Counterion effect in cobaltate-catalyzed alkene hydrogenation. Angewandte Chemie, International Edition. 2024;**63**:e202315381

[21] Zhou J, Xiong Y, Shi Y. Antioxidant consumption kinetics and shelf-life prediction for biodiesel stabilized with antioxidants using the Rancimat method. Energy & Fuels. 2016;**30**(12):10534-10542

[22] Nishimura S. Handbook of Heterogeneous Catalytic Hydrogenation for Organic Synthesis. New York: Wiley; 2001

[23] Hossain M, Israt SS, Muntaha N, Jamal MS. Effect of antioxidants and blending with diesel on partially hydrogenated fish oil biodiesel to upgrade the oxidative stability. Bioresource Technology Reports. 2022;**17**:100938

[24] Jones MD, Raja R, Thomas JM, Johnson BF, Lewis DW, Rouzaud J, et al. Enhancing the enantioselectivity of novel homogeneous organometallic hydrogenation catalysts. Angewandte Chemie International Edition. 2003;**42**(36):4326-4331

[25] Fache F, Schulz E, Tommasino ML, Lemaire M. Nitrogen-containing ligands for asymmetric homogeneous and heterogeneous catalysis. Chemical Reviews. 2000;**100**(6):2159-2232

[26] Lapointe S, Pandey DK, Gallagher JM, Osborne J, Fayzullin RR, Khaskin E, et al. Cobalt complexes of bulky PNP ligand: H2 activation and catalytic two-electron reactivity in hydrogenation of alkenes and alkynes. Organometallics. 2021;**40**(21):3617-3626

[27] Poitras AM, Knight SE, Bezpalko MW, Foxman BM, Thomas CM. Addition of H_2 across a cobalt-phosphorus bond. Angewandte Chemie, International Edition. 2018;**57**:1497-1500

[28] Yuwen J, Chakraborty S, Brennessel WW, Jones WD.

Additive-free cobalt-catalyzed hydrogenation of esters to alcohols. ACS Catalysis. 2017;7:3735-3740

[29] Olaoye OE, Oyetunji O, Makhubela BC, Kumar G, Darkwa J. Hydrogenation of biodiesel catalysed by pyrazolyl nickel (ii) and palladium (ii) complexes. RSC Sustainability. 2023;**1**(7):1814-1825

[30] Liu W, Xu L, Lu G, Zhang H. Selective partial hydrogenation of methyl linoleate using highly active palladium nanoparticles in polyethylene glycol. ACS Sustainable Chemistry & Engineering. 2017;**5**(2):1368-1375

[31] Carvalho MS, Lacerda RA, Leao JP, Scholten JD, Neto BA, Suarez PA. In situ generated palladium nanoparticles in imidazolium-based ionic liquids: A versatile medium for an efficient and selective partial biodiesel hydrogenation. Catalysis Science & Technology. 2011;**1**(3):480-488

[32] Cheng H, Dowd MK, Easson MW, Condon BD. Hydrogenation of cottonseed oil with nickel, palladium and platinum catalysts. Journal of the American Oil Chemists' Society. 2012;**89**:1557-1566

[33] Sejie FP, Oyetunji OA, Makhubela BC, Darkwa J, de Leeuw NH. Evaluation of cobalt, nickel, and palladium complexes as catalysts for the hydrogenation and improvement of oxidative stability of biodiesel. Catalysts. 2024;**14**(9):653

[34] Olaoye OE, Oyetunji O, Makhubela BC, Muyaneza A, Kumar G, Darkwa J. Catalytic hydrogenation of sorbic acid using pyrazolyl palladium (II) and nickel (II) complexes as precatalysts. South African Journal of Chemistry. 2021;**74**:50-56

[35] Sejie FP, Oyetunji OA, Darkwa J, Beas IN, Makhubela BC, Dzade NY, et al. The transfer hydrogenation of cinnamaldehyde using homogeneous cobalt (II) and nickel (II)(E)-1-(Pyridin-2-yl)-N-(3-(triethoxysilyl) propyl) methanimine and the complexes anchored on Fe_3O_4 support as pre-catalysts: An experimental and In silico approach. Molecules. 2023;**28**(2):659

[36] Anferov SW, Filatov AS, Anderson JS. Cobalt-catalyzed hydrogenation reactions enabled by ligand-based storage of dihydrogen. ACS Catalysis. 2022;**12**(16):9933-9943

[37] Liu W, Jiang Y, Dostert K-H, O'Brien CP, Riedel W, Savara A, et al. Catalysis beyond frontier molecular orbitals: Selectivity in partial hydrogenation of multi-unsaturated hydrocarbons on metal catalysts. Science Advances. 2017;**3**(7):e1700939

Chapter 7

Individual Transition Single-Atom Catalysts for Selective Oxidation Reaction

Shiyong Zhao, Jinfeng Xu, Jincheng Li and Chang Liu

Abstract

The selective transformation of chemicals, such as alcohols and ethylbenzene (EB) into their desired products, is a highly valuable process in the field of organic chemistry. Extensive efforts have been focused on developing efficient reaction strategies to achieve this goal. In contrast to homogeneous oxidative reactions that utilize stoichiometric toxic or expensive oxidants like hydrogen peroxide, heterogeneous catalysis has recently gained more attention due to its advantages of easy separation/recycling and promotion of environmentally friendly reaction processes. While precious metals like platinum, ruthenium, and palladium have demonstrated excellent catalytic activity for selective oxidation, their scarcity and high cost have limited their practical applications in the chemical industry. In comparison, transition metal single-atom catalysts (TM-SACs) are abundant in source and low in cost. More importantly, they possess high dispersion, controllable electronic configuration, and high stability, making them highly promising for practical applications. Unlike previous studies on homogeneous and heterogeneous metal catalysts, this chapter integrates the preparation of TM-SACs with high loading and high selectivity, and innovatively summarizes the key research and applications of TM-SACs in organic synthesis in recent years. This provides significant insights into exploring the impact of improving atom utilization efficiency on the selective oxidation of organic chemicals.

Keywords: TM-SACs, activity and stability, selective oxidation, organic synthesis, economical and efficient

1. Introduction

The research on the selective transformation of inexpensive and readily available hydrocarbons into higher-value chemicals is an extremely important process in the field of organic chemical synthesis [1]. In traditional homogeneous catalytic reactions, precise control of chemical selectivity in organic synthesis is often achieved through homogeneous organometallic complexes [2, 3]. This process requires the use of expensive and toxic solvents, oxidants, and organic ligands to regulate spatial and electronic parameters. However, this approach increases economic burden. It also leads to environmental pollution due to difficulties in separation and recycling [4]. Currently, various heterogeneous catalysts have been developed, among which noble metal catalysts have demonstrated good catalytic activity in selective oxidation [5]. Currently, various heterogeneous catalysts have been developed, among which noble metal catalysts have demonstrated good catalytic activity in selective oxidation. However, the low abundance and high cost of noble metals still pose significant challenges. Therefore, it is essential to develop non-precious metal catalysts with low cost, high activity, and good recyclability. Transition metals, such as Fe, Co, Ni, and Cu, are inexpensive and abundant, and they are widely used in various research fields. They also hold great potential in selective oxidation reactions for organic synthesis, especially in obtaining catalysts with immobilized transition metal active sites, which is an important approach to promote the development of heterogeneous catalytic reactions [6].

The dispersion and stability of the metal active sites determine the catalytic efficiency of transition metal catalysts [7]. In heterogeneous catalysis, the interaction between reactants and the catalyst occurs at the active sites on the surface of the solid catalyst. Increasing the surface atom ratio of the active metal component can directly enhance the catalytic efficiency [8, 9]. Reducing the metal active component to the single-atom scale is a way to provide catalysts with maximized metal utilization efficiency [10–12]. While the bulk materials are shrunk into nanoparticles, sub-nano clusters, and even single metal atoms, the specific activity and selectivity, stability and efficiency of catalyst will dramatically change, as shown in **Figure 1**. Through the size reduction, the amount of unsaturated coordination chemical environment will increase in the metal species [13]. Therefore, the surface free energy of the metal also significantly increases, which will make metal sites to be more active during the chemical reaction process. The enhanced chemical interactions between the metal

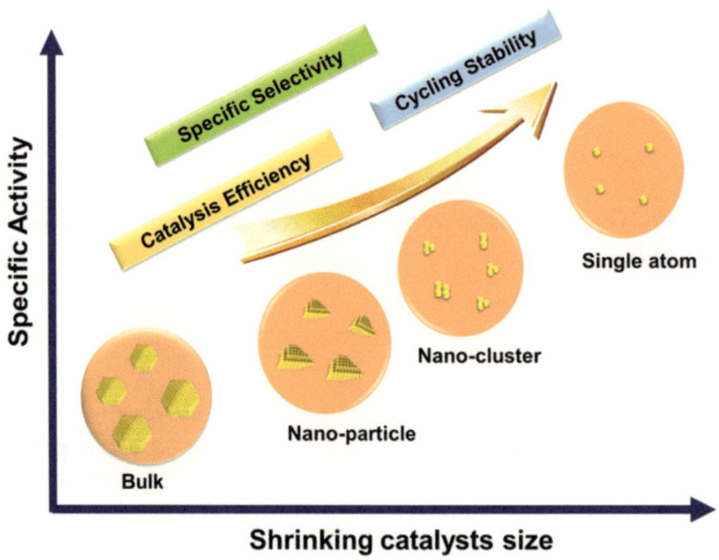

Figure 1.
Schematic illustrates the changes of specific activity, selectivity, stability, and efficiency per metal atom with catalyst size and the support effects.

and the support relate to the size of catalysts. When the metal catalysts are reduced to extreme single-atom catalysts (SACs), the quantum effects of electrons, the surface free energy will achieve the maximum due to the highly active valence electrons, which then results in the enhanced chemical activities of SACs for the catalytic process [14]. In the past few years, single-atom catalysts have garnered considerable interest due to the merged merits of homogeneous and heterogeneous catalysis, finding applications in diverse fields [15]. Therefore, it is crucial to integrate the high catalytic efficiency of transition metals with the high utilization and tunability of single-atom structures. This integration allows for the development of TM-SACs with excellent catalytic activity, selectivity, and stability. It represents an innovation in current catalytic technologies and is an effective way to achieve efficient, green, and sustainable organic chemical synthesis. By adjusting the metal center and coordination environments within SACs, it is possible to modulate the adsorption capacity of intermediates and the charge transfer, thereby enhancing catalytic performance.

TM-SACs have already seen extensive applications and developments, with several review articles summarizing the optimization strategies for TM-SACs. Therefore, this review will not repeat these discussions. Specifically, we focus on summarizing the significant syntheses of highly active and selective TM-SACs to date, as well as the progress achieved in their applications for organic chemical synthesis. This includes their design and fabrication methods, catalytic mechanisms, and application examples in various selective oxidation reactions. These insights will provide valuable guidance for the design of catalysts in various organic chemical syntheses. Finally, we discuss the challenges and opportunities that TM-SACs face in practical selective catalysis of organic chemicals.

2. Design and fabrication of transition metal single-atom catalysts

2.1 Selection of support materials

Different support materials influence the catalytic performance and stability of TM-SACs and are based on the large surface area of supports to absorb the metal atoms following the thermal, photo-catalytic, and electric treatments to form stable chemical bonding between metal atoms and supports [16], as can be seen in **Figure 2**. For better controllable synthesis of SACs, many efforts have been devoted to the dispersion of metal on different supports to achieve the single metal atoms.

To obtain high-loading active sites, we tried a series of strategies to prepare high-loading TM-SACs catalysts on substrates of different dimensions. In the field of 1D materials, Wang et al. successfully designed and synthesized a Cu-N_3 single-atom catalyst on nitrogen-doped carbon nanotubes (CNTs) with high curvature, with a Cu loading of 2.5 wt%, as depicted in **Figure 3a** [17]. Besides this, we successfully synthesized a class of TM-SACs with very high metal loading, which are based on nitrogen-doped carbon nanotubes [20]. In particular, the Ni single-atom catalyst supported on nitrogen-doped carbon nanotubes achieved a high metal loading of up to 20 wt%. This may be attributed to the self-constraining effect of the nanotube structure on metal single atoms, which results in the presence of not only a large number of single atoms on the carbon surface but also within the bamboo-like structure of the nanotubes. Based on 2D materials, we also reported a versatile strategy to synthesizing single-atom catalysts with high loading on various 2D substrate materials [18]. This method can effectively synthesize metal single atoms such as Ni,

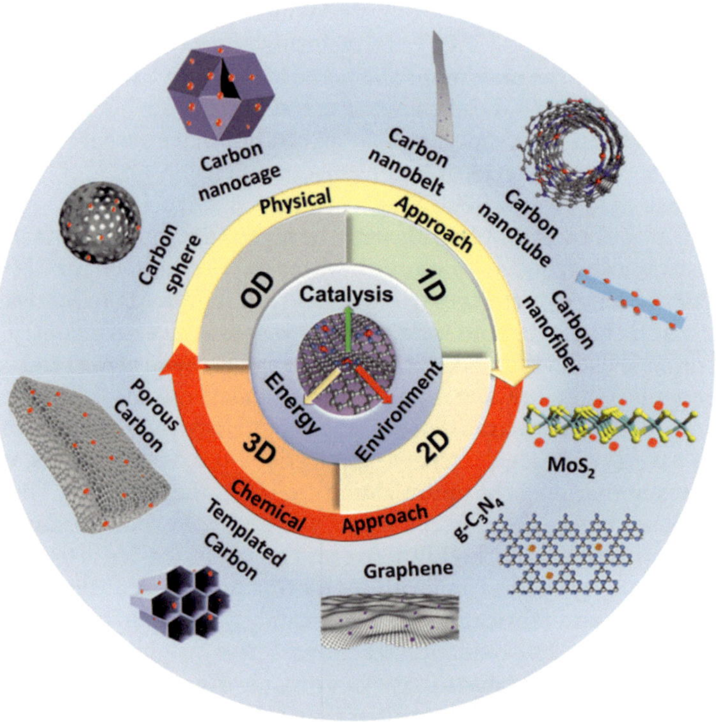

Figure 2.
Schematic illustration of the different substrates for the synthesis of SACs. The zero-dimensional (0D) carriers include nanocarriers, such as nanocarbon spheres and nanocages, which are at the nanoscale in all dimensions. One-dimensional (1D) carriers consist of nanoribbons, nanotubes, and nanofibers, among others. Two-dimensional (2D) carriers mainly include graphene, 2D sulfides, and carbon nitrides. Three-dimensional (3D) carriers mainly include templated carbons and integrated 3D structures. By combining chemical and physical methods, single atoms are deposited on the carriers to form single-atom catalysts.

Co, and Fe on 2D materials, achieving SAC mass loadings of approximately 3–7 wt% (**Figure 3b**). Additionally, as **Figure 3c** exhibits, we propose a universal strategy for obtaining TM-SACs (Mn, Cu, Cr, etc.…) supported on 3D carbon fiber foam (CF) using nitrogen coordination. The metal-doped graphitic carbon nitride (gCN) is thermally treated on CF, thereby generating stable and abundant single atoms [19]. Compared to the SAC supported on 1D materials such as CNTs, SACs supported on 2D/3D materials will have a much higher utilization efficiency. The synergistic combination of 2D/3D materials and single metal atom seeds makes it possible to achieve SACs with tunable mass loading and superior catalytic performance that cannot be otherwise achieved by using conventional synthesis method.

In addition to carbon-based materials, such as nitrogen-containing organic compounds, metal-organic frameworks (MOFs) and their derivatives, nitrogen-doped graphene, high-nitrogen-content g-C_3N_4, and carbide-derived carbon, common support materials also include metal oxides, such as titanium dioxide (TiO_2) and alumina. Among them, nitrogen-doped carbon materials are widely used for the preparation of TM-SACs. Due to nitrogen-doped carbon materials being a class of porous, nitrogen-rich carbon-based supports that have inherent nitrogen elements, very low density, large surface area, and ordered pore structures [21]. Typically, nitrogen atoms with lone pairs of electrons act as Lewis bases and can form strong metal-nitrogen (M-N)

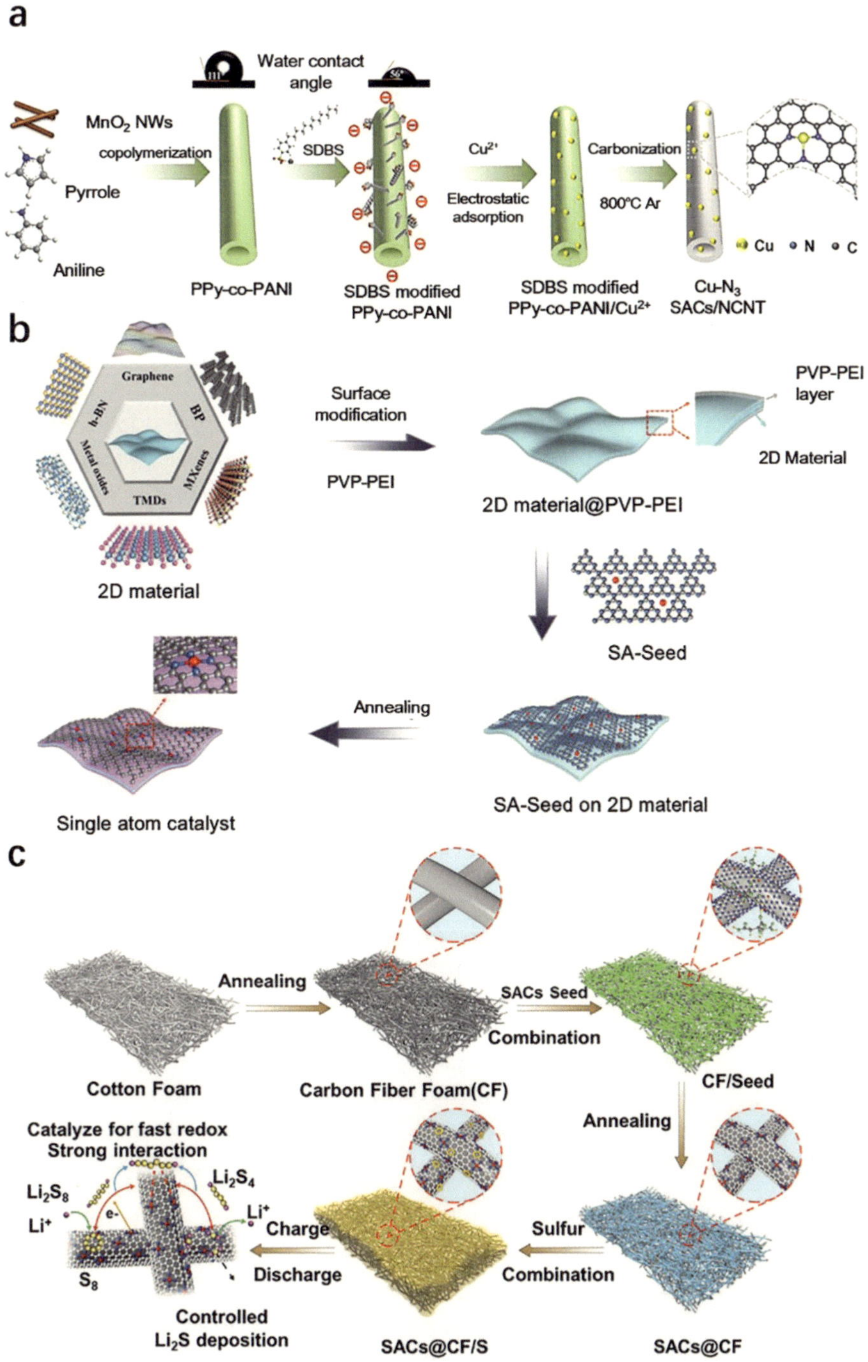

Figure 3.
Examples of preparation methods for anchoring transition metal single atoms on (a) 1D [17], (b) 2D [18], and (c) 3D [19] support materials. As shown in the figure, the methods for loading dispersed metal single atoms on 1D, 2D, and 3D carriers, respectively, are demonstrated. The main idea is to modify the surfaces of materials in three different dimensional substrates and then immobilize the dispersed metal ions or atoms on the carriers through chemical bonds. Because of their different dimensions, these materials can each offer unique advantages in catalytic reactions, providing a solid material basis for various catalytic processes.

coordination bonds with metal species, thereby greatly enhancing the thermal and chemical stability of the materials.

2.2 Synthesis methods

2.2.1 Impregnation method

The impregnation method is a traditional and commonly used approach for preparing single-atom catalysts. Its fundamental principle involves immersing the support material in a solution containing metal salts, allowing metal ions to be loaded onto the support surface through physical adsorption or ion exchange. Subsequently, through drying and calcination processes, the metal ions are reduced to a single-atom state and stably anchored onto the support [22].

The advantages of the impregnation method are its simplicity and low cost, making it suitable for large-scale preparation. However, the method has certain limitations in controlling the dispersion and loading amount of metal atoms, which can easily lead to metal atom aggregation, especially at high loadings [23]. To improve the dispersion of metal atoms, it is usually necessary to finely control the impregnation conditions, such as the concentration of the metal salt solution, impregnation time, temperature, and the selection of appropriate reducing agents.

2.2.2 Atomic layer deposition

Atomic layer deposition (ALD) is a self-limiting cyclic process based on the reaction of gaseous chemical precursors with the surface of a solid substrate, which can be used to prepare SACs in recent years [24]. By selecting appropriate metal precursors (such as metal halides or organometallic compounds) and introducing them into the reaction chamber in the gas phase, reactions and self-limiting growth can occur on suitable carrier materials, such as oxides, carbon-based materials, or metal-organic frameworks (MOFs). The precursors chemically adsorb onto the carrier surface to form a monolayer of atoms. By controlling the number of reaction cycles, metal atoms are deposited step by step. Additionally, by adjusting parameters such as deposition temperature, number of cycles, and precursor concentration, the uniform dispersion and high loading of metal atoms can be achieved. Finally, the single-atom sites can be further stabilized by methods such as thermal treatment or chemical reduction. As an example, Zhang et al. embedded Cu single atoms on hollow g-C_3N_4 nanospheres using ALD technology to construct a Cu-N_3 structure [25].

TM-SACs prepared by ALD have highly uniform metal atom distribution and precisely controllable loading amounts, enabling the precise construction of the interface structure between metal single atoms and the support at the atomic level. However, the high cost of ALD equipment, the complex preparation process, and the low yield limit its large-scale industrial application [26].

2.2.3 Photochemical method

The photochemical method employs metal precursors, including metal nitrates and chlorides. Under ultraviolet (UV) irradiation, photogenerated electrons are utilized to reduce metal ions, thereby forming single-atom sites on the surface of the carrier. Finally, acid washing is conducted to remove unbound metal ions, ensuring the uniform distribution of single atoms [27]. During the photochemical reaction process, the deposition of metal atoms can be effectively controlled by selecting appropriate light sources, photocatalysts, and reaction systems. Lee et al. proposed a general method for synthesizing atomically dispersed catalysts using photochemical defect engineering. This method can be applied to various catalytically active transition metals, including single atoms of Pt, Ir, and Cu dispersed on metal oxide supports [28]. The photochemical method has the advantages of mild reaction conditions and a simple and rapid preparation process, which allows for the loading of metal single atoms without compromising the structure of the support material. Moreover, the photochemical method can precisely control the loading amount and dispersion of metal atoms by adjusting parameters such as the wavelength, intensity, and exposure time of light. However, the method currently faces challenges, such as the selection and recovery of photocatalysts and relatively low reaction efficiency, which require further research and improvement.

2.2.4 Other preparation methods

In addition to the aforementioned common preparation methods, several other approaches have also been employed for the preparation of transition metal single-atom catalysts, such as wet-chemical methods, electrochemical methods, and high-temperature pyrolysis. Wet-chemical methods involve a series of chemical reactions in solution to achieve the precipitation and loading of metal atoms [29]. Butburee et al. synthesized various types of metal single atoms doped simultaneously in zinc oxide quantum dot (ZnO QD) structures through a simple one-step wet-chemical method [30]. Electrochemical methods utilize the principle of electrochemical deposition to deposit metal atoms onto a support material on the electrode surface [31]. Specifically, Kosar et al. prepared single-atom catalysts of various transition metals (such as Co, Cu, Zn, Ni, and Fe) using the method of electrochemical deposition [32]. High-temperature pyrolysis involves the thermal decomposition of a mixture containing metal precursors and supports at elevated temperatures, allowing metal atoms to decompose on the support surface and form single-atom catalysts [33]. In recent years, a method has been proposed to directly anchor metal atoms by pre-fabricating boron nitride (BvBN) with B vacancies, successfully constructing a series of boron nitride-based single-atom catalysts with well-defined M-N_3 sites [34]. Different synthetic methods each have their own advantages and disadvantages. The choice of an appropriate method should be based on specific application requirements, cost constraints, and experimental conditions. For example, the impregnation method is suitable for large-scale production but has relatively poor dispersion, while atomic layer deposition and photochemical methods, although offering better dispersion, are more costly. The future trend may involve combining the strengths of multiple methods to develop more efficient and cost-effective synthetic strategies.

2.3 Coordination environment regulation

Regulating the coordination environment of the metal center is key to enhancing the catalytic performance of TM-SACs. Common methods for regulation include the introduction of non-metal elements (such as boron (B), sulfur (S), phosphorus (P), etc.) and the regulation of the electronic structure of the metal center through defect engineering.

Chai et al. modified the Fe single-atom catalyst (Fe-NC) by introducing phosphorus (P) to alter its electronic structure, thereby enhancing the activation capability of peroxymonosulfate (PMS) (**Figure 4a**). Compared with the conventional Fe-NC, the Fe-PNC exhibited a degradation rate constant for bisphenol A (BPA) that was approximately three times higher, demonstrating that the P-induced Fe active sites possess excellent performance in activating PMS for BPA degradation [35]. This may be attributed to its unique coordination environment (Fe-N_4P_2) and high-spin state.

Zhang et al. obtained a sulfur-doped iron single-atom catalyst by co-pyrolyzing the prepared hierarchical porous structured network composite (Fe@NC-Ac) with thiourea, as depicted in **Figure 4b**. The catalyst exhibited remarkable catalytic activity in the oxygen reduction reaction (ORR), with higher onset and half-wave potentials compared to commercial Pt/C (platinum on carbon). When applied in practical batteries, the S-doped catalyst achieved the highest peak power density [36]. This work provides a simple synthetic strategy for designing non-metal atom-modified catalysts with ideal ORR performance.

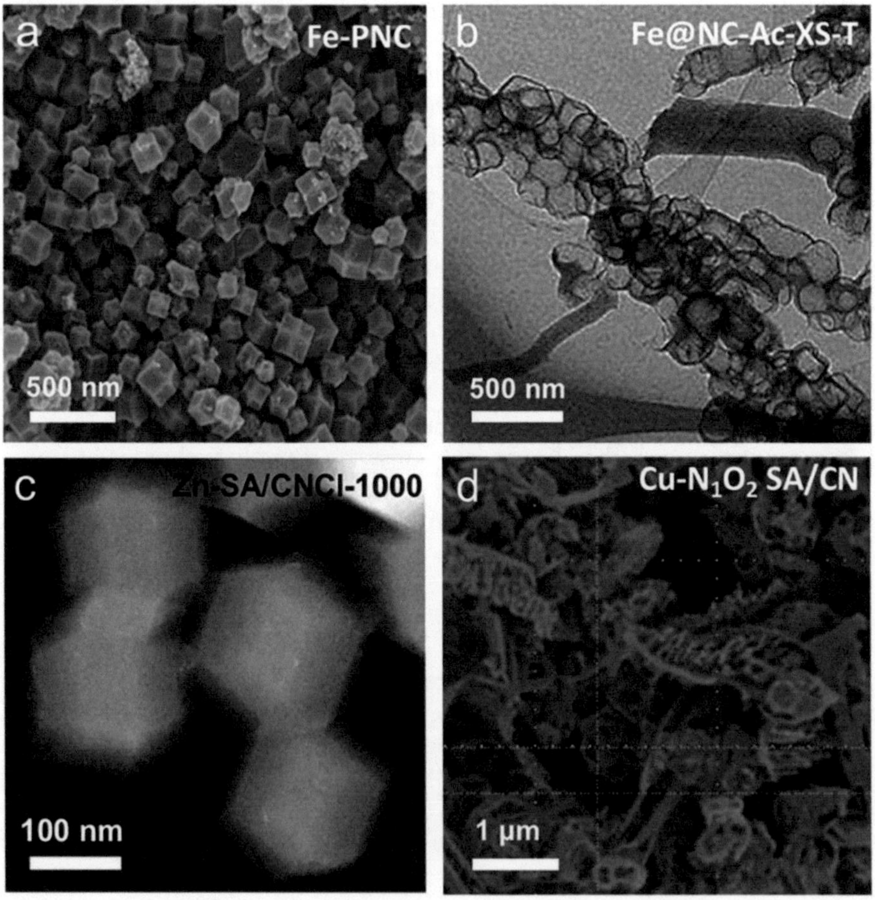

Figure 4.
Morphology of heteroatom-doped transition metal single-atom catalysts: (a) scanning electron microscopy (SEM) image of Fe-PNC [35], (b) transmission electron microscopy (TEM) image of Fe@NC-Ac-XS-T [36], (c) scanning transmission electron microscopy (STEM) image of Zn-SA/CNCl-1000 [37], and (d) TEM image of Cu-N_1O_2SA/CN [38].

Wei et al. used sodium chloride (NaCl) as the chlorine source and introduced chlorine (Cl) during the pyrolysis of zeolitic imidazolate framework 8 (ZIF-8), transforming the traditional Zn-N_4 sites into asymmetric Zn-N_3 sites (**Figure 4c**). Meanwhile, chlorine atoms were captured by carbon atoms in the plane. This design enhanced the selectivity and catalytic activity for CO_2. Among them, the optimal catalyst Zn-SA/CNCl-1000 achieved a high Faraday efficiency of 97 ± 3% for CO at a large current density of 200 mA cm^{-2}, with no significant degradation observed during stability testing [37]. This work provides important insights for designing transition metal single-atom catalysts with asymmetric active sites and for improving the catalytic performance of CO_2 reduction.

Zhang et al. have reported a Cu-N_1O_2 single-atom catalyst supported on carbon nitride (Cu-N_1O_2 SA/CN), which was prepared via a water-mediated pre-assembly pyrolysis method and single Cu sites with unsaturated N, O coordination structure (**Figure 4d**). The selective oxidation of benzene to phenol (SOBP) with low hydrogen peroxide consumption has been effectively enhanced [38]. The high benzene conversion rate and high phenol selectivity provide a simple and efficient method for the selective oxidation of benzene to produce phenol by enabling the tunable electronic configuration of single atoms.

3. Catalytic mechanism of transition metal single-atom catalysts in organic synthesis

The active sites of transition metal single-atom catalysts (TM-SACs) are the core of their catalytic reactions. Accurate identification and understanding of the structure and mechanism of action of these active sites are crucial for in-depth studies of catalytic reaction mechanisms. In TM-SACs, the interaction between the metal single atoms and the support, as well as the presence of surrounding coordinating atoms, endows the active sites with unique electronic structures and chemical properties.

By employing a variety of advanced characterization techniques, such as high-resolution transmission electron microscopy (HRTEM), X-ray absorption fine structure spectroscopy (XAFS), and X-ray photoelectron spectroscopy (XPS), researchers are able to precisely characterize the structure of active sites. In organic synthesis, the catalytic mechanism of TM-SACs mainly involves electron transfer and the adsorption and activation of reactants. TM-SACs can activate oxidants (such as peroxymonosulfate) to generate reactive species like radicals or singlet oxygen ($^1O_2$), thereby achieving efficient catalytic oxidation processes. For instance, in the selective oxidation of alcohols, the metal single-atom active sites first adsorb and activate the alcohol molecules by forming hydrogen bonds or coordination interactions with the hydroxyl groups of the alcohol molecules. Simultaneously, the activation of molecular oxygen by the active sites also plays a crucial role. Common reactive oxygen species (ROS) include sulfate radicals ($SO_4^{\cdot -}$), c), and $^1O_2$ in the activation processes of PMS and peroxydisulfate (PDS) [39] (**Figure 5a**). As depicted in **Figure 5b**, the obtained apparent rate of the Co-SAs is much higher than that of Co-NPs. Therefore, SACs, with their catalytic centers based on unique electronic structures and the high surface area and porous structures provided by different carbon supports, can expose more active sites, which are more conducive to the diffusion of PMS and electrons [40]. As shown in **Figure 5c**, in the PMS activation system, different transition metal-based single-atom catalysts can selectively generate $^1O_2$ to varying extents, which are used for the oxidation synthesis of aromatic alcohols. In addition, theoretical methods,

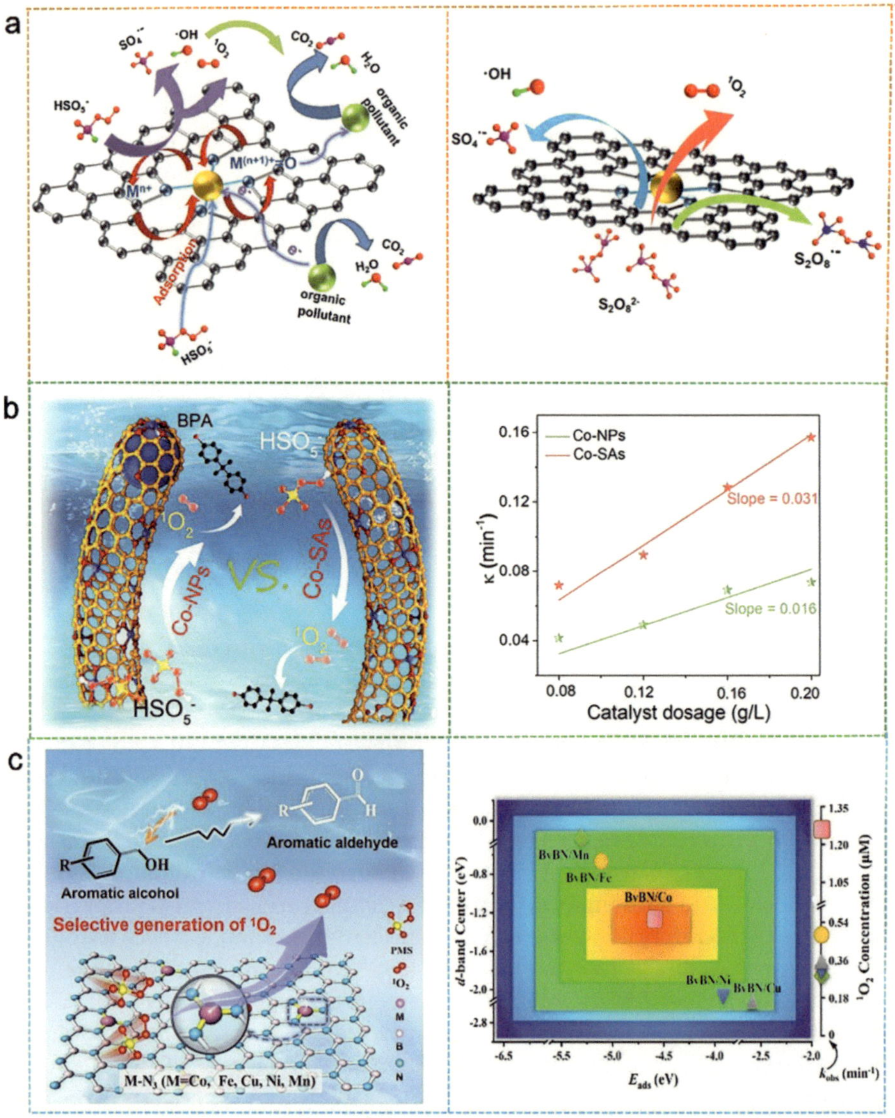

Figure 5.
The possible catalytic mechanisms (a) of SACs in the activation processes of PMS and PDS [39]. The possible degradation mechanisms using Co-NPs and Co-SAs as catalysts (b) for BPA degradation and comparison of the reaction rate constants [40]. The generation mechanism (c) of singlet oxygen ($^1O_2$) in the BvBN/M - PMS system [34].

such as density functional theory (DFT) calculations, can be used to gain a deeper understanding of the interaction mechanisms between the metal center and the reactants [34]. All the fabricated BvBN/M catalysts possess well-defined $M - N_3$ sites, which facilitate the self-decomposition of two PMS molecules to generate $^1O_2$ with high selectivity. Among them, BvBN/Co exhibits superior catalytic activity due to its moderate adsorption energy and optimally positioned d-band center.

Studies have shown that the type, number, and coordination mode of the atoms surrounding the active sites of metal single-atom catalysts significantly influence the electronic density and oxidation state of the metal atoms, thereby altering their ability to adsorb and activate reactant molecules. By adjusting the type and number of the coordinating atoms, the electronic structure of the metal atoms can be altered, which in turn optimizes their catalytic performance. Moreover, changes in the coordination mode can also affect the oxidation state and electronic density of the metal atoms, thereby influencing their adsorption capacity for reactants and catalytic activity. These findings provide an important theoretical basis for the design and optimization of metal single-atom catalysts [41].

4. Application examples

The application of transition metal single-atom catalysts in the selective oxidation of hydrocarbons, oxygen-containing compounds, nitrogen-containing compounds, sulfur-containing compounds, and biomass-related compounds not only enhances the efficiency and selectivity of reactions but also enables the efficient utilization of resources and sustainable environmental development, holding significant scientific and practical value.

4.1 Application in the selective oxidation of hydrocarbons

4.1.1 Activation and oxidation of alkanes

The selective oxidation of alkanes is highly challenging due to the high stability of C-H bonds. As a cheap yet efficient catalyst, Ni-based catalysts have attracted interest because of its ability to activate alkane. A desired propylene selectivity can be achieved by SACs, as a result of weaker propylene-π adsorption mode on single metal atoms, which contributes to fast propylene desorption.

Zhang et al. reported a type of anatase TiO_2-supported nickel single-atom catalyst (Ni SAC), which demonstrated excellent intrinsic activity and propylene selectivity in propane dehydrogenation (PDH). By comparing Ni SACs with nickel nanoparticles (Ni NPs), they showed that the propylene formation rate of nickel single atoms was approximately 65 times higher than that of nickel nanoparticles, and also brought better stability [42]. And there is a comparison of the performance with those of conventional catalysts and TM-SACs in **Table 1**. As can be seen from the table, the selectivity and Turnover Frequency (TOF) values of metal single-atom catalysts are generally higher than those of particulate catalysts.

4.1.2 Epoxidation of alkenes

In the selective oxidation of alkenes to epoxides, transition metal single-atom catalysts also perform remarkably well. Single-site titanium on metal oxide supports plays an important role in achieving high reactivity and selectivity in propylene epoxidation. Lu et al. successfully embedded single-atom titanium in a SiO_2 framework using ALD for propylene epoxidation. They demonstrated that the isolated titanium atoms on the support play a crucial role in achieving high reactivity and selectivity for propylene epoxidation. It was also determined that under similar propylene conversion, the selectivity for propylene oxide (PO) is closely related to the Ti-O coordination number and bond length. DFT calculations revealed that the partial charge of

Catalysts	T (°C)	Sel. (%)	TOF (h^{-1})	Ref.
Ni1/A-TiO_2	580	89.6	128	[42]
Ni_{NP}/A-TiO_2-H600	580	87.1	2.3	[42]
5%NiGa/Al_2O_3	600	94	98	[43]
2.6%Ni-P/SiO_2	600	90	2.9	[44]
Ga(III)/SiO_2	550	94.3	20.4	[45]
1.12Co-MFI	580	97	124	[46]
Co/S-1	550	95.3	138	[47]
Co@MCM-41	590	95	263	[48]
ZnO@NC/S-1(1.0)	600	94	65.1	[49]

Table 1.
Summary of some representative catalysts for the catalytic PDH.

individual Ti-SiO_2 sites varies linearly with the coordination number. The single-site, 4-fold coordinated Ti sites favors high PO selectivity and improving hydrogen efficiency, while the single-site, 6-fold coordinated Ti sites prefers the formation of undesired byproducts, such as propanal and carbon dioxide (CO_2) [50].

4.1.3 Application in the selective oxidation of aromatic hydrocarbons

TM-SACs catalysts possess unique electronic structures and active sites, which enable them to precisely activate specific chemical bonds in aromatic hydrocarbon molecules, thereby achieving highly selective oxidation reactions.

As illustrated in **Figure 6a**, we achieved a cobalt single-atom catalyst (SACo@g-C_3N_4) with a Co content of 3.17% on a graphitic carbon nitride support [51]. This catalyst was applied to the selective oxidation of ethylbenzene (EB) to acetophenone (AcPO). Under relatively low reaction temperatures, the EB conversion rate and AcPO selectivity of the SACo@g-C_3N_4/PMS system were both remarkably high, exceeding 95%. The selective conversion of EB to AcPO is attributed to the oxidative radicals generated from the decomposition of PMS molecules through electron transfer between Co atoms and PMS. During the oxidation of EB, sulfate radicals ($SO_4^{\cdot -}$) play a more significant role in activating C-H bonds. Additionally, the stable chemical bonding between Co atoms and N atoms endows SACo@g-C_3N_4 with excellent stability in acidic oxidative environments. This work reveals the potential of SACo@g-C_3N_4 to efficiently and environmentally benignly activate PMS for the oxidation of hydrocarbons.

4.2 Application in the selective oxidation of oxygen-containing compounds

The oxidation of alcohols to produce carbonyl compounds, such as aldehydes and ketones, is an important organic synthesis reaction. The loading of graphitic carbon nitride (gCN) with transition metals has received significant attention for efficient light-driven catalysis. Refer to **Figure 6b**, Sudrajat et al. further revealed that the single Cu atoms, which are coordinatively less saturated, are effective active sites to promote photocatalytic selective oxidation [52]. Single Cu atoms increase the

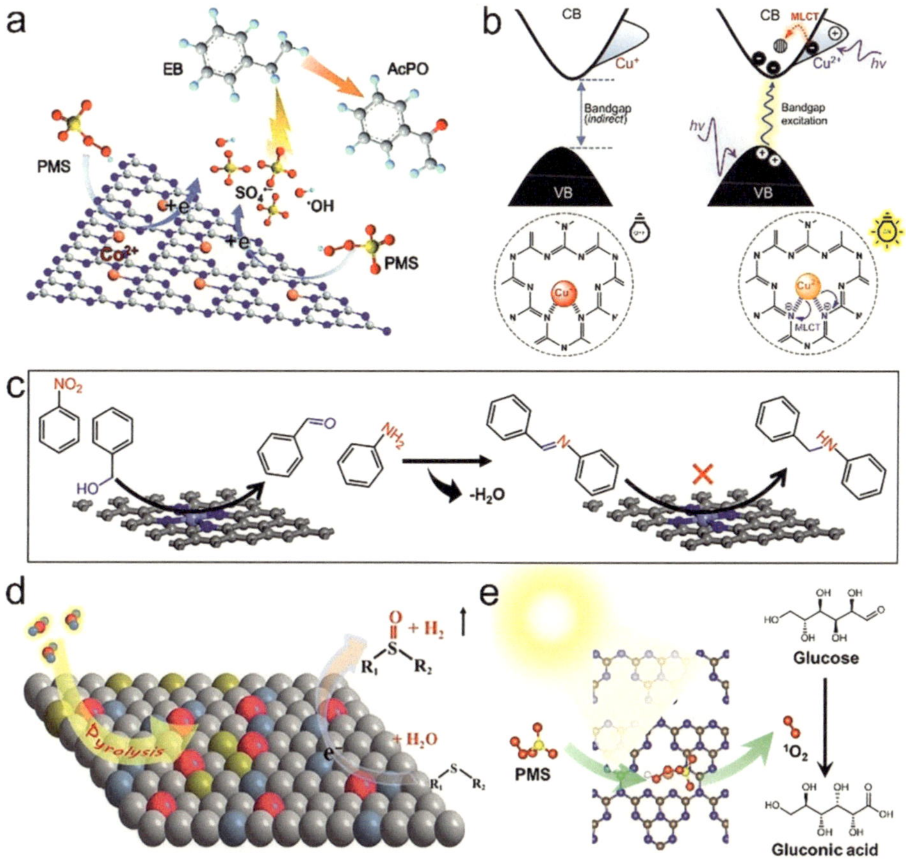

Figure 6.
Applications of transition metal single-atom catalysts in: (a) selective oxidation of aromatic hydrocarbons [51], (b) selective oxidation of oxygen-containing compounds [52], (c) selective oxidation of nitrogen-containing compounds [53], (d) selective oxidation of sulfur-containing compounds [54], and (e) selective oxidation of biomass-related compounds [55].

steady-state electron population in gCN by up to 8.1 times under UV light and 4.6 times under visible light, significantly enhancing the partial oxidation of benzyl alcohol to benzaldehyde. Using the most active photocatalyst, benzyl alcohol is converted to benzaldehyde in acetonitrile with 25% efficiency within 3 h of light irradiation at 465 nm, achieving nearly 100% selectivity, and revealed that tuning the coordination of single metal atoms to be less saturated is a promising approach for the selective valorization of biomass-derived compounds.

4.3 Application in the selective oxidation of nitrogen-containing compounds

4.3.1 Oxidation of amines

The selective oxidation of amines to produce imines, nitriles, and other compounds is of great significance in organic synthesis. In the oxidation of aniline to imine, Zhang et al. proposed single-atom Co-N_4 sites on nitrogen-doped carbon (Co SA/NC) for the synthesis of imines [56]. Specifically, compared to Co foil and cobalt oxide (CoO) counterparts, the isolated Co-N_4 sites exhibit an upshifted d-band center. This upshift in the d-band center of Co-N_4 sites facilitates the desorption of aniline and N-benzylideneaniline, thereby accelerating the spontaneous condensation coupling of aniline and benzaldehyde. As a result, an impressive yield of 97% was achieved in imine synthesis, demonstrating excellent catalytic performance. Additionally, the unique electronic structure of Co SA/NC enables high activity and selectivity for the transfer hydrogenation coupling of nitroarenes, benzyl alcohols, and various functionalized imines. This work provides an efficient Co single-atom catalyst for imine synthesis and offers a new pathway for designing single-atom catalysts with different metal active sites for selective hydrogenation reactions.

4.3.2 Reduction-oxidation of nitro compounds

In the transformation of nitro compounds, transition metal single-atom catalysts can participate in reduction-oxidation processes. For example, Wei et al. synthesized cobalt-nitrogen sites (CoN_x-NMCs) with a high loading of up to 1.7 wt% on nitrogen-doped mesoporous carbon [53] (**Figure 6c**). The CoN_x-NMCs feature a high surface area and abundant active sites. Under mild reaction conditions, CoN_x-NMCs exhibit good activity for the hydrogenation of a series of nitro compounds into amines, with high conversion rates and selectivity. The study also verified and revealed the crucial role of low-coordinated CoN_2 sites in chemoselective hydrogenation. This work may provide a general method and insights for the development of applications involving highly dispersed single-atom catalysts.

4.4 Application in the selective oxidation of sulfur-containing compounds

The selective oxidation of thiols and thioethers is crucial in organic synthesis and petrochemical fields. Transition metal single-atom catalysts can effectively catalyze the oxidation of thiols to disulfides and thioethers to sulfoxides or sulfones. As **Figure 6d** illustrates, Zhu et al. constructed a cobalt single-atom catalyst (Co-SAs/NSC) with nitrogen and sulfur coordination on a carbon support, achieving a 10 wt% loading [54]. The Co-S_2N_4 coordination structure of Co-SAs/NSC enabled excellent selective electro-oxidation of thioethers in acidic solution and demonstrated robust stability. Under low-potential conditions, Co-SAs/NSC achieved a thioether conversion rate and selectivity as high as 99.7 and 100%, respectively, for the production of sulfoxides. This performance outperformed those of most reported catalysts and commercial platinum catalysts. This superior performance is attributed to the Co-S_2N_4 structure, which significantly promotes the desorption of products, thereby endowing the cobalt single atoms with outstanding electro-oxidation activity. This work provides an important strategy for constructing single-atom catalysts with high catalytic performance and for the resourceful utilization of sulfur-containing pollutants.

4.5 Application in the selective oxidation of biomass-related compounds

4.5.1 Oxidation of sugars

Sugar compounds in biomass can be converted into a variety of high-value chemicals through selective oxidation. For example, in the oxidation of glucose to gluconic acid, Xia et al. developed a photocatalytic Fenton-like system with a high loading of atomic Fe catalyst (14 wt%) on carbon nitride (Fe_1CN), using singlet oxygen ($^1O_2$) as the primary active species to selectively oxidize glucose to gluconic acid (**Figure 6e**) [55]. Moreover, the individual Fe atoms loaded on carbon nitride not only significantly enhanced light absorption and simultaneously promoted the separation of photogenerated carriers but also rapidly activated PMS to generate various reactive oxygen species (ROS). Based on this, the Fe_1CN catalyst achieved a glucose conversion rate of 36.3% and a gluconic acid selectivity as high as 91.6%, which is superior to those of previously reported catalytic systems. This provides a reliable approach for the activation of PMS in photocatalytic Fenton-like systems and for selective oxidation systems, which is crucial for biomass transformation.

4.5.2 Depolymerization and oxidation of lignin

Lignin is an important component of biomass, and its depolymerization and selective oxidation are key to the full utilization of biomass. Meng et al. designed a molybdenum-based single-atom catalyst combined with aluminum Lewis acid sites on a magnesium oxide support (Mo_1Al/MgO) to form a multi-site synergistic catalyst for the depolymerization of Eucalyptus lignin via β-aryl ether bond cleavage [57]. Under an inert nitrogen atmosphere, Mo_1Al/MgO achieved a monophenol yield close to the theoretical value of 46%, with a selectivity of 92%. The catalyst also demonstrated good cycling durability. This study provides a feasibility basis for designing high-performance catalysts with synergistic multi-active sites and offers a safer and more economical catalytic system for the efficient conversion of lignin into high-value-added products.

5. Challenges and prospects

TM-SACs catalysts have demonstrated great potential in selective oxidation reactions for organic synthesis. Through rational design and fabrication, it is possible to achieve precise control over the reactivity and selectivity of the reactions. In recent research, it has been revealed that through the precise manipulation of atomic configuration and electronic structure, these catalysts can exhibit remarkable activity and selectivity. However, the development of TM-SACs still faces many challenges that require further research and exploration.

5.1 Challenges faced

First and foremost, there is a pressing need to enhance technologies for large-scale production. Despite the fact that current preparation methods are capable of generating high-performance single-atom catalysts, the process is often hindered by challenges, such as elevated preparation costs, suboptimal yields, and intricate preparation procedures during mass production. These factors significantly impede their widespread industrial application. Consequently, a key research focus at present is the innovation of preparation technologies that are not only efficient and cost-effective but also readily scalable. Second, the durability and service life of catalysts are critical barriers to their practical implementation. In actual reaction settings, TM-SACs can encounter influences from reactants, products, and the surrounding reaction environment, which may result in the clustering of individual metal atoms, the depletion of metal atoms, or the contamination of active sites. These phenomena can significantly curtail the stability and lifespan of the catalysts. Therefore, enhancing the stability and lifetime of TM-SACs is a challenge. It requires development from various perspectives, such as catalyst surface protection, optimization of carrier materials, introduction of catalyst regeneration technologies, and dynamic monitoring and precise control. Furthermore, the comprehension of the catalytic reaction mechanisms and structure-activity correlations of transition metal single-atom catalysts within intricate organic synthesis frameworks is not sufficiently profound. The bulk of current research is concentrated on straightforward model reaction systems, with comparatively fewer studies delving into multi-step reactions, intricate reactant systems, and the interplay between catalysts and reactants in actual organic synthesis processes. A more profound exploration into the operational mechanisms of catalysts in complex systems is of paramount importance for the further optimization of catalyst performance and the development of novel catalytic reaction systems.

5.2 Future development trends

Looking ahead, TM-SACs are poised to make significant strides in the realm of selective oxidation for organic synthesis. Initially, as materials science, nanotechnology, and catalytic science continue to advance, it is anticipated that innovative preparation methods and technologies will emerge, facilitating the large-scale and cost-effective production of SACs. Additionally, through meticulous research and precise design of support materials and the coordination environment of metal single atoms, the development of new types of single-atom catalysts with enhanced activity, selectivity, and stability is within reach.

Subsequently, the convergence of multidisciplinary approaches will yield fresh perspectives and methodologies for the study of TM-SACs. For instance, the integration of theoretical computational chemistry, in situ characterization techniques, and artificial intelligence (AI) technologies can provide deeper insights into catalytic mechanisms and guide the rational design of catalysts. In particular, AI and machine learning (ML) are capable of processing and analyzing vast amounts of experimental data, thereby swiftly identifying promising catalysts. AI algorithms can analyze complex reaction conditions and employ iterative optimization techniques to determine the optimal reaction parameters, such as temperature, pressure, and catalyst selection. This enables the application of catalysts in organic synthesis to be more efficient and precise. Furthermore, through AI models, researchers can simulate the stability of catalysts under various reaction conditions, thereby predicting their service life in advance. This is conducive to the development of more stable and longer-lasting single-atom catalysts. Thus, this interdisciplinary synergy is expected to unlock new avenues for understanding and optimizing the performance of TM-SACs in complex organic synthesis systems.

Finally, with the rapid development of TM-SACs, their large-scale application in the future faces many considerations in terms of strategy and economy. Developing simple and efficient synthesis pathways is the key to the large-scale application of TM-SACs. By targeting research and development (R&D) investment and promoting collaborative innovation between industry, academia, and research institutions, costs can be controlled. Moreover, TM-SACs combine the advantages of homogeneous and heterogeneous catalysis, offering higher recyclability and lower waste generation, and thus hold great market potential for the future.

Acknowledgements

This work was supported by the National Natural Science Foundation of China (Grants 52130209, 52402065, and 52472055) and the Research Funding from Institute of Metal Research Grants E351L501A1. The authors declare that they have no known competing financial interests or personal relationships that could have appeared to influence the work reported in this review.

Conflict of interest

The authors declare no conflict of interest.

Author details

Shiyong Zhao[1*], Jinfeng Xu[2], Jincheng Li[2] and Chang Liu[1]

1 Shenyang National Laboratory for Materials Science, Institute of Metal Research, Chinese Academy of Sciences, Shenyang, China

2 Faculty of Chemical Engineering, Yunnan Provincial Key Laboratory of Energy Saving in Phosphorus Chemical Engineering and New Phosphorus Materials, Kunming University of Science and Technology, Kunming, China

*Address all correspondence to: zhaosy@imr.ac.cn

IntechOpen

© 2025 The Author(s). Licensee IntechOpen. This chapter is distributed under the terms of the Creative Commons Attribution License (http://creativecommons.org/licenses/by/4.0), which permits unrestricted use, distribution, and reproduction in any medium, provided the original work is properly cited.

References

[1] Cheng X, Fu C, Chen BB, et al. Asymmetric relay catalysis enables unreactive allylic alcohols to participate in 1,3-dipolar cycloaddition of Azomethine Ylides. Journal of the American Chemical Society. 2025;**147**(6):5014-5024. DOI: 10.1021/jacs.4c14761

[2] Absalan Y, Gholizadeh M, Kim EB, et al. Recent progress on organic metal compound/MOF hybrids: From controllable synthesis to potential catalytic applications. Coordination Chemistry Reviews. 2024;**515**:215972. DOI: 10.1016/j.ccr.2024.215972

[3] Shi W, Li J, Gao F, et al. Strongly coordinating mediator enables single-step resource recovery from heavy metal-organic complexes in wastewater. Nature Communications. 2024;**15**(1):10828. DOI: 10.1038/s41467-024-55174-1

[4] Chen L, Zhang S, Liu X, et al. Recent advances in water-mediated multiphase catalysis. Current Opinion in Colloid & Interface Science. 2023;**65**:101691. DOI: 10.1016/j.cocis.2023.101691

[5] Peng M, Ge Y, Gao R, et al. Thermal catalytic reforming for hydrogen production with zero CO2 emission. Science. 2025;**387**(6735):769-775. DOI: 10.1126/science.adt0682

[6] Lu Y, Knezevic M, Prescimone A, et al. Site-selective C(sp3)–H oxidation of alkyl substrates devoid of functional handles. Chem. 2025;**11**(3):102442. DOI: 10.1016/j.chempr.2025.102442

[7] Bates JS. Progress and pitfalls in designing heterogeneous catalysts with molecular precision. Nature Chemistry. 2025;**17**:1-7. DOI: 10.1038/s41557-024-01731-6

[8] Chen BWJ, Mavrikakis M. Modeling the impact of structure and coverage on the reactivity of realistic heterogeneous catalysts. Nature Chemical Engineering. 2025;**2**:1-17. DOI: 10.1038/s44286-025-00179-w

[9] Zhu A, Qiao L, Liu K, et al. Rational design of precatalysts and controlled evolution of catalyst-electrolyte interface for efficient hydrogen production. Nature Communications. 2025;**16**(1):1880. DOI: 10.1038/s41467-025-57056-6

[10] Yuan C, Zhang S, Zhang J. Oxygen reduction electrocatalysis: From conventional to single-atomic platinum-based catalysts for proton exchange membrane fuel cells. Frontiers in Energy. 2024;**18**(2):206-222. DOI: 10.1007/s11708-023-0907-3

[11] Yu Y, Zhu Z, Huang H. Surface engineered single-atom systems for energy conversion. Advanced Materials. 2024;**36**(16):2311148. DOI: 10.1002/adma.202311148

[12] Song ZY, Li PH, Li YY, et al. High-throughput screening of selective bimetallic atomic catalysts for self-adaptive matched electrochemical reduction. Nano Letters. 2025;**25**(10):3947-3955. DOI: 10.1021/acs.nanolett.4c06524

[13] Zhu ZS, Wang Y, Wang P, et al. Multidimensional engineering of single-atom cobalt catalysts for ultrafast Fenton-like reactions. Nature Water. 2025;**3**(2):211-221. DOI: 10.1038/s44221-024-00382-8

[14] Zhao SH, Pan Y. Principles of coordination structure design of single-atom catalysts in electrocatalytic oxygen reduction reaction. Rare Metals.

2025;**44**:2900-2920. DOI: 10.1007/s12598-024-03085-x

[15] Song W, Xiao C, Ding J, et al. Review of carbon support coordination environments for single metal atom electrocatalysts (SACS). Advanced Materials. 2024;**36**(1):2301477. DOI: 10.1002/adma.202301477

[16] Liu S, Lin C, Chen J, et al. Finely tailoring metal–support interactions via transient high-temperature pulses. ACS Applied Materials & Interfaces. 2025;**17**(8):12328-12338. DOI: 10.1021/acsami.4c22469

[17] Wang Y, Zhang W, Wen W, et al. Atomically dispersed unsaturated Cu-N3 sites on high-curvature hierarchically porous carbon nanotube for synergetic enhanced nitrate electroreduction to ammonia. Advanced Functional Materials. 2023;**33**(46):2302651. DOI: 10.1002/adfm.202302651

[18] Zhao S, Chen G, Zhou G, et al. A universal seeding strategy to synthesize single atom catalysts on 2D materials for electrocatalytic applications. Advanced Functional Materials. 2020;**30**(6):1906157. DOI: 10.1002/adfm.201906157

[19] Han Z, Zhao S, Xiao J, et al. Engineering d-p orbital hybridization in single-atom metal-embedded three-dimensional electrodes for Li–S batteries. Advanced Materials. 2021;**33**(44):2105947. DOI: 10.1002/adma.202105947

[20] Cheng Y, Zhao S, Johannessen B, et al. Atomically dispersed transition metals on carbon nanotubes with ultrahigh loading for selective electrochemical carbon dioxide reduction. Advanced Materials. 2018;**30**(13):1706287. DOI: 10.1002/adma.201706287

[21] Huang J, Ajmal S, Kumar A, et al. Atomically precise M-N-C electrocatalysts for oxygen reduction: Effects of inter-site distance, metal-metal interaction, coordination environment, and spin states. Journal of Energy Chemistry. 2025;**101**:132-155. DOI: 10.1016/j.jechem.2024.08.068

[22] Ge X, Shao Y, Gan Y, et al. Thermally-driven fabrication of single-atom catalysts from bulk metal foil for oxygen reduction. New Journal of Chemistry. 2025;**49**:4806-4812. DOI: 10.1039/D4NJ05417B

[23] Zhang ZQ, Duan PJ, Zheng JX, et al. Nano-island-encapsulated cobalt single-atom catalysts for breaking activity-stability trade-off in Fenton-like reactions. Nature Communications. 2025;**16**(1):115. DOI: 10.1038/s41467-024-55622-y

[24] Liu B, Han Z, Bendavid A, et al. Atomic-layer deposition of the single-atom Pt catalyst on vertical graphene for H2 sensing. ACS Applied Nano Materials. 2024;**7**(19):22605-22616. DOI: 10.1021/acsanm.4c03416

[25] Zhang L, Liao J, Li Y, et al. Cu single atoms embedded on hollow g-C3N4 nanospheres with enhanced charge transfer and separation for efficient photocatalysis. Chinese Chemical Letters. 2024;**35**(2):108568. DOI: 10.1016/j.cclet.2023.108568

[26] Oke JA, Jen TC. Atomic layer deposition thin film techniques and its bibliometric perspective. International Journal of Advanced Manufacturing Technology. 2023;**126**(11):4811-4825. DOI: 10.1007/s00170-023-11478-y

[27] Lin Z, Yu X, Zhao Z, et al. Controlling crystallization in covalent organic frameworks to facilitate photocatalytic hydrogen

production. Nature Communications. 2025;**16**(1):1940. DOI: 10.1038/s41467-025-57166-1

[28] Lee C, Lee B, Park S, et al. Photochemical tuning of dynamic defects for high-performance atomically dispersed catalysts. Nature Materials. 2024;**23**:552-559. DOI: 10.1038/s41563-024-01799-y

[29] Hasan IM, Xu N, Liu Y, et al. Noble and non-noble metal based catalysts for electrochemical nitrate reduction to ammonia: Activity, selectivity and stability. Electrochemical Energy Reviews. 2024;**7**(1):36. DOI: 10.1007/s41918-024-00236-7

[30] Butburee T, Prasert A, Rungtaweevoranit B, et al. Engineering Lewis-acid defects on ZnO quantum dots by trace transition-metal single atoms for high glycerol-to-glycerol carbonate conversion. Small. 2024;**20**(44):2403661. DOI: 10.1002/smll.202403661

[31] Gong X, Bao J, Wang X, et al. Dry-solid-electrochemical synthesis: A general method for large scale synthesis of single-atom catalysts with high metal loadings. Angewandte Chemie International Edition. 2025;**64**:e202419374. DOI: 10.1002/anie.202419374

[32] Kosar N, Mahmood T, Arshad M, et al. Scrutinization of late first-row transition metals decorated octagonal boron (B8) ring complexes as single-atom catalysts for green hydrogen and oxygen production. RSC Advances. 2025;**15**(9):6863-6874. DOI: 10.1039/d4ra07274j

[33] Han J, Meng X, Lu L, et al. Single-atom Fe-N-C as an efficient electrocatalyst for zinc–air batteries. Advanced Functional Materials. 2019;**29**(41):1808872. DOI: 10.1002/adfm.201808872

[34] Zhen J, Sun J, Xu X, et al. M−N3 configuration on boron nitride boosts singlet oxygen generation via peroxymonosulfate activation for selective oxidation. Angewandte Chemie International Edition. 2024;**63**(26):e202402669. DOI: 10.1002/anie.202402669

[35] Chai Y, Dai H, Duan X, et al. Elucidation of the mechanistic origin of spin-state-dependent P-doped Fe single-atom catalysts for the oxidation of organic pollutants through peroxymonosulfate activation. Applied Catalysis B: Environmental. 2024;**341**:123289. DOI: 10.1016/j.apcatb.2023.123289

[36] Zhang J, Deng Z, Bai S, et al. Fe, N, S co-doped carbon network derived from acetate-modified Fe-ZIF-8 for oxygen reduction reaction. Journal of Colloid and Interface Science. 2024;**658**:373-382. DOI: 10.1016/j.jcis.2023.12.047

[37] Wei S, Zhu J, Chen X, et al. Planar chlorination engineering induced symmetry-broken single-atom site catalyst for enhanced CO2 electroreduction. Nature Communications. 2025;**16**(1):1652. DOI: 10.1038/s41467-025-56271-5

[38] Zhang T, Sun Z, Li S, et al. Regulating electron configuration of single Cu sites via unsaturated N,O-coordination for selective oxidation of benzene. Nature Communications. 2022;**13**(1):6996. DOI: 10.1038/s41467-022-34852-y

[39] Zhai C, Chen Y, Huang X, et al. Recent progress on single-atom catalysts in advanced oxidation processes for water treatment. Environmental Functional Materials. 2022;**1**(3):219-229. DOI: 10.1016/j.efmat.2022.11.001

[40] Gao Y, Yang C, Zhou M, et al. Transition metal and metal–N Co

doped MOF-derived fenton-like catalysts: A comparative study on single atoms and nanoparticles. Small. 2020;**16**(50):2005060. DOI: 10.1002/smll.202005060

[41] Pei S, Wang S, Lu Y, et al. Application of metal-based catalysts for fenton reaction: From homogeneous to heterogeneous, from nanocrystals to single atom. Nano Research. 2024;**17**(11):9446-9471. DOI: 10.1007/s12274-024-6973-y

[42] Zhang Q, Jiang X, Su Y, et al. Catalytic propane dehydrogenation by anatase supported Ni single-atom catalysts. Chinese Journal of Catalysis. 2024;**57**:105-113. DOI: 10.1016/S1872-2067(23)64584-X

[43] He Y, Song Y, Cullen DA, et al. Selective and stable non-Noble-metal intermetallic compound catalyst for the direct dehydrogenation of propane to propylene. Journal of the American Chemical Society. 2018;**140**(43):14010-14014. DOI: 10.1021/jacs.8b05060

[44] Zhang G, Yang C, Miller JT. Tetrahedral nickel(II) phosphosilicate single-site selective propane dehydrogenation catalyst. ChemCatChem. 2018;**10**(5):961-964. DOI: 10.1002/cctc.201701815

[45] Searles K, Siddiqi G, Safonova OV, et al. Silica-supported isolated gallium sites as highly active, selective and stable propane dehydrogenation catalysts. Chemical Science. 2017;**8**(4):2661-2666. DOI: 10.1039/C6SC05178B

[46] Hu ZP, Qin G, Han J, et al. Atomic insight into the local structure and microenvironment of isolated co-motifs in MFI zeolite frameworks for propane dehydrogenation. Journal of the American Chemical Society. 2022;**144**(27):12127-12137. DOI: 10.1021/jacs.2c02636

[47] Li Y, Zhang Q, Fu S, et al. Active species and fundamentals of their creation in co-containing catalysts for efficient propane dehydrogenation to propylene. Chemical Engineering Journal. 2023;**460**:141778. DOI: 10.1016/j.cej.2023.141778

[48] Gao Y, Peng L, Long J, et al. Hydrogen pre–reduction determined co–silica interaction and performance of cobalt catalysts for propane dehydrogenation. Microporous and Mesoporous Materials. 2021;**323**:111187. DOI: 10.1016/j.micromeso.2021.111187

[49] Zhao D, Li Y, Han S, et al. ZnO nanoparticles encapsulated in nitrogen-doped carbon material and silicalite-1 composites for efficient propane dehydrogenation. iScience. 2019;**13**:269-276. DOI: 10.1016/j.isci.2019.02.018

[50] Lu Z, Liu X, Zhang B, et al. Structure and reactivity of single site Ti catalysts for propylene epoxidation. Journal of Catalysis. 2019;**377**:419-428. DOI: 10.1016/j.jcat.2019.07.051

[51] Li J, Zhao S, Yang SZ, et al. Atomically dispersed cobalt on graphitic carbon nitride as a robust catalyst for selective oxidation of ethylbenzene by peroxymonosulfate. Journal of Materials Chemistry A. 2021;**9**(5):3029-3035. DOI: 10.1039/D0TA11503G

[52] Sudrajat H, Wella SA, Phanthuwongpakdee J, et al. Atomistic understanding of enhanced selectivity in photocatalytic oxidation of benzyl alcohol to benzaldehyde using graphitic carbon nitride loaded with single copper atoms. Nanoscale. 2024;**16**(31):14813-14830. DOI: 10.1039/D4NR01610F

[53] Wei X, Hu Z, Li C, et al. High-density atomically dispersed CoNx

catalysts supported on nitrogen-doped mesoporous carbon materials for efficient hydrogenation of nitro compounds. Catalysis Today. 2022;**405-406**:92-100. DOI: 10.1016/j.cattod.2022.06.028

[54] Zhu P, Shi WX, Wang Y, et al. Fabrication of N,S co-doped carbon immobilized high-density co single atoms toward electro-oxidation of organic sulfides with water as feedstock. Nano Research. 2023;**16**(5):6593-6600. DOI: 10.1007/s12274-023-5381-z

[55] Xia T, Ju M, Qian H, et al. Photocatalytic fenton-like system with atomic Fe on carbon nitride boost selective glucose oxidation towards gluconic acid. Journal of Catalysis. 2024;**429**:115257. DOI: 10.1016/j.jcat.2023.115257

[56] Zhang T, Zhao W, Li S, et al. Single-atom co sites with upshifted d-band center efficiently boost transfer hydrogenation for selective imines synthesis. Chemical Engineering Journal. 2024;**484**:149614. DOI: 10.1016/j.cej.2024.149614

[57] Meng G, Lan W, Zhang L, et al. Synergy of single atoms and Lewis acid sites for efficient and selective lignin disassembly into monolignol derivatives. Journal of the American Chemical Society. 2023;**145**(23):12884-12893. DOI: 10.1021/jacs.3c04028

Chapter 8

Magnetic Spinel Ferrite Nanomaterials: Fabrication and Application

Hua Li, Wenqin Li, Jacques Robichaud and Yahia Djaoued

Abstract

Spinel ferrites represented crystallographically as $\left[M_{1-j}^{2+}Fe_j^{3+}\right]_A\left[M_j^{2+}Fe_{2-j}^{3+}\right]_B O_4$, where the inversion degree j defines the fraction of tetrahedral (A) or octahedral (B) sites occupied by Fe^{3+} or M^{2+} cations, exhibit properties that are highly sensitive to cation distribution. This inversion degree j, reflecting the distribution of cations between A and B sites, has a profound impact on their physical properties, particularly magnetic and electrical characteristics. Tailoring these properties often requires precise control over cation distribution, achieved through various synthetic approaches. Unlike traditional high-temperature synthesis methods, such as ceramic processing, spinel ferrite nanoparticles synthesized at lower temperatures often exhibit non-equilibrium cation distributions. This results in physical properties markedly distinct from their bulk counterparts. This chapter begins by introducing generic ferrimagnetic systems and nanoscale magnetic phenomena before focusing on spinel ferrite nanomaterials. The discussion covers their crystal structure, diverse synthesis methods—from conventional techniques to novel approaches—and strategies for tuning their magnetic properties. Additionally, their applications in magnetism, environmental remediation, biomedicine, and hyperthermia are explored. The chapter concludes with an overview of current scientific challenges and future research directions in the field.

Keywords: spinel ferrtites, nanomaterials, ferrimagnetism, synthesis, application

1. Introduction

Ferrimagnetism, a term introduced by Néel, describes a magnetic phenomenon exhibited by certain materials (e.g., ferrites) that have unbalanced antiparallel-aligned sublattice magnetic moments below a specific critical temperature [1, 2]. Unlike ferromagnets, such as transition metals, which possess unpaired electrons aligned in parallel spins leading to spontaneous and strong permanent magnetism, ferrimagnets demonstrate a weaker net magnetization. This reduced magnetization results from the unequal contributions of opposing magnetic moments: stronger moments in one direction are only partially canceled by weaker moments in the opposite direction. Ferrimagnets also differ from antiferromagnets, such as hematite, which have completely balanced antiparallel spins, resulting in a net-zero magnetic moment [3, 4]. Beyond their magnetic characteristics, ferrimagnetic insulators—predominantly transition-metal oxides—stand out for their electrical insulating properties. These materials, with their low dielectric losses and reduced eddy current generation, are invaluable in applications involving microwaves, spin-waves, and other advanced technologies [5–9].

This chapter focuses on ferrimagnetic ferrites, particularly spinel ferrites. It provides a comprehensive discussion of their structural and magnetic properties, as well as the strategies used to tune these properties. Topics include various synthesis methods for spinel ferrites, emphasizing nanoscale and doping approaches and a detailed exploration of their magnetic-related applications in both traditional and emerging technological domains.

2. Nanoscale ferrimagnetism

2.1 General considerations

Magnetism arises from the motion of charges and the intrinsic magnetic properties of electrons. The macroscopic magnetic behavior of materials stems from the magnetic moments associated with individual electrons. At the atomic level, there are two primary types of electronic motion that contribute to magnetism: the *orbital motion* of electrons around the nucleus and the *spin motion* of electrons. The orbital motion is analogous to the current flowing in a loop of wire, while in an isolated electron, the total magnetic moment is primarily due to its spin (**Figure 1**) [10]. When atoms have uncompensated (or unpaired) electrons, magnetism arises because these unpaired electrons behave like tiny magnets due to their spin and movement around the nucleus. The term 'exchange interactions' refers to the quantum mechanical forces between these unpaired electrons in neighboring atoms. These interactions determine whether the atomic 'mini-magnets' align in the same direction (ferromagnetism), in opposite directions (antiferromagnetism), or adopt more complex arrangements, which ultimately define the material's magnetic properties.

Magnetic materials can be influenced by an external magnetic field, which can either attract or repel them. Based on their response to this field, materials are

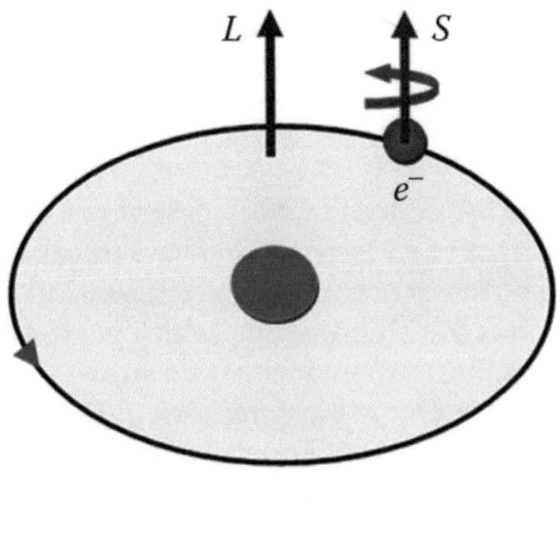

Figure 1.
Origin of magnetism [10].

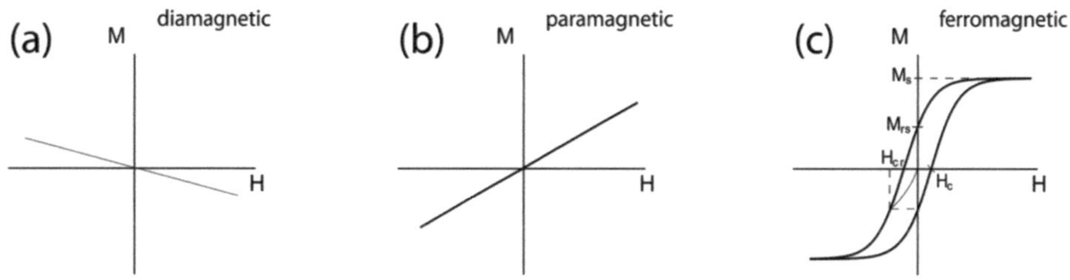

Figure 2.
Schematic behavior of the magnetic susceptibility of three types of magnetic materials. (a) Diamagnetic, (b) paramagnetic, and (c) ferromagnetic [11].

generally classified into three main categories (**Figure 2**): *diamagnetic*, *paramagnetic*, and *ferromagnetic*, according to their magnetic susceptibility (χ) [11]. Magnetic susceptibility (χ) is a dimensionless parameter that quantifies how a material becomes magnetized in response to an external magnetic field. It is expressed mathematically as:

$$M = \chi H \tag{1}$$

where M is the induced magnetization (A/m), and H is the applied magnetic field (A/m).

In diamagnetic materials, such as quartz, the atoms have no permanent magnetic moment. When an external magnetic field is applied, these materials develop a weak magnetization in the opposite direction to the applied field. This effect is due to the induced currents that oppose the change in the magnetic field. Diamagnetic materials have a small negative susceptibility, typically around $\chi \approx -10^{-5}$.

In paramagnetic materials, such as those containing transition-metal ions, the atoms have randomly oriented magnetic moments in the absence of an external field. When an external field is applied, these moments align partially in the direction of the field, producing a weak net magnetization. Paramagnetic materials exhibit a small positive susceptibility, typically in the range of $\chi \approx 10^{-5}$ to 10^{-3} [12].

Ferromagnetic materials, characterized by uncompensated electron spins, exhibit strong exchange interactions that result in a net magnetization even in the absence of an external magnetic field, provided the temperature is below their magnetic ordering temperature. These materials can be classified into different types based on the configuration of their magnetic sublattices: ferromagnetic, ferrimagnetic, antiferromagnetic, and spin-canted antiferromagnetic [11]. The relationship between magnetization and an external field is nonlinear, as illustrated by the hysteresis loop of magnetization versus field (**Figure 2c**) [13].

In ferromagnets, strong positive exchange interactions align atomic magnetic moments in a parallel configuration (**Figure 3a**). This results in significant magnetic susceptibilities, as observed in metals like Fe, Co, and Ni, when the temperature is below the Curie temperature. Above this temperature, thermal energy randomizes the magnetic moments, causing spontaneous magnetization to vanish and the material to exhibit paramagnetic behavior.

Antiferromagnetism, in contrast, displays weak overall magnetism due to the antiparallel alignment of spin magnetic moments, driven by exchange interactions between neighboring atoms (**Figure 3b**). This leads to a macroscopically balanced state with minimal net magnetization. Similar to ferromagnets, antiferromagnets possess a critical temperature, called the Néel temperature. Above this temperature,

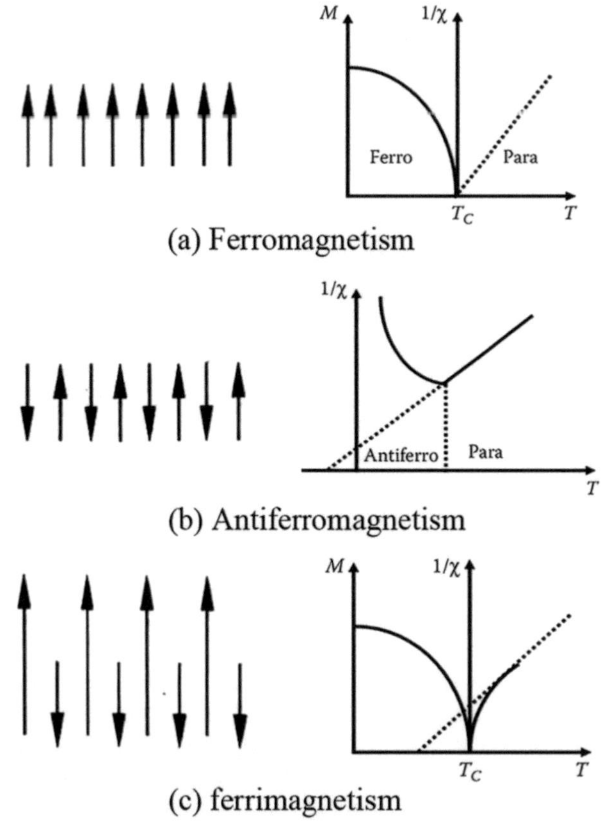

Figure 3.
Illustration of three simple exchange models of spontaneous spin alignment (left) and corresponding magnetic behaviors (right): for a single parallel magnetic lattice, ferromagnetism (a), for two equal antiparallel sublattices, antiferromagnetism (b), and two unbalance sublattices, ferrimagnetism (c) [10].

thermal agitation disrupts the exchange interactions, and the material exhibits susceptibility similar to that of paramagnets.

Ferrimagnetism, initially described in ferrites, refers to materials with a small net magnetization arising from the antiferromagnetic coupling of inequivalent sublattice moments (**Figure 3c**). Unlike antiferromagnets, the sublattice moments in ferrimagnets are unbalanced, resulting in a net magnetic moment equal to the difference between the opposing sublattice moments. Like ferromagnets, ferrimagnets lose their spontaneous magnetization above the Curie temperature and behave paramagnetically. Ferrimagnets share many properties with ferromagnets, including high permeability, significant saturation magnetization, and the presence of hysteresis.

2.2 Magnetic parameters

Magnetic properties are universally characterized using the following parameters: magnetic flux (φ_m), magnetization vector (M), magnetic permeability (μ), and magnetic susceptibility (χ)

2.2.1 Magnetic flux (φ_m)

In a constant field, magnetic flux is defined as the product of the magnetic field and the area it penetrates, given by:

$$\varphi_m = \oint_s \overline{B.ds} \tag{2}$$

Here, B is the magnetic field (or magnetic flux density) and *ds* is an infinitesimal area element.

Since isolated magnetic poles do not exist in nature, the net magnetic flux through a closed surface is always zero. This is mathematically expressed by Gauss's law for magnetism:

$$\oint_s \overline{B.ds} = 0 \tag{3}$$

2.2.2 Magnetization vector ($\bar{M}$)

Under an external magnetic field, the dipoles within a material partially align, creating a net magnetic moment in the direction of the field. The magnetization (M) is defined as the magnetic moment per unit volume of the material (M = magnetic moment/volume) [14]. The tendency of dipoles to align depends on the applied field strength, referred to as the *intensity of magnetization*. The unit of magnetization is amperes per meter (A/m), as the magnetic moment is measured in A m^2.

2.2.3 Magnetic permeability (μ)

Magnetic permeability quantifies a material's ability to support the formation of magnetic flux in response to an applied field (H). It is defined as:

$$\mu = \frac{Magnetic\ flux\ density}{Magnetic\ field} = \frac{B}{H} \tag{4}$$

The unit of magnetic permeability is Henries per m (H/m).

The *relative permeability* (μ_r) is the ratio of the permeability of a specific medium (μ) to the permeability of free space (μ_0):

$$\mu_r = \frac{\mu}{\mu_0} \tag{5}$$

Here, $\mu_0 = 4\pi \times 10^{-7}$ H/m. The *relative permeability* μ_r is dimensionless.

2.2.4 Magnetic susceptibility (χ)

Magnetic susceptibility (χ) is a measure of how easily a material becomes magnetized in an external magnetic field. It is defined as the ratio of the magnetization vector (M) to the applied magnetic field (H): χ = M/H. Since M and H have the same units (A/m), susceptibility is dimensionless.

2.2.5 Relation between magnetic parameters

The relationships among the magnetic parameters are summarized as follows:

a. Magnetic flux density (B) is related to permeability and the applied field:

$$B = \mu_r \mu_0 H \text{ or } B = \mu H \tag{6}$$

b. Magnetization (M) is related to susceptibility (χ) and magnetic flux density (B):

$$M = \frac{\chi}{\mu} B \tag{7}$$

c. The relationship between permeability, relative permeability, and susceptibility is:

$$\frac{\mu}{\mu_0} = \mu_r = 1 + \chi \text{ or } \mu = \mu_0 + \mu_0 \chi \tag{8}$$

2.3 Nanoscale magnetic phenomena

2.3.1 Magnetic domains

Ferromagnetic materials in a demagnetized state exhibit no net magnetic moment despite their magnetic nature. This is because they spontaneously divide into small regions known as *magnetic domains*, each of which is magnetized in a different direction to minimize magnetostatic energy. These domains typically have a volume of about 10^{-9} cm^3, depending on the material and various anisotropy energy contributions. The opposing directions of adjacent domains often form closed-loop configurations, effectively canceling the magnetostatic energy (**Figure 4**).

The boundaries between these domains are called *domain walls*, which serve as transitional regions where the magnetization gradually changes from one domain to another. Domain walls increase the material's total energy. Their thickness, usually around 10^{-5} cm, depends on the balance between *exchange energy* and *magnetic anisotropy energy*:

- Narrow walls reduce exchange energy.
- Thicker walls reduce magnetic anisotropy energy.

When an external magnetic field is applied, the material's magnetization occurs primarily through two mechanisms: *domain wall movement* and *domain rotation*, as discussed below.

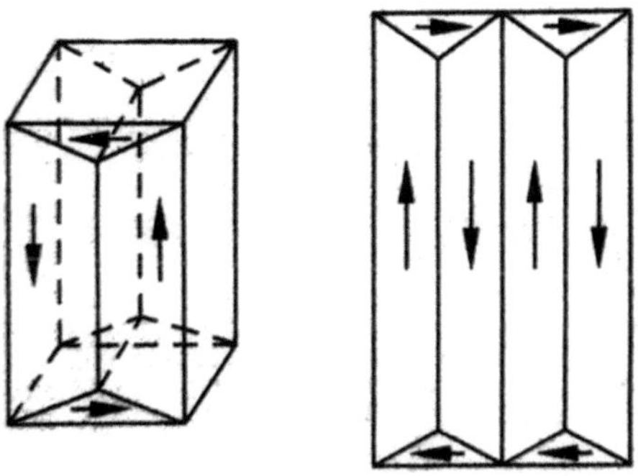

Figure 4.
Illustration of close circle of domains in a ferromagnet.

2.3.2 Hysteresis

The magnetization process under an external magnetic field can be described by the following mechanisms (**Figure 5**).

1. *Domain wall movement* (**Figure 5b**)

    - At weak external fields (H), domain walls move along the field direction, expanding domains that align favorably with the field and shrinking those that do not.

2. *Domain rotation* (**Figure 5c**)

    - At strong external fields, the magnetization mechanism shifts to the physical rotation of domains, aligning them parallel to the field direction.

Hysteresis describes the behavior of magnetization (M) as a function of the applied magnetic field (H) well below the material's Curie temperature (T_c). **Figure 6** illustrates this phenomenon.

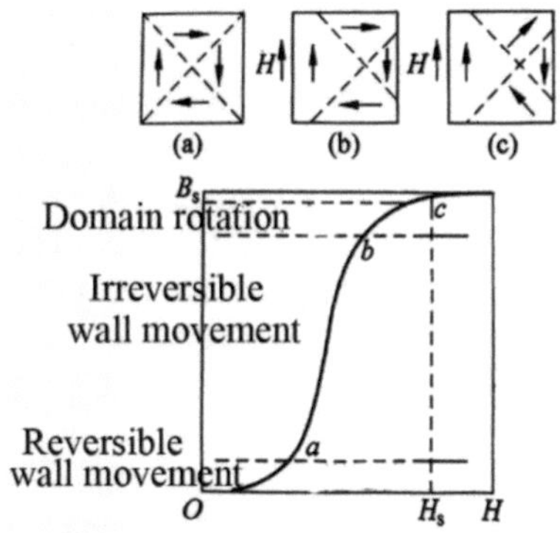

Figure 5.
Domain and wall changes with magnetization process. (a) Demagnetized state; (b) domain wall movement state; (c) domain rotation state.

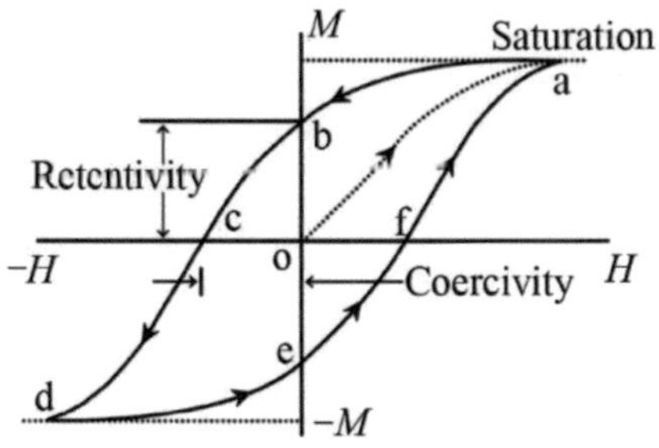

Figure 6.
The magnetization curve [10].

- Initially, M initially increases linearly with H and then transitions to nonlinear behavior due to domain wall movement and rotation.
- At high enough H, magnetization reaches a maximum (saturation magnetization, M_s) when all domains align with the field.
- When H is gradually reduced, M decreases but does not retrace the initial path. Instead, the material retains some magnetization even after H = 0. This remaining magnetization is called *remanent magnetization* (M_r).
- To fully demagnetize the material, H must be reversed to a value called the *coercive field* (H_c).

As H continues to increase in the negative direction, the material becomes saturated again but in the opposite orientation (point d). When H is subsequently reduced and reversed, the material follows a new path through points e and f and back to saturation at a.

This complete cycle forms the *hysteresis loop*, which characterizes the material's magnetic behavior. The area enclosed by the loop represents the energy loss during the magnetization process. This energy is dissipated as heat, indicating inefficiencies in magnetic applications.

2.3.3 Nanoscale magnetic phenomena

The alignment of sublattices *via* exchange interactions primarily determines the type of magnetism in a material. However, other interactions, such as anisotropy, dipolar interactions, and magnetoelastic effects, significantly influence magnetic behavior by forming and stabilizing magnetic domains. The magnetization process, therefore, depends on the response of these domains and their boundaries (domain walls) to external magnetic fields.

When the size of a magnetic material is reduced from the microcrystalline to the nanocrystalline scale, two primary effects emerge:

1. *Domain size limitations and broken translation symmetry*: These changes alter the fundamental magnetic system.
2. *Increased interfacial atoms and defects*: Nanoparticles exhibit a high surface-to-volume ratio, significantly impacting magnetic properties due to interfacial

atomic behavior and an increased number of defects, further complicating the magnetism of the system [15, 16].

2.3.3.1 Single-domain particles

As discussed earlier (Section 2.3.1 on domains), ferromagnetic materials are spontaneously divided into domains to minimize internal energy. However, when the size of a particle becomes smaller than or comparable to the critical single-domain diameter (typically tens of nanometers), the inter-particle and intra-particle interactions together dictate the magnetic properties [17].

In single-domain particles, domain wall movement is absent. Instead, magnetization is reversed solely through domain rotation, leading to reduced energy stability. A uniformly magnetized single-domain particle can be represented by a *superspin*, where the entire particle behaves as a single magnetic entity (**Figure 7**) [18].

2.3.3.2 Superparamagnetism and blocking temperature

Magnetic anisotropy, which arises from spin-orbit interactions, is proportional to the particle's volume. As the particle size decreases, its magnetic anisotropy energy can become comparable to or smaller than the thermal energy. When this occurs, the magnetic moments of particles can freely rotate, and the material loses its magnetism in the absence of an external field.

The temperature at which this transition occurs is termed the *blocking temperature* (T_B), which depends on several factors, including particle size, the effective anisotropy constant, the applied magnetic field, dipolar interactions, and the experimental measurement time.

- When $T > T_B$: The particles enter a *superparamagnetic state*, where thermal energy causes their magnetic moments to fluctuate rapidly around the easy axis of magnetization. Despite their large individual magnetic moments, these fluctuations prevent remanent magnetization or coercivity, but they allow for a rapid response to external magnetic fields [19–21].

- $T < T_B$: The thermal energy is insufficient to overcome the anisotropy energy, causing the moments to freeze. In this state, hysteresis and coercivity are observed during magnetization cycles.

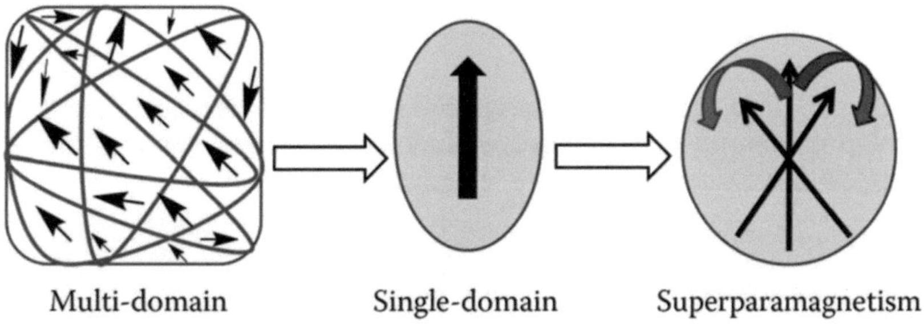

Figure 7.
Formation of superparamagnetic structures [10].

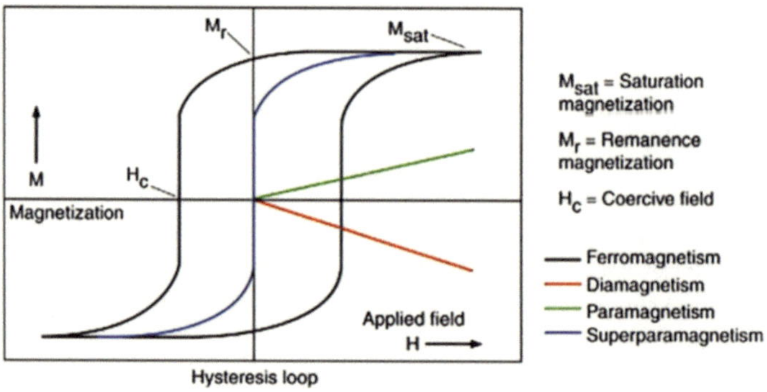

Figure 8.
Hysteresis loops for various types of magnetic materials [23].

Type of magnetic material	Coercivity	Remanence magnetization	Saturation magnetization	Examples	Ref.
Ferromagnetic	Very high	Very high	Very high	Ni, Fe, Co	[24]
Ferrimagnetic	High at low T	High at low T	High	Fe_3O_4, $MnFe_2O_4$	[25]
Antiferromagnetic	low	low	low	Hematite ($\alpha-Fe_2O_3$)	[3]
superparamagnetic	Zero above T_B	Zero above T_B	High	Fe_3O_4, $NiFe_2O_4$ below 10 nm	[26]

Table 1.
Summary of the different types of magnetic behavior exhibited by ferrites [4].

2.3.3.3 Particle size and coercivity

Particle size also significantly impacts magnetic coercivity.

- For particles larger than the critical single-domain diameter, coercivity increases as size decreases because domain wall movement becomes less efficient in responding to the external magnetic field.

- Below the critical diameter, coercivity decreases as thermal energy facilitates the rotation of single-domain particles, eventually reaching zero. At this point, the material transitions into a superparamagnetic state [22].

Figure 8 and **Table 1** compare the hysteresis loops for various types of magnetic behavior. Superparamagnetic materials are characterized by negligible or zero coercivity at high temperatures, a key feature that distinguishes them from other magnetic states.

3. Magnetic spinel ferrites nanomaterials

Ferrimagnets are a class of magnetic materials characterized by unbalanced antiparallel-aligned sublattice moments, as previously described. They typically encompass two main categories: *ferrimagnetic insulators* and *ferrimagnetic alloys*. The former includes ferrites, while the latter is composed primarily of rare-earth-transition-metal (RE-TM) alloys. Ferrites stand out due to their *high specific electrical resistance*, which minimizes eddy current losses, their *customizability in magnetic properties*, and their exceptional performance in applications involving microwaves, spin-waves, and other advanced technologies [27–29].

The term *ferrite* is derived from *ferrum*, the Latin word for iron. While the practical use of ferrites dates back over 2000 years—pioneered by Ancient Greeks for tool-making—the modern scientific understanding of ferrites began to emerge in the 1930s, driven by advances in physics and chemistry [30–33]. Ferrites are broadly classified into four subcategories based on their crystal structures: *garnets*, *orthoferrites*, *hexagonal ferrites*, and *spinels* (**Figure 9**).

- *Garnets*: These have the composition $M_3Fe_5O_{12}$, where M is a trivalent rare-earth element, such as yttrium. Garnets are typically *hard magnetic materials*, exhibiting high coercivity and significant remanent magnetization.

- *Orthoferrites*: With the composition $MFeO_3$, where M is a rare-earth element, orthoferrites are generally classified as *soft magnetic materials*, known for their low coercivity.

- *Hexaferrites*: Characterized by the formula $MFe_{12}O_{19}$, hexaferrites are *hard magnetic materials* with pronounced magnetic anisotropy, making them ideal for high-frequency and permanent magnet applications.

- *Spinel ferrites*: Perhaps the most widely studied class, spinel ferrites have the general formula MFe_2O_4, where M is a divalent cation such as Mn^{2+}, Fe^{2+}, or Ni^{2+}. These cations can be partially substituted with trivalent ions like Mn^{3+} or Co^{3+} [35],

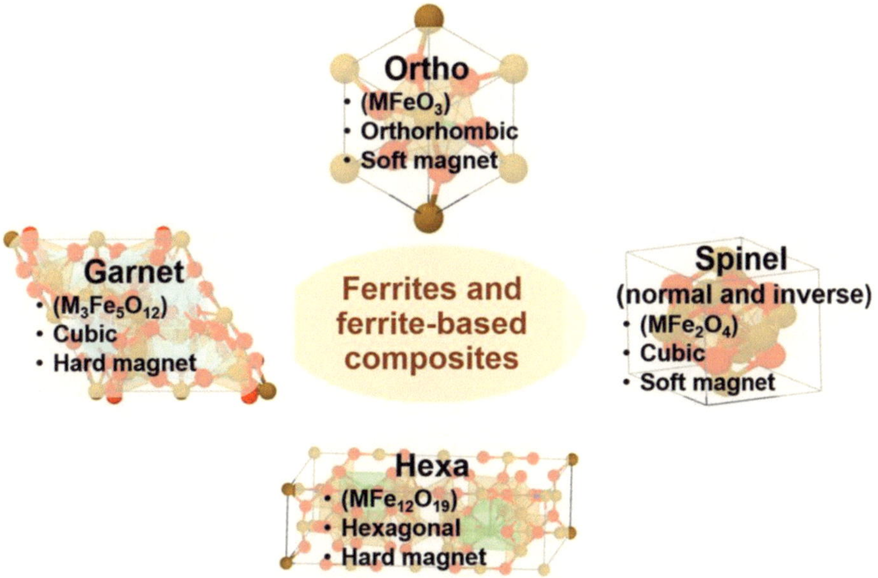

Figure 9.
The four classes of ferrites: garnet, ortho, hexagonal, and spinel [34].

allowing for precise tuning of their physical and magnetic properties. The distribution of cations within the lattice significantly influences the material's overall magnetic characteristics.

Given their versatility and unique properties, *spinel ferrites* will be the primary focus of this discussion, particularly in the context of their structural features and magnetic-related applications.

3.1 Structure of spinel ferrites

Spinel ferrites possess a face-centered cubic (FCC) crystal structure, classified under the space group Oh7-F3dm. The structure comprises a close-packed arrangement of oxygen (O^{2-}) ions, creating two types of interstitial sites (**Figure 10**):

- *Tetrahedral sites (A sites)*: Formed by coordination with four nearest O^{2-} ions.
- *Octahedral sites (B sites)*: Formed by coordination with six nearest O^{2-} [1, 3].

In this configuration, 12.5% of the tetrahedral A sites and 50% of the octahedral B sites are occupied by metal cations.

3.1.1 Cation distribution and inversion degree

Both M^{2+} (divalent) and Fe^{3+} (trivalent) ions can occupy either the A or B sites. To describe their distribution, the inversion degree (j) is introduced [36]. This parameter quantifies the fraction of tetrahedral (A) or octahedral (B) sites occupied by Fe^{3+}or M^{2+}, respectively, and is expressed as:

$$\left[M_{1-j}^{2+}Fe_j^{3+}\right]_A\left[M_j^{2+}Fe_{2-j}^{3+}\right]_B, \tag{9}$$

where j ranges between 0 and 1:

- *Normal spinel* (j = 0): All M^{2+} ions occupy A sites, and all Fe^{3+} ions occupy B sites. Examples include $ZnFe_2O_4$ and $CdFe_2O_4$ (**Figure 10b**).
- *Inverse spinel* (j = 1): All M^{2+} ions occupy B sites, and Fe^{3+} ions are evenly distributed between A and B sites. Examples include $NiFe_2O_4$ (**Figure 10c**).

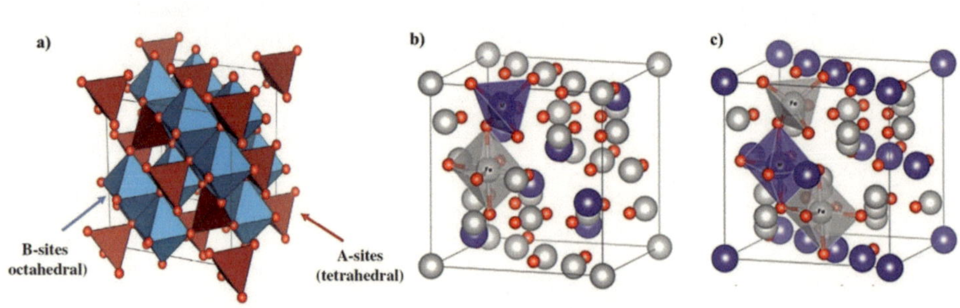

Figure 10.
Schematic representation of spinel structure: Face-centered cubic lattice, (a) octahedral (A sites) and tetrahedral (B sites) [36]; (b) normal spinel ferrite; (c) inverse spinel ferrite [12].

- *Partial inverse spinel* ($0 < j < 1$): Cation distribution is shared between A and B sites, as seen in $MnFe_2O_4$ with $j = 0.2$.

The degree of inversion is influenced by the material's synthesis conditions, including temperature, cooling rate, and cation radii [37–44]. These factors significantly affect the cation arrangement, which in turn impacts the magnetic and electrical properties of the spinel ferrite.

3.2 Magnetization properties of spinel ferrite nanomaterials

The magnetization in spinel ferrites arises from the super-exchange interactions between cations occupying the tetrahedral (A) and octahedral (B) sites in the crystal lattice. There are three types of super-exchange pathways: J_{AA} (A-O-A), J_{BB} (B-O-B), and J_{AB} (A-O-B). Among these, the strength of the J_{AB} (A-O-B) interaction primarily determines the overall magnetization of spinel ferrites [36, 45]. In most *normal spinels*, the net magnetization is zero due to the specific arrangement of magnetic moments, except in rare cases such as $CuCr_2O_4$ and Mn_3O_4. Conversely, *inverse* or *partially inverse spinels* exhibit ferrimagnetism [1]. For example, $ZnFe_2O_4$ is antiferromagnetic below its Néel temperature (10 K) because the magnetic moments originate solely from Fe^{3+} ions, which all occupy the B sites, leading to antiparallel alignment. However, a redistribution of cations (where $j \neq 0$) can alter its magnetic properties [46–49].

The saturation magnetization (M_s) at absolute zero temperature (T = 0) is given by:

$$M_s = \frac{N \cdot d}{M_M}\left[\sum_B n_{B,B} - \sum_A n_{A,B}\right] \cdot \mu_B = \frac{N \cdot d}{M_M}\mu_{eff} \tag{10}$$

Here $n_{B,i}$ represents the number of Bohr magnetons (μ_B) associated with i site in the unit cell, M_M is the molar mass of the ferrite, d is the density, and N is Avogadro's number.

According to Eq. (9), $MnFe_2O_4$ achieves the largest value of saturation magnetization.

As discussed in Section 3.2.2 'Nanoscale phenomena', when size of particle decreases below a critical radius that corresponds to a single domain, it acts as a superspin, by showing superparamagnetic state. Such critical diameter for spinel ferrite nanoparticle is between 15 and 100 nm above the blocking temperature (T_B) [37, 50].

In addition to particle size, the *surface/interface effect* plays a critical role in the magnetization of nanoparticles. The breaking of translational symmetry at the surface results in *magnetic frustration* due to changes in super-exchange angles and bond lengths. This leads to *spin canting* (non-collinear spin alignment), which primarily occurs at the particle surface [37]. Consequently, the surface magnetization differs from the core, reducing the overall magnetization of the nanoparticle. For instance, Kodama et al. observed reduced magnetization in nickel ferrite ($NiFe_2O_4$) nanoparticles due to disordered surface ions and disrupted exchange interactions [22, 51–53].

3.2.1 Magnetocrystalline anisotropy

Magnetocrystalline anisotropy determines the most energetically favorable direction for spontaneous magnetization within a crystal. For spinel ferrites, the magnetization axis typically aligns along the [54] direction of the close-packed structure, except for cobalt ferrites.

In *bulk spinel ferrites*, the anisotropy primarily depends on:

1. The *metal ion anisotropy* (originating from spin-orbit coupling).

2. The *symmetry of the interstitial site*, which depends on the chemical composition and cation distribution [55–57].

At the nanoscale, however, magnetocrystalline anisotropy is significantly influenced by the *particle shape* and *surface effects*. For example:

- *Spherical particles* do not exhibit shape anisotropy because their demagnetizing factors are isotropic.

- *Acicular (elongated) particles*, in contrast, have smaller demagnetizing fields along their long axis, making this axis easier to magnetize compared to the short axis [58, 59].

Additionally, studies have shown that *magnetic anisotropy*—the tendency of a material's magnetization to prefer certain directions—tends to increase as the size of particles decreases. This is because smaller particles have a higher proportion of atoms at the surface, where magnetic properties are influenced differently than in the bulk interior of the material. The effective anisotropy constant (K_{eff}) is a measure of this directional preference and can be calculated using the equation:

$$K_{eff} = K_V + \frac{6}{D} K_S \tag{11}$$

In this formula:

- K_V represents the *volume anisotropy constant*, which accounts for the internal structure of the material and its directional preferences.

- K_s is the *surface anisotropy constant*, quantifying how the material's surface contributes to directional magnetic behavior.

- D is the diameter of the nanoparticle.

This equation shows that as the particle diameter D decreases, the contribution of the surface anisotropy constant (K_s) becomes more significant. Experimental studies on materials such as copper ferrite ($CuFe_2O_4$) and nickel ferrite ($NiFe_2O_4$) nanoparticles confirm this relationship (see **Figure 11**).

3.3 Synthetic methods of nanoparticles

The synthesis of uniform nanoparticles is of paramount importance, as their *magnetic*, *electrical*, and *optical properties* are highly correlated with their size and morphology [60]. For example, Hua Li et al. demonstrated that the coercivity (*Hc*)

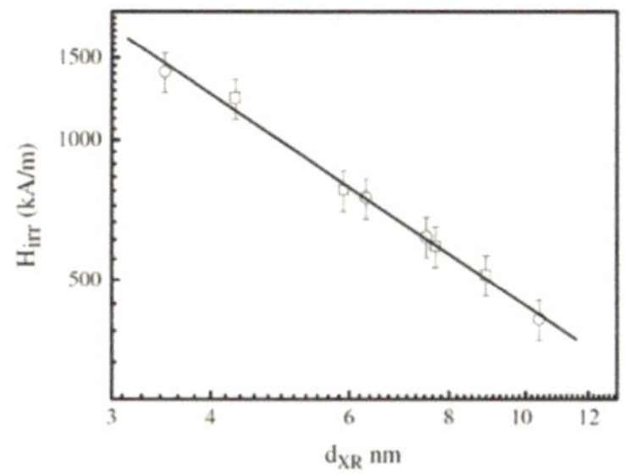

Figure 11.
Irreversible field dependence with the nanoparticle diameter for $CuFe_2O_4$ (□) and (○) $NiFe_2O_4$ [36].

of $NiFe_2O_4$ nanoparticles synthesized through hydrothermal treatment is directly proportional to particle size (**Figure 12**) [61]. Therefore, significant efforts have been dedicated to optimizing techniques for synthesizing ferrite-based nanomaterials.

In general, nanoparticle synthesis can be categorized into two main approaches: *top-down* and *bottom-up* (**Figure 13**).

1. *Top-down approach*: In the top-down method, bulk materials are mechanically or physically broken down into smaller nanoscale particles. This technique primarily relies on *physical processes* and includes:

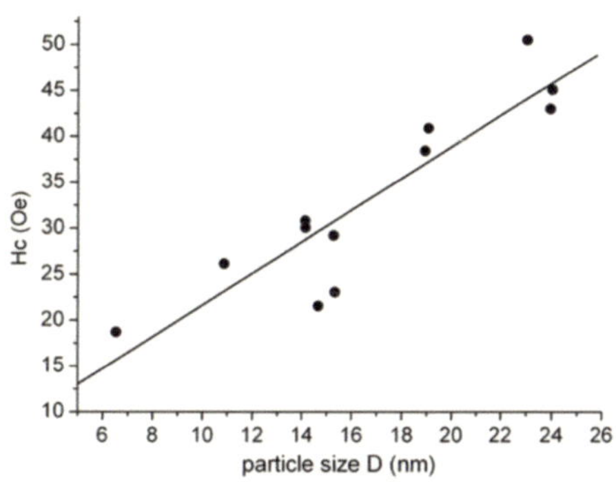

14 / 51

Figure 12.
The relation between the particle size and the coercivity [61].

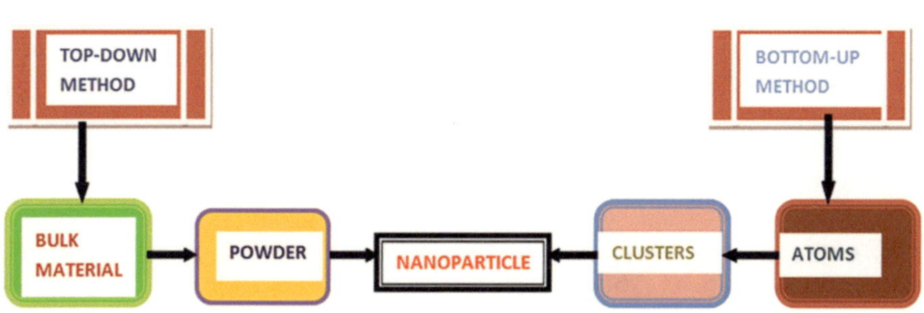

Figure 13.
Schematic flow chart of the synthesis techniques includes bottom-up ones and top-down ones [62].

- Pulsed layer deposition [63],
- Ionic layer adsorption reaction method [64],
- Mechanosynthesis (e.g., ball milling) [65].

2. *Bottom-up approach*: The bottom-up method involves the *assembly of atoms or molecules* into nanoscale structures, typically in solution or gas phases. This category predominantly includes *chemical routes*, such as:

    - *Co-precipitation*: A simple and cost-effective method widely used for ferrite synthesis [66].
    - *Thermal decomposition*: Produces highly crystalline nanoparticles under controlled temperature [67].
    - *Electrochemical methods*: Enables precise size control by tuning reaction parameters [68].
    - *Sol-gel method*: Offers uniform particles and allows for doping with other elements [69].
    - *Micro-emulsion*: Produces monodispersed nanoparticles in a controlled environment [70].
    - *Hydrothermal/solvothermal methods*: Utilizes high temperature and pressure in a liquid medium for the controlled growth of nanoparticles [71].
    - *Solid-state method*: Involves reactions between solid precursors to form nanoparticles [72].

3. Emerging techniques: In addition to conventional methods, several *newly developed techniques* have gained attention in recent years due to their advantages, such as mild reaction conditions, scalability, and improved product selectivity. These include:

    a. *Microwave-assisted synthesis*: Offers rapid heating, uniform energy distribution, and reduced reaction time [73]·

    b. *MOF precursor method*: Utilizes metal-organic frameworks (MOFs) as templates to produce highly structured nanoparticles [74].

    c. *Sonochemical methods*: Employs ultrasonic waves to accelerate chemical reactions and control nanoparticle size [75].

    d. *Steam-assisted methods*: Enhances the crystallinity and uniformity of synthesized nanoparticles [76].

    e. *Biological methods*: Leverages microorganisms, plants, or enzymes for environmentally friendly nanoparticle synthesis [77].

In this section, we will first introduce some conventional synthesis methods, followed by an overview of the emerging techniques mentioned above.

3.3.1 Conventional synthesis

Physical methods were the first approaches explored for synthesizing spinel ferrites, with many relying on *top-down processes* to reduce bulk materials to nanoscale particles. Techniques such as electrical wire explosion [78], mechanical milling [79, 80], and laser target evaporation [81, 82] are widely recognized examples. Ball milling, for instance, involves grinding precursor materials like oxides and carbonates in high-energy mills, followed by calcination at elevated temperatures to achieve the desired spinel ferrite phase. While physical methods can achieve high production yields—sometimes up to 200 grams per hour—they come with significant challenges. Controlling particle shape and size distribution is particularly difficult, and these approaches are often energy- and time-intensive. Furthermore, the efficiency of magnetic nanoparticle production remains relatively low, meeting only a fraction of industrial demands [83–85].

In contrast, chemical synthesis methods offer significant advantages in terms of cost, control, and scalability. The *ceramic method*, a solid-state reaction technique, remains the most industrially mature approach. It involves grinding a stoichiometric mixture of metal oxides or carbonates, followed by prolonged annealing at temperatures often exceeding 1000°C [85, 86]. While straightforward, this method struggles to produce uniform crystallites and requires extensive processing time. High-energy input and prolonged milling can also introduce impurities into the final product, limiting its application [87–92].

The co-precipitation method is among the most popular chemical approaches, known for its simplicity, scalability, and cost-effectiveness. Here, metal salts such as chlorides, sulfates, or nitrates are mixed with a precipitating agent like a base or oxalic acid to form a solid precipitate. The resulting precipitate is then annealed to form the final spinel ferrite phase. Take $ZnFe_2O_4$ as an example; the whole process is schematically illustrated in **Figure 14**. The process is influenced by key factors such as pH, temperature, reaction time, and precursor ratios, all of which determine particle size, shape, and purity [93, 94]. While co-precipitation allows for large-scale synthesis under mild conditions, controlling morphology and particle size distribution remains challenging, particularly due to the need for post-annealing at high temperatures [83, 95, 96].

For applications requiring well-crystallized spinel ferrite nanoparticles with precise morphology and high purity, *hydrothermal and solvothermal syntheses* have

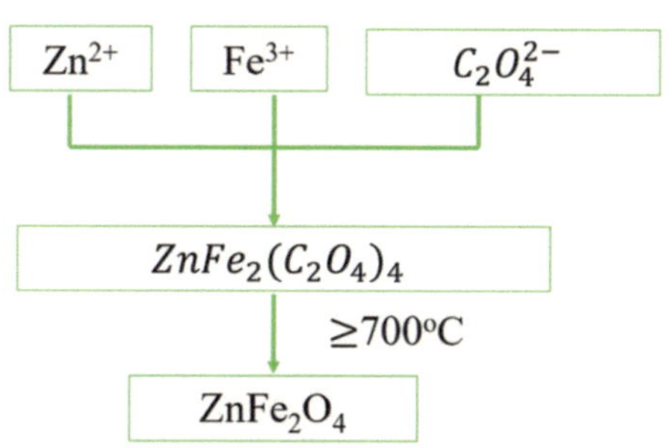

Figure 14.
Schematic illustration of $ZnFe_2O_4$ prepared by co-precipitation route.

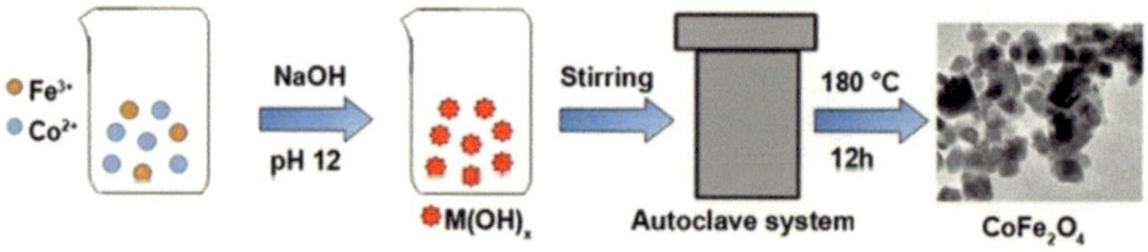

Figure 15.
Schematic representation of the hydrothermal reaction procedure [97]·

proven to be excellent alternatives. **Figure 15** schematically illustrates the hydrothermal synthesis of $CoFe_2O_4$. Both methods rely on operating in sealed systems at elevated temperatures (100–300°C) and pressures (1–30 MPa), allowing ions in solution to gain sufficient energy to form highly crystalline nanoparticles [98–100]. Hydrothermal synthesis employs water as a reaction medium, while solvothermal synthesis uses organic solvents, which offer additional advantages in controlling particle dispersion and morphology. Solvothermal methods are particularly useful when precursors are sensitive to oxidation or hydrolysis in water. Reaction parameters such as temperature, heating rate, stirring speed, and reaction time play a critical role in determining the size and shape of the final nanoparticles [101].

The *sol-gel method* is another widely used chemical approach that offers control over particle size and shape [102]. In this process, precursors undergo hydrolysis or polymerization to form a *sol*, which condenses into a gel upon evaporation of the solvent. Further calcination converts the gel into the final spinel ferrite product. Compared to the ceramic method, sol-gel synthesis operates at significantly lower annealing temperatures, reducing particle aggregation [78]. However, post-annealing remains necessary to achieve the desired purity and crystallinity [103].

An advanced variation of this process is the *sol-gel auto-combustion method*, which combines so-–gel techniques with combustion synthesis (**Figure 16**) [104]. This method involves preparing xerogels from aqueous solutions containing metal salts and organic fuels like urea, glycine, or citric acid [104]. Upon ignition, an exothermic redox reaction generates flames with temperatures ranging from 600 to 1350°C, rapidly producing uniform nanoparticles [105, 106]. The auto-combustion

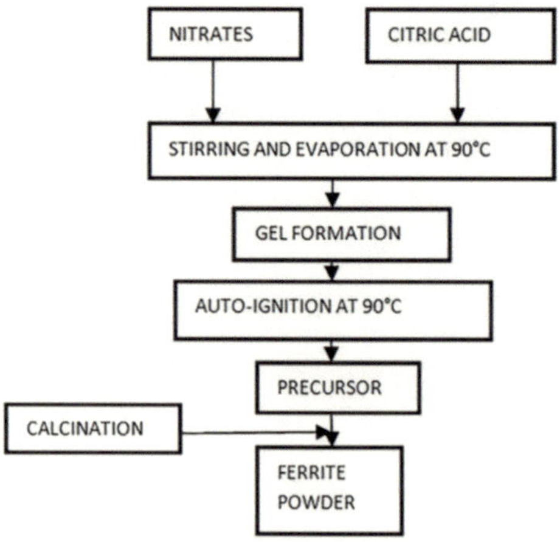

Figure 16.
Flowchart for the preparation of ferrite powder by sol-gel auto-combustion [62].

method enables fine control over particle morphology, producing structures such as nanospheres, hollow nanocages, and nanorods. For example, mixed ferrites such as $Co_xM_yCu_zFe_2O_4$ (M = Ni, Zn, Mg) have been synthesized using this technique, demonstrating excellent crystallinity after sintering at 1000°C [107].

The *micro-emulsion method* offers another route for synthesizing highly uniform and well-dispersed nanoparticles. In this approach, water-in-oil or oil-in-water nanodroplets act as confined reactors, restricting particle growth to the nanoscale range. With droplet sizes typically between 1 and 100 nanometers, the resulting spinel ferrite particles exhibit narrow size distributions and excellent dispersion properties [104].

Additional techniques, including *thermal decomposition* and *electrodeposition*, have also been explored for synthesizing spinel ferrites [108–110]. Thermal decomposition methods produce highly crystalline particles under precisely controlled heating conditions, while electrodeposition provides an efficient route for creating thin films and nanoscale particles. Although conventional methods remain integral to the synthesis of spinel ferrites, recent advancements in synthesis techniques have opened new possibilities for improved control, efficiency, and scalability. The following section will discuss these emerging methods.

3.3.2 Microwave-assisted synthesis

Microwaves, electromagnetic waves with frequencies between 0.3 and 300 GHz and wavelengths ranging from 1 mm to 1 m, have emerged as a powerful tool in nanomaterial synthesis [111]. These waves can effectively heat materials with polar molecules or conducting ions, commonly referred to as *microwave absorbers*. When exposed to an oscillating electric field, these substances undergo rotational motion, friction, and collision, generating internal heating due to dielectric loss. The two primary mechanisms responsible for this loss are ionic conduction and dipole polarization [112]. As a result, the efficiency of microwave absorption by a material depends on its dielectric constant at the relevant microwave frequency. If the reaction mixture contains at least one species with a strong permanent dipole [4], rapid and efficient heating can occur, with localized temperature increases reaching rates as high as 10°C per second [54].

Microwave-assisted synthesis has been recognized as an environmentally friendly and efficient method for producing spinel ferrite nanomaterials. This approach offers several distinct advantages [83, 113]:

1. *Versatility in producing magnetic nanoparticles with different phases*: The method enables the synthesis of ferrite nanoparticles with various phases and surface coatings, which is especially valuable for applications requiring compatibility with non-polar or weakly dielectric phases, such as in biological systems.

2. *Stability of aqueous colloids*: Magnetic nanoparticles often aggregate in aqueous solutions due to their intrinsic magnetization. Microwave-assisted synthesis mitigates this by producing stable aqueous colloids without the need for additional surfactants.

3. *Rapid reaction times*: Unlike conventional heating methods, where thermal gradients are unavoidable, microwaves generate homogeneous internal heating throughout the reaction system. This significantly accelerates chemical processes, with reactions often completing within seconds.

4. *Large-scale production of uniform nanoparticles*: Homogeneous heating ensures simultaneous nucleation throughout the reaction medium, enabling the large-scale synthesis of nanoparticles with uniform size and morphology [114].

Microwave heating is frequently combined with other synthesis techniques to enhance their efficiency. For instance, *microwave-assisted hydrothermal synthesis* merges the advantages of homogeneous microwave heating with the high-pressure environment of hydrothermal methods. This combination has been successfully employed to synthesize Ti-doped ferrites, resulting in nanoparticles with uniform sizes around 30 nm [115]. The experimental setup for this process is shown in **Figure 17a**.

Similarly, *microwave-assisted auto-combustion synthesis* has been used to prepare Co-doped $NiFe_2O_4$ nanoparticles [116]. This method leverages the rapid heating and energy efficiency of microwaves alongside the exothermic reactions of auto-combustion. For example, $Co_{0.9}Ni_{0.1}$ ferrite nanoparticles synthesized using this approach demonstrated a significant improvement in their saturation magnetization (*Ms*), increasing from 41.5 emu/g in pure $NiFe_2O_4$ to 62.8 emu/g.

In summary, microwave-assisted synthesis provides a robust platform for producing high-quality ferrite nanomaterials with diverse applications, combining speed, uniformity, and environmental benefits. As advancements in microwave technology continue, this method is expected to play an increasingly pivotal role in nanomaterial research and development.

3.3.3 Biological methods in magnetic nanoparticle synthesis

Biological synthesis, or biosynthesis, offers an environmentally friendly approach to producing magnetic nanoparticles by leveraging bacteria or plant extracts as stabilizing and reducing agents. This green methodology relies on the reducing properties of biological molecules found in plants, such as alkaloids, flavonoids, phenols, tannins, quinones, and terpenoids. These molecules facilitate the reduction of metal ions into nanoparticles and play an active role in the formation of spinel ferrites [115, 117]. Various plant sources, including seaweed, *Sesbania grandiflora* leaves, *Rubus glaucus*, pineapple, and lemon, have been successfully used in this process.

For instance, *Limonia acidissima* juice has been utilized as a reducing agent to synthesize $ZnFe_2O_4$ nanoparticles [118], while tea extracts were employed to produce $CuFe_2O_4$ magnetic nanoparticles with an average size of 8.78 nm [119]. Similarly, maghemite nanoparticles with superparamagnetic properties have been synthesized using actinobacteria in aerobic environments, highlighting the potential of microbial systems for magnetic nanoparticle production.

The properties of nanoparticles synthesized *via* biological methods, such as structure, morphology, particle size, and magnetic characteristics, are influenced by

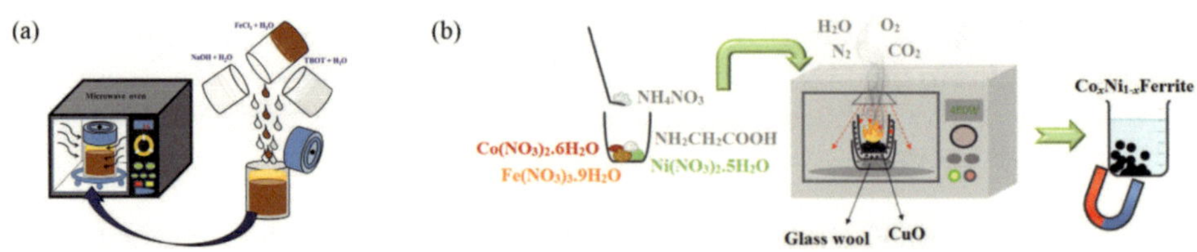

Figure 17.
Schematic experimental setup used for the synthesis of spinel ferrite nanoparticles using microwave-assisted hydrothermal technology (a) [113] and microwave-assisted auto-combustion technology [116].

Microorganisms	Culturing temperature	Size (nm)	Shape	Location	Reference
Shewanella oneidensis	28	40–50	Rectangular, rhombic, hexagonal	Extracellular	[121]
Recombinant AMB-1	28	20	Cubo-octahedral	Intracellular	[122]
Yeast cell	36	n/a	Nanopowders	Extracellular	[123]
Shewanella oneidensis MR-1	25	30–43	Pseudohexagonal/ irregular or rhombohedral	Intracellular	[124]

Table 2.
Metal ferrite nanoparticles synthesized by biosynthesis [84].

several factors. These include the type of bacteria or plant extract used, the reaction conditions (aerobic or anaerobic environments), and parameters like the concentration of metal salts, reducing agent quantity, reaction time, pH, and temperature [84, 120]. **Table 2** provides examples of ferrite nanoparticles produced through biosynthesis, showcasing its versatility [84].

This method offers several advantages, including shorter processing times, energy efficiency, and environmental benefits. However, challenges remain, particularly in controlling particle morphology, achieving consistent saturation magnetization, scaling up production, and addressing the presence of organic contaminants. These difficulties are partly due to the limited understanding of the underlying mechanisms of biosynthesis [118, 120, 125–127].

To enhance the efficiency of biosynthesis, it is often combined with other technologies. For example, Naik et al. developed nickel-doped cobalt ferrite nanoparticles using a microwave-assisted bioengineered method. *Andrographis paniculata* plant extract served as the biological agent, yielding nanoparticles with uniform sizes of 20–30 nm and a 67% chemical yield. These ferrite nanoparticles demonstrated promising photocatalytic activity against various dyes under visible light irradiation, emphasizing their potential for environmental remediation applications [128].

Biological methods, with their inherent sustainability and adaptability, are an exciting avenue in magnetic nanoparticle synthesis. Continued research and technological integration are expected to overcome current limitations and unlock their full potential.

3.3.4 Sonochemical methods for nanoparticle synthesis

Ultrasound radiation. This technique employs sound waves with frequencies exceeding 20 kHz, using vibrational energy to promote chemical reactions [129–133]. The core principle of sonochemistry is acoustic cavitation, where ultrasonic energy accumulates within pulsating bubbles. These bubbles grow progressively until they collapse, releasing the stored energy. The rupture of these bubbles generates localized "hot spots," characterized by extreme conditions—temperatures as high as 5000 K and pressures reaching 1000 bar—within a fraction of a second. These conditions are intense enough to drive various chemical reactions effectively [134, 135].

The efficiency of this process depends significantly on the chemical reagents and solvents used, as they influence the pressure dynamics inside the bubbles [136]. Compared to other synthesis methods, sonochemical synthesis offers advantages such as energy efficiency and reduced reaction times [137].

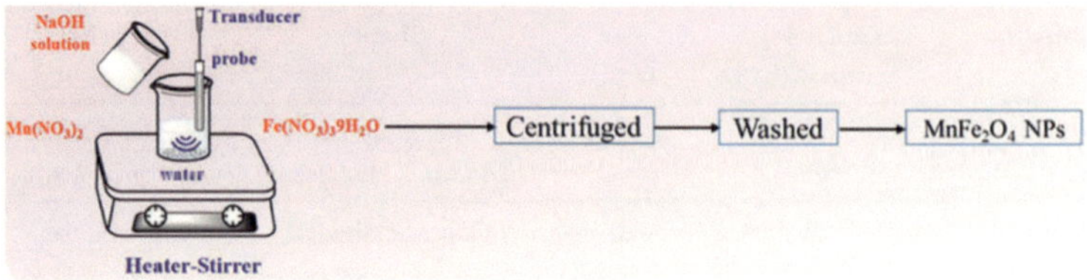

Figure 18.
Schematic diagram of sonochemical process [84].

For example, Amulya et al. synthesized nickel ferrite nanoparticles using a sonochemical approach, reacting iron (III) chloride and nickel (II) chloride with sodium hydroxide in a probe sonicator for 60 minutes [138]. Similarly, Yadav et al. demonstrated the synthesis of $MnFe_2O_4$ nanoparticles with crystallite sizes ranging from 1.8 nm to 22.1 nm by varying the sonication duration from 20 minutes to 80 minutes [139]. The experimental setup for sonochemical synthesis, often employing probe sonicators, is illustrated schematically in **Figure 18**.

Despite its advantages, the sonochemical method faces challenges that limit its scalability beyond laboratory settings. The nanoparticles produced through this technique often exhibit characteristics such as amorphous structures, porous morphologies, aggregation, and a broad size distribution, which can be difficult to control. These limitations underscore the need for further research and optimization to improve particle uniformity and control over the synthesis process, enabling broader industrial applications [83, 140].

Sonochemical synthesis remains a promising area of exploration in nanoparticle production, particularly for its rapid reaction rates and unique reaction conditions. With continued advancements, this method has the potential to overcome current limitations and emerge as a viable option for large-scale production.

3.3.5 Emerging methods for nanoparticle synthesis: MOF precursors and steaming-assisted techniques

3.3.5.1 MOF precursors for spinel ferrite synthesis

Metal-organic frameworks (MOFs) are crystalline materials made up of inorganic metal centers and organic linkers, offering high porosity and surface area. These attributes make MOFs advantageous precursors for fabricating porous spinel ferrites, which are particularly valuable in applications requiring high surface areas, such as electrochemical batteries [74, 141, 142]. The transformation of MOFs into spinel ferrites typically involves chemical etching, cation exchange, or direct thermal oxidation, enabling the creation of porous structures.

For example, nickel ferrites have been synthesized by annealing MOF precursors at 460°C, 500°C, and 550°C. The morphology of the resulting structures evolved with temperature, transitioning from thread-like to mesh-like or agglomerated grains. The mesh-like structure obtained at 500°C exhibited the best electrochemical performance, likely due to its optimized porosity [142, 143]. Despite its potential, this technique often requires post-calcination or pyrolysis at high temperatures, which can compromise the porous structure and result in increased production costs.

3.3.5.2 Steaming-assisted conversion technology

Steaming-assisted conversion is a novel synthesis method that integrates the strengths of hydrothermal and ceramic routes. This hybrid approach combines the low-temperature, high-pressure advantages of hydrothermal processes with the simplicity of solid-state ceramic methods. In this technique, solid reactants are kept separate from gaseous reactants in a pressure vessel (**Figure 19**). The metal ion precursors are isolated from the vapor in a sealed system, which is then heated in an oven. The vaporized reactants interact with the dry powder under controlled temperature and autogenous pressure, leading to the formation of well-crystallized ferrites. This method's liquid-free reaction environment and relatively low operating temperatures make it promising for specialized applications, such as the fabrication of magnetic ferrite films. By minimizing liquid-phase interactions and leveraging the advantages of vapor-based synthesis, steaming-assisted technology offers a unique avenue for developing advanced ferrite materials with precise control over crystallinity and morphology [76].

Both MOF precursor and steaming-assisted methods highlight innovative approaches to synthesizing spinel ferrites. They address challenges like energy efficiency, cost, and structural control while expanding their applicability across emerging fields. Further refinement and scaling of these techniques could unlock even greater potential in industrial and research contexts.

3.4 Tuning magnetic properties of spinel ferrites

The magnetic properties of spinel ferrite nanoparticles can be precisely tuned by altering the composition, structure, or synthesis parameters. These adjustments are crucial for tailoring spinel ferrites (MFe_2O_4) for specific applications.

The mechanism of magnetism in ferrites arises from the distribution of 3d electrons across tetrahedral (A) and octahedral (B) sites [144, 145]. The difference in magnetic moments at these sites determines the total magnetization, which is antiparallel due to the spin structure. The relationship is expressed as [146]:

$$M_{total} = |M_A - M_B| \tag{12}$$

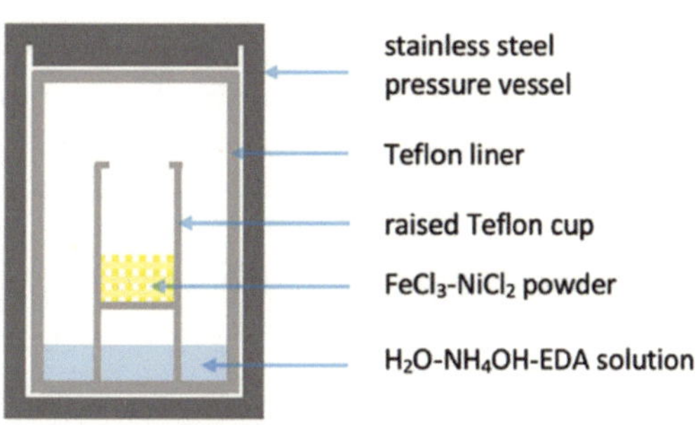

Figure 19.
Illustration of the pressure vessel with the setup for the alkaline steaming process [76].

The theoretical saturation magnetization (*Ms*) of a single-phase ferrite structure can be calculated using the formula [147]:

$$M_s = \frac{8n\mu_B}{10^3 a^3 \rho} \tag{13}$$

where n is the theoretical number of magnetic moments per formula unit, μ_B is the Bohr magneton, a is the lattice parameter, and ρ is the density.

Experimental *Ms* values are often lower than theoretical values due to several factors:

1. Imperfect cation distribution at A and B sites.
2. Deviations from Néel's two-sublattice theory.
3. The presence of spin canting, particularly at B sites, disrupting collinear alignment [12, 148].

The methods to Tune Saturation Magnetization (*Ms*) are:

1. *Substitution or doping*: Replacing ions with lower magnetic moments (e.g., Fe^{3+}:5 μB) by non-magnetic or weakly magnetic ions (e.g., Mg^{2+}: 0 μB) decreases Ms. In some cases, doping can enhance *Ms*, as seen in Ni-ferrites doped with Co^{2+}, which modifies the exchange interactions between A and B sublattices [149]. Larger dopant ionic radii compared to host ions reduce *Ms* [150]. Rare-earth ions occupying octahedral sites often decrease Ms. [150].
2. Variations in crystallite size influence *Ms* due to surface spin canting and incomplete magnetic alignment at smaller sizes [151–153].
3. *Addition of a second phase*: The effect on Ms. depends on the magnetic nature of the secondary phase compared to the primary ferrite phase [154].
4. *Use of reducing agents*: Reducing agents can enhance magnetic properties by modifying cation distributions and ferrite morphology [155].

Coercivity (Hc) and influencing factors. Coercivity depends on magnetocrystalline anisotropy, particle morphology, strain, and grain size. For example, Ni-Co ferrites show high *Hc* due to Co^{2+} ions at the B sites, which possess strong anisotropy. Sintering temperature, which affects grain size and porosity, also plays a significant role [146, 151, 156–159].

Squareness ratio (SQ). The squareness ratio is defined as:

$$SQ = \frac{M_r}{M_s} \tag{14}$$

The squareness ratio provides insights into the domain structure of ferrites, where values close to 0.5 indicate pseudo-single-domain behavior, while lower values suggest multi-domain particles. Magnetic anisotropy, particle size, and synthesis methods are critical in determining this parameter [151, 160]. For example, Co-doped Ni-ferrites with an SQ smaller than 0.1 are found to have multi-domain grains [161].

Curie temperature (T_C), which marks the transition from ferrimagnetic to paramagnetic behavior, decreases with the substitution of ions with lower magnetic moments. The Curie temperature decreases when ions with high magnetic moments are substituted by those with lower moments, weakening the A-B exchange interactions [162, 163]. For example, Curie temperature decreased in Ni-ferrites from 875.7 K of pure Ni-ferrites to 674.38 K for $NiCr_{0.6}Fe_{1.4}O_4$ ferrites when Cr^{3+} ions substitute Fe^{3+} ions [164].

In summary, through substitution, doping, particle size manipulation, and tailored synthesis processes, the magnetic properties of spinel ferrites can be customized for applications ranging from data storage to biomedical uses. Advances in understanding the interplay between these factors will further enhance their functionality.

4. Magnetic-related application of spinel ferrites

The production costs of non-functionalized spinel ferrite nanoparticles vary between $200/kg and $400/kg, depending on particle size, purity, and synthesis methods. While this cost is competitive for laboratory-scale production, it remains a significant challenge for large-scale industrial applications. Despite these concerns, spinel ferrite nanoparticles exhibit several advantages, including low cost, abundant availability, high chemical stability, biocompatibility, and superparamagnetic behavior, making them attractive for practical use in biomedical and environmental sectors.

Magnetic spinel ferrite nanoparticles are versatile materials known for their unique magnetic, electrical, and optical properties. These properties, influenced by factors such as cation type, particle size, morphology, and synthesis methods, make spinel ferrites suitable for applications in magnetic, environmental, biomedical, and energy-related fields (**Figure 20**).

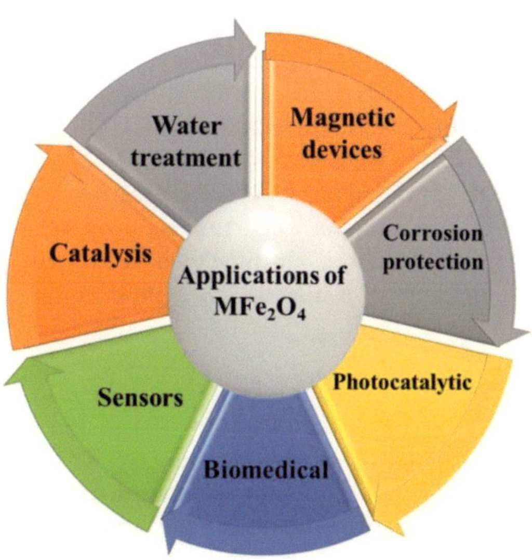

Figure 20.
Applications of MFe_2O_4 [84].

1. *Magnetic applications*: Due to their high saturation magnetization and low eddy current loss, spinel ferrite nanoparticles play a critical role in information storage technologies. They are integral to devices like high-density magnetic storage media and magnetic storage disks, enhancing the capacity and efficiency of data storage solutions such as magnetic tapes and hard drives. Their significant magnetoresistance under various temperatures, magnetic fields, and doping conditions also make them ideal for high-frequency electronic applications, including microwave-absorbing materials and advanced electronic devices.

2. *Environmental applications*: In the environmental sector, spinel ferrite nanoparticles serve as effective adsorbents for removing pollutants like dyes, heavy metals, pesticides, antibiotics, and aromatic compounds from wastewater. Their photocatalytic properties enable the efficient degradation of organic pollutants under light irradiation. Furthermore, their superparamagnetic nature allows for easy separation and recovery using external magnetic fields, making them reusable and sustainable for multiple cycles.

3. *Biomedical applications*: Spinel ferrite nanoparticles have revolutionized biomedicine with their multifunctional roles. As magnetic resonance imaging (MRI) contrast agents, they enhance imaging quality, aiding in the diagnosis and monitoring of diseases, including cancer. In cancer therapy, they facilitate targeted drug delivery by utilizing external magnetic fields to direct therapeutic agents precisely to tumor sites. Additionally, these nanoparticles are employed in magnetic hyperthermia therapy, where an alternating magnetic field induces localized heating to destroy cancer cells. Their surface can be functionalized to improve DNA extraction efficiency and enable cell labeling and tracking. Recent studies also highlight their antibacterial properties, offering promising applications in wound healing and infection prevention.

4. *Energy storage and conversion*: Spinel ferrites, particularly cobalt-iron oxide-based materials, are pivotal in energy storage and conversion technologies. They are widely used in lithium-ion batteries and supercapacitors due to their excellent electrochemical properties, stability, and capacity for efficient energy storage and discharge. These materials contribute to advancements in renewable energy technologies, supporting sustainable energy solutions.

In summary, the multifunctional nature of spinel ferrite nanoparticles makes them indispensable in various domains. Their tunable properties, combined with advancements in synthesis techniques, continue to expand their application potential, fostering innovation in technology, healthcare, and environmental sustainability.

4.1 Magnetic application

Spinel ferrite is a unique material with ferrimagnetism and semiconductor properties, which can be regarded as a magnetic semiconductor. They have excellent magnetic properties. Its magnetic properties depend on the nature of the divalent cations in the spinel structure (except for iron) and their distribution at tetrahedral and octahedral sites, synthesis methods, grain size, and morphology [165, 166].

Spinel ferrite materials are divided into soft magnetic, gyromagnetic, and moment magnetic ferrite according to magnetic properties. Mn-Zn series (Mn-Zn) and Ni-Zn series (Ni-Zn) are used as soft magnetic materials. They are used as magnetic heads and cores for audio, intermediate-frequency and high-frequency recording, video recording, communication, radio, television, and so on. The soft magnetic ferrite has low coercivity (usually less than 100 A/m), and it is easy to change its magnetization state under the action of an external magnetic field. Hard magnets contrast with this. The coercivity will vary according to the particle size. Generally, as the surface area increases and the grain size decreases, the coercivity and saturation magnetization will increase, so the nanoparticles have a larger surface area and smaller grain size and exhibit high magnetic strength [167]. It is widely used in high-frequency magnetic components, inductor cores, microwave devices, etc. [168]. There are room temperature Mg-Mn, Mn-Zn, Cu-Mn, Cd-Mn series, and wide temperature Li-Mn, Li-Ni, Li-Cu, and Li-Zn series composite ferrites. The wide-temperature-moment ferrite has high Curie temperature, good temperature coefficient stability, short switching time ($t \approx 195$ nms), and large switching coefficient ($S = 36.86$Aμs/m), which is used as a fast memory core for large-capacity electronic computers. Spin magnetic (microwave) ferrites include Mg-Mn, Mg-Mn-Al, Ni-Zn, Ni-Al, Li-Al, and Li-Ti composite ferrites. $MgMn_{0.1}Fe_{1.7}Al_{0.3}O_4$ has high saturation magnetization, high resistivity ($\rho = 1010\ \Omega\ cm^{-1}$), and low dielectric loss ($tg\delta \leq 10^{-4}$). It is suitable for the frequency range above the S-band and can be used to make microwave devices, which can isolate, circulate, phase shift, limit frequency, filter, delay, and amplify the signal.

4.1.1 Magnetic storage applications

Among all spinel ferrites, cobalt ferrite is one of the most attractive materials. $CoFe_2O_4$ is a ferrimagnetic spinel oxide structure with high magnetic anisotropy, high saturation magnetization, high wear resistance, and good chemical stability. Because of its unique characteristics, $CoFe_2O_4$ is widely used in magnetic recording equipment. High saturation magnetization means that saturation magnetization can be achieved at a relatively low external magnetic field, thereby increasing the storage density. The high magnetic anisotropy means that the magnetization direction is very sensitive to the change of the external magnetic field, which helps to improve the stability of the storage unit so that the magnetic information can be maintained for a long time, even under a small external magnetic field.

Cobalt ferrite has a cubic spinel structure with Co^{+2} substitution at the octahedral lattice site. At the same time, Fe^{+3} is evenly distributed between the octahedral and tetrahedral lattice sites and is arranged in a spin-polarized state through exchange interaction to form strong ferromagnetism [169]. The distribution of $CoFe_2O_4$ makes the overall magnetism of $CoFe_2O_4$ strongly stable, reducing the risk of information loss. Enhancing the magnetic moment and magnetocrystalline anisotropy constant is necessary for the device application of cobalt ferrite. Therefore, we can enhance the magnetism of nanocrystalline cobalt ferrite by doping different transition metals such as Cd, Cu, Zn, Mg, Ni, and Mn [170]. Maaz et al. [171] studied the synthesis of nickel-substituted cobalt ferrite magnetic nanoparticles by co-precipitation method and analyzed their crystal properties.

The magnetization hysteresis loop results show that the sample has high magnetic properties at room temperature. The coercivity and saturation magnetization of the samples decreases linearly with the increase of nickel concentration in cobalt ferrite, and the superparamagnetic barrier temperature decreases linearly with the increase of nickel concentration [172]. These results show that the synthesis of nickel-substituted cobalt ferrites can adjust their magnetic properties and provide valuable guidance for their applications in magnetic storage, sensors, and other fields.

4.1.2 Sensor application

The sensor is a detection device that can feel the measured information and transform the felt information into electrical signals or other required forms of information output according to specific rules to meet the requirements of information transmission, processing, storage, display, recording, and control. Spinel ferrite is widely used in sensors, which can be applied to electrochemical, gas, humidity, and other sensors. One of the most common uses of sensors is to detect changes in humidity. Humidity is the concentration of water vapor in the air. It is an important environmental parameter in preservation, storage, and transportation operations. It plays an important role in climatology, the medical industry, industrial processes, environmental monitoring, agriculture, and other fields. For meteorological applications, real-time collection and processing of data on ambient humidity is critical, which may help predict precipitation, dew, and fog [173].

Nano-sized polycrystalline spinel ferrite exhibits humidity characteristics. Rezlescu et al. studied the effect of multivalent cation (Sn and Mo) substitution in Mg ferrites. These materials were synthesized by the sol-gel auto-ignition method and reported that Sn ions enhanced the humidity sensitivity of magnesium ferrite [174]. Spinel ferrites achieve high sensitivity and selectivity to specific gases in gas sensors. Spinel ferrites have an open pore structure, allowing them to adsorb test gas analytes more quickly, accelerating the reaction. The larger pores provide more surface area, making it easier for gas molecules to contact and react with the sensor surface. Since Seiyama et al. first proposed the gas sensing properties of semiconductor metal oxide thin films in 1962, metal oxide semiconductor gas sensor materials have been deeply studied [175]. $ZnFe_2O_4$ nanofibers exhibit high sensitivity and selectivity to H_2S gas and can detect H_2S at sub-ppm level [176]. The grain size and particle size significantly affect the performance of channeling and drying. Generally, smaller grain sizes will have higher gas sensitivity. The spinel ferrite gas sensor principle is based on the semiconductor gas sensor mechanism. Oxygen is chemically adsorbed onto the metal oxide particles, trapping electrons to form an electron-depleted space charge layer. When gas molecules react with chemically adsorbed oxygen, electrons are released and oxidized five, thereby changing the conductivity of the space charge layer [177]. Spinel ferrite can also be used in temperature sensors due to its magnetic properties and dynamic response characteristics of heat input. Ferrite materials can directly detect temperature changes by measuring changes in their magnetic characteristics (such as saturation magnetic beam density, retention force, and magnetic permeability). Ni-Co magnetic ferrite materials have been studied for magnetostrictive ceramics. This material has potential magnetoelastic properties and is suitable for the application of magnetomechanical sensors. The Ni-Co spinel ferrite sample treated by ceramic method shows a magnetoelastic sensitivity close to that of $CoFe_2O_4$, and the hysteresis phenomenon is significantly reduced, indicating its application potential in mechanical stress sensing [178].

4.1.3 Microwave absorption and high-frequency applications

Ferrites are well-known materials for microwave absorption and high-frequency applications. The spinel structure is composed of a face-centered cubic lattice in which the lattice sublayers between the tetrahedral and octahedral sites are merged, resulting in antiparallel spins between the sites [179]. This structure leads to spinel ferrite's low coercivity and high magnetization properties.

Using nickel spinel ferrite as a nano-filler, combined with conductive polyaniline, a composite of magnetic nanoparticles and conductive polymers was successfully prepared. These materials exhibit excellent electromagnetic interference (EMI) shielding performance in the microwave region. Spinel ferrite has high resistivity, high dielectric constant, and medium magnetic properties, making it a leader in microwave materials. These characteristics make spinel ferrites exhibit excellent microwave absorption properties in a wide frequency range. Ferrite-based microwave-absorbing materials (MAM) absorb electromagnetic waves through specific mechanisms, effectively solving the problem of electromagnetic pollution. Its performance is affected by two factors: impedance matching and attenuation characteristics. The impedance matching determines the degree of electromagnetic wave entering the material, and the attenuation characteristics determine the loss capacity of the material. The moderate impedance matching and dielectric loss ability of spinel ferrite make it have application value in electromagnetic pollution prevention and radar stealth of military targets. It can be used to manufacture radar stealth coatings to effectively absorb or scatter incident radar waves, thereby reducing target radar reflection signals and achieving stealth effects.

The magnetic and electrical properties of nano-ferrites are enhanced by doping and optimizing the chemical composition of spinel ferrites. High crystalline Mn-Zn ferrite thin films suitable for high-frequency devices were prepared by a wet chemical process. The films prepared by dextran as a complexing agent are highly crystalline and have a large magnetic loss. It can be used as an electromagnetic compatibility material to effectively suppress the conducted noise in mobile computers and mobile phones [180]. Similarly, Ni-Zn ferrites have relatively low permeability and high resistivity in the high-frequency region. It can better replace the hexaferrite widely used in microwave device applications (such as circulators) because high saturation magnetization, medium permeability, dielectric constant, and low loss are the basic necessary parameters of materials used in circulators [181].

Spinel ferrites have strong magnetic properties, especially at the nanoscale. Their superparamagnetism and high permeability make them important magnetic materials. They can be applied in the medical field. Due to their superparamagnetism, nanoparticles can be controlled by an external magnetic field to achieve advanced drug delivery. Spinel ferrites also have good biocompatibility and mild toxicity. Their magnetic response can be used in medical, industrial detection, environment, and other fields to design highly sensitive magnetic sensors [182, 183].

4.2 Environmental application

4.2.1 Photocatalytic applications

In recent years, the study of ferrite as a visible light photocatalyst to degrade pollutants in water has attracted great interest. Photocatalysts are important materials to promote the use of solar energy in oxidation and reduction reactions. They have a variety of applications, including water decomposition to produce hydrogen, decomposition of harmful substances in the air, sewage treatment, carbon dioxide reduction, degradation of organic pollutants, and so on [184–187].

When selecting a photocatalyst, the material's band gap determines the wavelength at which light can be absorbed. At present, TiO_2, as one of the most commonly used photocatalysts, has strong photocatalytic activity, but it is only effective in the ultraviolet range, which limits its application efficiency in the solar spectrum [188]. Compared with TiO_2, some spinel ferrites ($ZnFe_2O_4$ and $NiFe_2O_4$) can also exhibit good catalytic performance under visible light [189]. The band gap of ferrite is smaller than that of other commonly used visible light catalysts, and a smaller band gap enhances their ability to absorb visible light. Due to the spinel crystal structure of the ferrite, its lattice has additional catalytic sites and a large surface area, thereby improving the catalytic efficiency. The effectiveness of the photocatalyst depends on various factors, such as the choice of photocatalyst, the size of the nanoparticles, the crystallinity, the accessibility of the active surface to the pollutants, and the resistance to the diffusion of organic pollutants. These parameters affect the photocatalytic activity. Small particle size and high crystallinity are important factors to improve the active sites. Ferrite is an excellent candidate material for photocatalysts due to its spinel structure and band gap that can absorb visible light. Spinel ferrite can be coupled with common photocatalysts, and various composite photocatalysts have improved photocatalytic performance [190]. Due to their good magnetic properties, they can be easily separated from the reaction mixture through an external magnetic field [191].

4.2.2 Wastewater treatment applications

Magnetic ferrites' excellent magnetic properties and high adsorption capacity have also attracted special attention and are widely used in water purification and wastewater treatment [186]. With the world's population growth, people's demand for clean water is increasing. It is inevitable that new and cost-effective water purification technologies will be found to provide sufficient and high-quality purified water. In particular, the application of nanomaterials has great potential in enhancing water purification [192]. Traditional wastewater treatment technologies such as precipitation, electrocoagulation, flotation, and ion exchange often face challenges, including sludge production, high equipment and disposal costs, low flow rates, and significant energy or pressure requirements [193]. Among these methods, adsorption is particularly promising due to its simplicity and cost-effectiveness. Unlike commonly used adsorbents like activated carbon [194], spinel ferrites provide a critical advantage: magnetic nanomaterials can be separated and desorbed efficiently using an external magnetic field, facilitating higher reusability [195]. In contrast, activated carbon often requires more complex and expensive recovery processes.

The application of magnetic spinel ferrite nanoparticles in water purification and wastewater treatment is mainly reflected in its efficient adsorption capacity and easy recovery. These nanoparticles are usually prepared by chemical synthesis and can be surface modified to enhance their adsorption capacity for specific contaminants. These nanoparticles usually have a core-shell structure, which can effectively remove pollutants in water, including heavy metal ions, organic matter and inorganic matter. Fe_3O_4 @ SiO_2 core-shell nanoparticles not only exhibit efficient removal of metal particles (such as Zn, Co, Cu, Ni), but also are easy to recycle due to their high magnetic saturation value [196]. Magnetic nanoparticles can be separated and recycled by external magnetic field. The gas-assisted magnetic separation combined with low magnetic field separation and flotation was used to treat Gd in wastewater, showing good adsorption capacity and reusability [197].

The specific mechanism of spinel ferrite nanoparticles (such as Fe_3O_4) in degrading organic pollutants mainly involves the redox reaction on their surface. These nanoparticles can effectively degrade organic pollutants through a variety of mechanisms. Magnetic nanoparticles can adsorb organic pollutants to their surface and promote the oxidative degradation of organic pollutants through a Fenton-like reaction in the presence of oxygen. The Fenton reaction is an advanced oxidation process (AOP) that emphasizes a mixture of ferrous ions (Fe^{2+}) and hydrogen peroxide (H_2O_2) to produce highly reactive hydroxyl radicals ($^{\cdot}OH$) [198]. The Fe^{+2} ions and H_2O_2 added in this reaction are easy to handle and harmless to the environment. Therefore, great efforts have been made to develop the classical Fenton reaction in recent years [199].

In practical applications, magnetic spinel ferrite nanoparticles have been used in various wastewater treatment scenarios. For example, they are used to remove heavy metal ions in industrial wastewater, and the wastewater produced by industrial dyes contains many non-biodegradable organic dyes. Untreated sewage negatively impacts not only humans and animals but also flora and fauna [200]. Heavy metals have a relatively high density of metal elements. Toxicity can be induced even at low levels of heavy metal exposure. Cd, Cr, Cu, Pb, Ni, and Zn are the most common heavy metal pollutants in industrial pollution. Heavy metals are a class of persistent toxic substances (PTS), which have acute toxicity and accumulation effects on organisms. Nanoparticles have great potential in removing heavy metals due to their large surface area and strong magnetism. Nanomaterials can not only adsorb heavy metal ions but also convert Cr(VI) into harmless Cr(III) by catalytic reduction to achieve harmless treatment of Cr(VI)-containing wastewater [201]. Rhodamine B (RhB) is a synthetic, highly toxic, water-soluble organic dye often occurring in wastewater. The use of $CoFe_2O_4$ for wastewater treatment is based on its high adsorption capacity, allowing the use of external magnetic fields to separate the magnetism of magnetic nanoparticles, and its low energy band gap, which can photocatalytically degrade RhB under visible light irradiation [202].

4.3 Biomedical application

For biomedical applications, nanoparticles must have these characteristics: (1) Magnetic nanoparticles are biocompatible and non-toxic. (2) Magnetic nanoparticles should be small enough (10–50 nm). (3) Magnetic particles must have high saturation magnetization. When the nanoparticles are small enough, if the magnetic interaction of the nanoparticles is reduced, the stability of the colloid will be preserved, and the aggregation will be resisted. The ratio of dipole-dipole interaction is r^6 (r is the particle's radius). Therefore, when the particle size becomes very small, the dipole interaction becomes very small. This will help minimize particle aggregation when the field is applied. Ferrite materials usually have good biocompatibility and can be degraded in vivo without significant toxic reactions, giving them broad application potential in the biomedical field. It is widely used in magnetic resonance imaging, targeted drug delivery, biosensors, magnetic separation, antibacterial and other fields [203–206].

4.3.1 MRI application

Imaging is one of the most important tools in cancer diagnosis. Magnetic nanoparticles have been used as magnetic resonance imaging (MRI) contrast agents in clinical practice, which is helpful for the identification and diagnosis of lesion areas. The advantage is that they can perform various functions triggered by invisible, tissue-penetrating magnetic fields. Two independent relaxation times, longitudinal and transverse relaxation (T1 and T2), are used to generate MR images [207]. Cobalt ferrite magnetic nanoparticles have excellent contrast. Among the three components studied, cobalt ferrite nanoparticles have the highest magnetization, initial susceptibility, and anisotropic energy, showing an r_2 proton relaxation rate, which is higher than that achieved by using commercial phase contrast agents [207]. The tissue contrast can be further enhanced by applying external CA, thereby shortening the longitudinal (T_1) and transverse (T_2) relaxation time [208]. The synthesis method of spinel ferrite and the size and shape of nanoparticles have an effect on the application of MRI. In clinical applications, spinel ferrites have been used in various MRI applications, including cell labeling, molecular imaging, and imaging of specific tissues such as the liver, spleen, and bone marrow. Spinel ferrite nanoparticles are used as contrast agents in MRI due to their excellent magnetic properties and biocompatibility. Spinel ferrite nanoparticles provide significant imaging advantages as contrast agents in MRI due to their excellent magnetic properties and biocompatibility. Through further research and development, these nanoparticles are expected to play a greater role in future medical diagnosis.

4.3.2 Drug delivery application

Due to its economy and simple operation, magnetic spinel ferrite nanoparticles have been widely studied in drug delivery applications. Unlike traditional drug delivery, magnetic nanoparticles are used as drug delivery agents, attached to the outer surface of the drug or dissolved in the coating. Using the superparamagnetism of ferrite for targeted delivery, all domains in the ferrite nanostructure are aligned along the direction of the applied external magnetic field, resulting in increased magnetic momentum so that the direction and position of the nanocarrier can be controlled without aggregation when the external magnetic field is removed. Nano-spinel ferrite has a specific load capacity, which can load drugs, genes, or other therapeutic molecules onto its surface or inside through surface modification. The precise regulation of drug release can be achieved by controlling the surface properties. For example, the anticancer drug doxorubicin (DOX) can be loaded onto magnetic spinel ferrite nanoparticles and transported to the target through a magnetic conductor [209]. It can carry drugs and circulate without leakage, easily move to the target tumor site with the help of an external magnetic field, and help provide effective treatment for cancer cells by bypassing normal cells [210]. Once delivered to the site of action, the drug will be released and provide its therapeutic effect [211]. The surface of these particles is usually modified by organic polymers and inorganic metals or oxides to make them biocompatible and suitable for further functionalization by connecting various bioactive molecules [212].

4.3.3 Biosensors application

Recently, biosensors have been widely used in the healthcare industry. The sensor device consists of a sensor (semiconductor material/nanomaterial), a biological device that can be a receptor (enzyme, antibody, cell, aptamer, or nucleic acid), and an electronic device consisting of a signal amplifier, a processor, and a display. It is a device that produces a detectable signal after the interaction of biomolecules with the target analyte. It plays an important role in detecting pathogens, metabolite monitoring, and disease diagnosis. Glucose biosensors are commonly used for blood glucose detection in diabetic patients. It can also be used to detect the presence of viruses, bacteria, or other pathogens, such as the detection of novel coronaviruses. Biosensors can also assist in cancer diagnosis and detection of tumor markers in the blood. In clinical studies, it is often used to detect the physiological state of patients, evaluate drug efficacy, or perform drug screening. In personal daily life, it is also widely used in portable health monitoring equipment, such as smart bracelets and smart watches. They can monitor physiological data such as body temperature, heart rate, exercise, and blood oxygen saturation to help users understand their health status in real time.

4.4 Hyperthermia application

Hyperthermia is a treatment method that heats human tissue to achieve therapeutic effects. In this treatment, the body tissue is exposed to slightly higher temperatures, damaging or killing cancer cells by causing cell death. The application of nano-spinel ferrite in hyperthermia is mainly reflected in its use as magnetic nanoparticles (MNPs) in magnetic hyperthermia (MH) [213]. Magnetothermal therapy is a treatment that uses the rotation of nanoparticles in an alternating magnetic field and the reversal of magnetic moments to generate heat to heat tumor tissue, thereby killing or inhibiting cancer cells. Nano-spinel ferrites have shown great potential in the biomedical field, especially in cancer treatment, due to their excellent magnetic properties, high stability, and good biocompatibility.

Nano-spinel ferrites (such as $CoFe_2O_4$ and $NiFe_2O_4$) have cubic spinel structures, showing superparamagnetism or ferromagnetism. This structure endows them with high resistance and permeability, which is very suitable for electromagnetic wave propagation because of its negligible eddy current loss [214].

The biocompatibility and low toxicity of nano-spinel ferrite are very important considerations in biomedical applications. Compared with other nanomaterials, nano-spinel ferrites show better biocompatibility under specific conditions (such as coating and focusing state). For example, citric acid-coated nano-spinel ferrites have less effect on cell mitochondrial activity at high concentrations [215].

The biocompatibility and low toxicity of nano-spinel ferrites mainly depend on their surface modification and treatment methods, as well as their morphology and size. Nano-spinel ferrites exhibit certain toxicity without heat treatment. Heat treatment can remove some precursors (solutions or surfactants) used in the synthesis process, thereby reducing the toxicity of nanomaterials [216]. The surface properties of nano-spinel ferrite can be significantly changed by functionalizing biological surface activators. For example, chitosan and polyethylene glycol (PEG) encapsulation prolong the circulation time of nanoparticles *in vivo* [217]. The size and morphology of nanoparticles also have an important influence on their properties. Spherical nanoparticles have high aggregation and dispersion, while cubic nanoparticles show different magnetic and hyperthermia properties. By adjusting the synthesis conditions, such as temperature, reaction time, and solvent type, the size distribution of nanoparticles can be precisely controlled, thereby improving their magnetothermal induction ability and biocompatibility [218].

In thermomagnetic therapy, spinel ferrite nanoparticles generate heat through the magnetic loss mechanism (such as Néel relaxation, Brownian motion relaxation, and hysteresis loss) under an external oscillating magnetic field, thereby achieving local heating of the tumor [219]. Spinel ferrite nanoparticles such as cobalt ferrite ($CoFe_2O_4$) and zinc ferrite ($ZnFe_2O_4$) exhibit a high specific absorption rate (SAR) in aqueous solution and can be heated to the optimal temperature of 41–46°C in a short time to remove cancer tissues [220]. Magnetic hyperthermia can be combined with other treatments to achieve a synergistic therapeutic effect [221]. $CoFe_2O_3$@DMSA/DOX NPs improve the efficiency of their therapeutic applications by surface functionalization and coupling with the anticancer agent doxorubicin [222].

However, one main challenge of hyperthermia is reducing the loss of normal tissues nearby. Tumor cells are more sensitive to heat under certain conditions (such as low pH or metabolically restricted environment) [223]. Tumors have tortuous blood vessels that feed them and cannot dissipate heat like other blood vessels that feed normal cells. High temperatures above 41°C can push cancer cells toward acidosis and reduce cell pH), thereby decreasing cell viability and transplantation ability and changing the function of many enzymes and structural proteins that affect cell growth and differentiation, which may lead to apoptosis [223].

Significant differences exist in the application effect of nano-spinel ferrite *in vitro* and *in vivo* magnetothermal therapy [224]. *In vitro* experiments, magnetic ferrite nanoparticles usually show high heat generation efficiency. Studies have shown that using highly charged dextran-coated magnetic nanofluids can quickly increase the temperature *in vitro* experiments. *In vivo* applications, although nano-spinel ferrite can also effectively produce thermal effects, its efficiency may be affected by the complex environment in the organism, and its distribution, stability, and potential impact on surrounding healthy tissues need to be considered.

Nano-spinel ferrites show good biocompatibility and versatility in hyperthermia, but there are still some challenges and limitations, for instance, the long-term fate and toxicity of nanoparticles in the human body, especially those non-biodegradable nanomaterials containing toxic elements [63]. Another major challenge is the targeting and penetration of nanoparticles in solid tumors. The complexity of the tumor microenvironment, such as uneven vascular leakage and dense interstitial structure, makes it difficult for nanoparticles to be effectively delivered to tumor cells [225]. In addition, non-specific protein adsorption on the surface of nanoparticles also affects their targeting ability [226]. Future research will focus on developing biodegradable materials to reduce the risk of toxicity in long-term applications. And optimize the physical and chemical properties of nanoparticles.

However, the scalability of nanoparticle production remains a critical issue. Cost-effective and scalable methods, such as sol-gel techniques, are being explored to meet industrial requirements [227, 228]. Additionally, long-term stability and reusability in real-world applications demand further investigation to ensure consistent performance over extended periods. Addressing these aspects is essential to bridge the gap between laboratory research and practical deployment.

5. Scientific challenges and future perspectives

The field of nanomagnetism is bursting with day-by-day enhancement and cross-disciplinary innovation. In particular, nano-spinel ferrites, featuring flexibility in tuning their magnetic properties, have been growing interest in recent years. Therefore, this chapter has covered some important concepts, such as magnetic order, superparamagnetism, and domain, which help understand the concept of nanomagnetic-related applications. The synthesis methods of nano-spinel ferrites are introduced, and some newly developed methods are discussed in detail. Owing to their distinctive characteristics, including stability in chemical and thermal environments, high electrical resistance, and most importantly, tunable magnetic properties and corresponding minimal loss of eddy current, nano-spinel ferrites are showing their versatile potential applications from microwave devices and sensors to MRI applications.

Although great progress has been achieved, numerous efforts are still needed. We propose the subsequent scope studies as follows:

1. Advanced new chemical synthesis or techniques are needed for the preparation of nano-sized spinel-type ferrite.

   First, the particle size and distribution, shape, crystallinity, and purity depend on the synthesis method, which plays key roles in the application-related properties.

2. The mechanism of how dopant affects their magnetic behavior needs more investigation.

   The uncertainty in the mechanism and corresponding uncontrollable tuning in magnetism results in the difficulty of a large-scale synthesis, which is required for practical application.

3. In some special applications, such as biomedical application, the efficient methods for structural and surface modification need more exploration.

Author details

Hua Li[1,2], Wenqin Li[1], Jacques Robichaud[2] and Yahia Djaoued[2]*

1 Department of Materials Chemistry, School of Materials Science and Engineering, Jingdezhen Ceramic University, Jingdezhen, Jiangxi, P.R. China

2 Laboratoire de Recherche en Matériaux et Micro-spectroscopies Raman et FTIR, Université de Moncton-Campus de Shippagan, Shippagan, NB, Canada

*Address all correspondence to: yahia.djaoued@umoncton.ca

IntechOpen

© 2025 The Author(s). Licensee IntechOpen. This chapter is distributed under the terms of the Creative Commons Attribution License (http://creativecommons.org/licenses/by/4.0), which permits unrestricted use, distribution, and reproduction in any medium, provided the original work is properly cited.

References

[1] Wolf WP. Ferrimagnetism. Reports on Progress in Physics. 1961;**24**(1):212. DOI: 10.1088/0034-4885/24/1/306

[2] Kim SK, Beach GSD, Lee KJ, et al. Ferrimagnetic spintronics. Nature Materials. 2022;**21**(1):24-34. DOI: 10.1038/s41563-021-01139-4

[3] Shyam K. Gore, Santosh S. Jadhav. Basics of ferrites. In: Spinel Ferrite Nanostructures for Energy Storage Devices. Amsterdam, Netherlands: Elsevier; 2020. pp. 1-11. DOI: 10.1016/B978-0-12-819237-5.00001-8

[4] Jadhav VV, Shirsat SD, Tumberphale UB, Mane RS. Properties of ferrites. In: Spinel Ferrite Nanostructures for Energy Storage Devices. Amsterdam, Netherlands: Elsevier; 2020. pp. 35-50. DOI: 10.1016/B978-0-12-819237-5. 00003-1

[5] Jadhav VV, Shirsat SD, Tumberphale UB, et al. Properties of ferrites/Spinel Ferrite Nanostructures for Energy Storage Devices. Amsterdam, Netherlands: Elsevier; 2020. pp. 35-50. DOI: 10.1016/B978-0-12-819237-5. 00003-1

[6] Dionne GF. A review of ferrites for microwave applications. Proceedings of the IEEE. 1975;**63**(5):777-789. DOI: 10.1109/PROC.1975.9827

[7] Serga AA, Chumak AV, Hillebrands B. YIG magnonics. Journal of Physics D: Applied Physics. 2010;**43**(26):264002. DOI: 10.1088/0022-3727/43/26/264002

[8] Sharma R, Thakur P, Sharma P, et al. Mn 2+ doped Mg–Zn ferrite nanoparticles for microwave device applications. IEEE Electron Device Letters. 2018;**39**(6):901-904. DOI: 10.1109/LED.2018.2829926

[9] Andrade RGD, Veloso SRS, Castanheira EMS. Shape anisotropic iron oxide-based magnetic nanoparticles: Synthesis and biomedical applications. International Journal of Molecular Sciences. 2020;**21**(7):2455. DOI: 10.3390/ijms21072455

[10] Shrivastava N, Sarveena MS, Sharma SK. The basis of Nanomagnetism: An overview of exchange bias and spring magnets. Exchange Bias. Standalone publication; 2017:1-45. DOI: 10.1201/9781351228459

[11] Martín-Hernández F, Ferré EC. Separation of paramagnetic and ferrimagnetic anisotropies: A review. Journal of Geophysical Research: Solid Earth. 2007;**112**(B3):B03105. DOI: 10.1029/2006JB004340

[12] Narang SB, Pubby K. Nickel spinel ferrites: A review. Journal of Magnetism and Magnetic Materials. 2021;**519**:167163. DOI: 10.1016/j.jmmm.2020.167163

[13] Dunlop DJ, Özdemir Ö. 1997. Rock Magnetism. Fundamentals and Frontiers. In: Magnetism Series. Cambridge, New York, Port Chester, Melbourne, Sydney: Cambridge University Press; Geological Magazine. 1998;**135**(2):287-300. DOI:10.1017/S0016756898218437

[14] Givord D. Magnetism in nanomaterials. In: Nanomaterials and Nanochemistry. Berlin, Heidelberg: Springer Berlin Heidelberg; 2007. pp. 101-134. DOI: 10.1007/978-3-540-72993-8

[15] Guimaraes AP. Principles of Nanomagnetism. In: NanoScience and Technology Series. 2nd ed. Berlin, Germany: Springer; 2009. DOI: 10.1007/978-3-319-59409-5

[16] Getzlaff M. Fundamentals of Magnetism. Berlin, Germany: Springer

Science and Business Media; 2008. DOI: 10.1007/978-3-540-31152-2

[17] Rao C, Thomas P, Kulkarni G. Nanocrystals: Synthesis, Properties and Applications. In: Springer Series in Materials Science. Berlin, Germany: Springer; 2007. DOI: 10.1007/978- 3-540-68752-8

[18] Ortega D. Structure and magnetism in magnetic nanoparticles. In: Magnetic Nanoparticles. Boca Raton, US: CRC Press; 2012. DOI:10.1201/B11760-3

[19] Muraca D, Zélis PM, Javed Y, et al. Surface and interface interplay on the oxidizing temperature of iron oxide and Au–iron oxide core–shell nanoparticles. RSC Advances. 2016;**6**(74):70394-70404. DOI: 10.1039/C6RA15610J

[20] Coey JMD. Magnetism and Magnetic Materials. Cambridge, England: Cambridge University Press; 2010

[21] Cahen D, Kahn A. Electron energetics at surfaces and interfaces: Concepts and experiments. Advanced Materials. 2003;**15**(4):271-277. DOI: 10.1002/adma.200390065

[22] Issa B, Obaidat IM, Albiss BA, et al. Magnetic nanoparticles: Surface effects and properties related to biomedicine applications. International Journal of Molecular Sciences. 2013;**14**(11):21266-21305. DOI: 10.3390/ ijms141121266

[23] Arruebo M, Pacheco R, Ibarra MR, Santamaría J. Magnetic nanoparticles for drug delivery. Nano Today. 2007;**2**(3):22-32. DOI: 10.1016/ S1748-0132(07)70084-1

[24] Weiss P. The magnetic properties of the alloys of the ferromagnetic metals: Iron-nickel, nickel-cobalt, cobalt-iron. Transactions of the Faraday Society. 1912;**8**(October):149-156. DOI: 10.1039/ tf9120800149

[25] Kozlenko DP, Dubrovinsky LS, Kichanov SE, et al. Magnetic and electronic properties of magnetite across the high pressure anomaly. Scientific Reports. 2019;**9**(1):4464. DOI: 10.1038/ s41598-019-41184-3

[26] Lu AH, Salabas EL, Schüth F. Magnetic nanoparticles: Synthesis, protection, functionalization, and application. Angewandte Chemie International Edition. 2007;**46**(8):1222-1244. DOI: 10.1002/anie.200602866

[27] Rashad MM, Rayan DA, Turky AO, et al. Effect of Co^{2+} and Y^{3+} ions insertion on the microstructure development and magnetic properties of $Ni_{0.5}Zn_{0.5}Fe_2O_4$ powders synthesized using co-precipitation method. Journal of Magnetism and Magnetic Materials. 2015;**374**:359-366. DOI: 10.1016/j. jmmm.2014.08.031

[28] Mund HS, Ahuja BL. Structural and magnetic properties of Mg doped cobalt ferrite nano particles prepared by sol-gel method. Materials Research Bulletin. 2017;**85**:228-233. DOI: 10.1016/j. materresbull.2016.09.027

[29] Moradmard H, Shayesteh SF, Tohidi P, et al. Structural, magnetic and dielectric properties of magnesium doped nickel ferrite nanoparticles. Journal of Alloys and Compounds. 2015;**650**:116-122. DOI: 10.1016/j.jallcom.2015.07.269

[30] Hilpert S. Correspondence as to structure and origin in magnetic properties of ferrite and iron oxide. Berichte der Deutschen Chemischen Gesellschaft. 1909;**42**:2248-2261. DOI: 10.1002/cber.190904202121

[31] Forestier H. Magnetic transformation of the sesquioxid of iron, of its solid solution and of its ferromagnetic combinations (in Fr.). Annales de Chimie et de Physique. 1928;**9**:316

[32] Sugimoto M. The past, present, and future of ferrites. Journal of the American Ceramic Society. 1999;**82**(2):269-280. DOI: 10.1111/j.1551-2916.1999.tb20058.x

[33] Snelling EC. Soft Ferrites - Properties and Applications. London: Iliffe Books Ltd; 1969. p. 411

[34] Jain A, Ong SP, Hautier G, et al. Commentary: The materials project: A materials genome approach to accelerating materials innovation. APL Materials. 2013;**1**(1):011002. DOI: 10.1063/1.4812323

[35] Momma K, Izumi F. VESTA 3 for three-dimensional visualization of crystal, volumetric and morphology data. Journal of Applied Crystallography. 2011;**44**(6):1272-1276. DOI: 10.1107/S0021889811038970

[36] Da Silva FG, Depeyrot J, Campos AF, et al. Structural and magnetic properties of spinel ferrite nanoparticles. Journal of Nanoscience and Nanotechnology. 2019;**19**(8):4888-4902. DOI: 10.1166/jnn.2019.16877

[37] Peddis D. In: Trohidou KN, editor. Magnetic Properties of Spinel Ferrite Nanoparticles: Influence of the Magnetic Structure, Magnetic Nanoparticle Assemblies. Singapore: Pan Stanford Publishing; 2014. pp. 978-981. DOI: 10.1201/b15657-6

[38] Peddis D, Mansilla MV, Mørup S, et al. Spin-canting and magnetic anisotropy in ultrasmall $CoFe_2O_4$ nanoparticles. The Journal of Physical Chemistry B. 2008;**112**(29):8507-8513. DOI: 10.1021/jp8016634

[39] Chinnasamy CN, Yang A, Yoon SD, et al. Size dependent magnetic properties and cation inversion in chemically synthesized $MnFe_2O_4$ nanoparticles. Journal of Applied Physics. 2007;**101**(9):09M509-1-09M509-3. DOI: 10.1063/1.2710218

[40] Tang ZX, Sorensen CM, Klabunde KJ, et al. Size-dependent Curie temperature in nanoscale $MnFe_2O_4$ particles. Physical Review Letters. 1991;**67**(25):3602. DOI: 10.1103/PhysRevLett.67.3602

[41] Van der Zaag PJ, Noordermeer A, Johnson MT, et al. Comment on "Size-dependent Curie temperature in nanoscale $MnFe_2O_4$ particles". Physical Review Letters. 1992;**68**(20):3112. DOI: 10.1103/PhysRevLett.68.3112

[42] Tang ZX, Sorensen CM, Klabunde KJ, et al. Size-dependent magnetic properties of manganese ferrite fine particles. Journal of Applied Physics. 1991;**69**(8):5279-5281. DOI: 10.1063/1.348048

[43] Vamvakidis K, Katsikini M, Sakellari D, et al. Reducing the inversion degree of $MnFe_2O_4$ nanoparticles through synthesis to enhance magnetization: Evaluation of their 1 H NMR relaxation and heating efficiency. Dalton Transactions. 2014;**43**(33):12754-12765. DOI: 10.1039/C4DT00162A

[44] Cannas C, Musinu A, Piccaluga G, et al. Magnetic properties of cobalt ferrite–silica nanocomposites prepared by a sol-gel autocombustion technique. The Journal of Chemical Physics. 2006;**125**(16):164714. DOI: 10.1063/1.2354475

[45] Néel L. Théorie du paramagnétisme constant; application au manganèse. Comptes Rendus de l'Académie des Sciences. 1936b;**203**:304-306. Available from: https://hal.science/hal-02878478

[46] Gomes JA, Azevedo GM, Depeyrot J, et al. Structural, chemical, and magnetic investigations of core–shell zinc ferrite nanoparticles. The Journal of Physical Chemistry C. 2012;**116**(45):24281-24291. DOI: 10.1021/jp3055069

[47] Blanco-Gutiérrez V, Torralvo-Fernández MJ, Sáez-Puche R. Magnetic behavior of $ZnFe_2O_4$ nanoparticles: Effects of a solid matrix and the particle size. The Journal of Physical Chemistry C. 2010;**114**(4):1789-1795. DOI: 10.1021/jp908395v

[48] Yelenich OV, Solopan SO, Kolodiazhnyi TV, et al. Magnetic properties and high heating efficiency of $ZnFe_2O_4$ nanoparticles. Materials Chemistry and Physics. 2014;**146**(1-2):129-135. DOI: 10.1016/j.matchemphys.2014.03.010

[49] Virumbrales-del OM, Delgado-Cabello A, Andrada-Chacón A, et al. Effect of composition and coating on the interparticle interactions and magnetic hardness of MFe_2O_4 (M= Fe, Co, Zn) nanoparticles. Physical Chemistry Chemical Physics. 2017;**19**(12):8363-8372. DOI: 10.1039/C6CP08743D

[50] Frenkel J, Doefman J. Spontaneous and induced magnetisation in ferromagnetic bodies. Nature. 1930;**126**(3173):274-275. DOI: 10.1038/126274a0

[51] Coey JMD. Noncollinear spin arrangement in ultrafine ferrimagnetic crystallites. Physical Review Letters. 1971;**27**(17):1140. DOI: 10.1103/PhysRevLett.27.1140

[52] Kodama RH. Magnetic nanoparticles. Journal of Magnetism and Magnetic Materials. 1999;**200**(1-3):359-372. DOI: 10.1016/ S0304-8853(99)00347-9

[53] Kodama RH, Berkowitz AE, McNiff EJ Jr, et al. Surface spin disorder in $NiFe_2O_4$ nanoparticles. Physical Review Letters. 1996;**77**(2):394. DOI: 10.1103/PhysRevLett.77.394

[54] Mirzaei A, Neri G. Microwave-assisted synthesis of metal oxide nanostructures for gas sensing application: A review. Sensors and Actuators B: Chemical. 2016;**237**:749-775. DOI: 10.1016/j.snb.2016.06.114

[55] Vasilakaki M, Trohidou KN, Nogués J. Enhanced magnetic properties in antiferromagnetic-core/ferrimagnetic-shell nanoparticles. Scientific Reports. 2015;**5**(1):9609. DOI: 10.1038/srep09609

[56] Nogués J, Sort J, Langlais V, et al. Exchange bias in nanostructures. Physics Reports. 2005;**422**(3):65-117. DOI: 10.1016/j.physrep.2005.08.004

[57] Dino F. Surface effects in magnetic nanoparticles. Berlin, Germany: Springer Science & Business Media; 2005. DOI: 10.1007/b136494

[58] Prozorov T, Prozorov R, Koltypin Y, et al. Sonochemistry under an applied magnetic field: Determining the shape of a magnetic particle. The Journal of Physical Chemistry B. 1998;**102**(50):10165-10168. DOI: 10.1021/jp982453k

[59] Prozorov R, Yeshurun Y, Prozorov T, et al. Magnetic irreversibility and relaxation in assembly of ferromagnetic nanoparticles. Physical Review B. 1999;**59**(10):6956. DOI: 10.1103/PhysRevB.59.6956

[60] Baykal A, Karaoglu E, Sözeri H, et al. Synthesis and characterization of high catalytic activity magnetic Fe_3O_4 supported Pd nanocatalyst. Journal of Superconductivity and Novel Magnetism. 2013;**26**:165-171. DOI: 10.1007/s10948-012-1702-2

[61] Li H, Wu H, Xiao G. Effects of synthetic conditions on particle size and magnetic properties of $NiFe_2O_4$. Powder Technology. 2010;**198**(1):157-166. DOI: 10.1016/j.powtec.2009.11.005

[62] Thakur P, Taneja S, Chahar D, et al. Recent advances on synthesis,

characterization and high frequency applications of Ni-Zn ferrite nanoparticles. Journal of Magnetism and Magnetic Materials. 2021;**530**:167925. DOI: 10.1016/j.jmmm.2021.167925

[63] Chen J, Ling Y, Qu D, et al. Enhanced electrocatalysis of NiMnIn Heusler alloy films for hydrogen evolution reaction by magnetic field. Journal of Alloys and Compounds. 2021;**877**:160271. DOI: 10.1016/j.jallcom.2021.160271

[64] Majumder S, Quang ND, Hien TT, et al. Effect of SILAR-anchored $ZnFe_2O_4$ on the $BiVO_4$ nanostructure: An attempt towards enhancing photoelectrochemical water splitting. Applied Surface Science. 2021;**546**:149033. DOI: 10.1016/j.apsusc.2021. 149033

[65] Šepelák V, Bergmann I, Feldhoff A, et al. Nanocrystalline nickel ferrite, $NiFe_2O_4$: Mechanosynthesis, nonequilibrium cation distribution, canted spin arrangement, and magnetic behavior. The Journal of Physical Chemistry C. 2007;**111**(13):5026-5033. DOI: 10.1021/jp067620s

[66] Chaibakhsh N, Moradi-Shoeili Z. Enzyme mimetic activities of spinel substituted nanoferrites (MFe_2O_4): A review of synthesis, mechanism and potential applications. Materials Science and Engineering: C. 2019;**99**:1424-1447. DOI: 10.1016/j.msec.2019.02.086

[67] Urbain F, Du R, Tang P, et al. Upscaling high activity oxygen evolution catalysts based on $CoFe_2O_4$ nanoparticles supported on nickel foam for power-to-gas electrochemical conversion with energy efficiencies above 80%. Applied Catalysis B: Environmental. 2019;**259**:118055. DOI: 10.1016/j.apcatb.2019.118055

[68] Karimzadeh I, Aghazadeh M, Ganjali MR, et al. Saccharide-coated superparamagnetic Fe_3O_4 nanoparticles (SPIONs) for biomedical applications: An efficient and scalable route for preparation and in situ surface coating through cathodic electrochemical deposition (CED). Materials Letters. 2017;**189**:290-294. DOI: 10.1016/j.matlet.2016.12.010

[69] Vivekanandhan S, Venkateswarlu M, Satyanarayana N. Effect of ethylene glycol on polyacrylic acid based combustion process for the synthesis of nano-crystalline nickel ferrite ($NiFe_2O_4$). Materials Letters. 2004;**58**(22-23):2717-2720. DOI: 10.1016/j.matlet.2004.02.030

[70] Kale A, Gubbala S, Misra RDK. Magnetic behavior of nanocrystalline nickel ferrite synthesized by the reverse micelle technique. Journal of Magnetism and Magnetic Materials. 2004;**277**(3):350-358. DOI: 10.1016/j.jmmm.2003.11.015

[71] Xiang S, Wang Y, Xiang Y, et al. Megahertz magneto-dielectric properties of nanosized NiZnCo ferrite from CTAB-assisted hydrothermal process. Transactions of Nonferrous Metals Society of China. 2009;**19**(6):1588-1592. DOI: 10.1016/S1003-6326(09)60075-3

[72] Su H, Zhang H, Tang X, et al. Complex permeability and permittivity spectra of polycrystalline Ni–Zn ferrite samples with different microstructures. Journal of Alloys and Compounds. 2009;**481**(1-2):841-844. DOI: 10.1016/j.jallcom.2009.03.133

[73] Mayakkannan M, Murugan A, Shameem A, et al. Investigations on ternary transition metal ferrite: $NiCoFe_2O_4$ as potential electrode for supercapacitor prepared by microwave irradiation method. Journal of Energy Storage. 2021;**44**:103257. DOI: 10.1016/j.est. 2021.103257

[74] Yin W, Shen Y, Zou F, et al. Metal–organic framework derived ZnO/ Zn Fe_2O_4/C nanocages as stable cathode material for reversible lithium–oxygen batteries. ACS Applied Materials & Interfaces. 2015;**7**(8):4947-4954. DOI: 10.1021/ am509143t

[75] Moller L, Thauer E, Ottmann A, Deeg L, Ghunaim R, Hampel S, et al. $CoFe_2O_4$-filled carbon nanotubes as anode material for lithium-ion batteries. Journal of Alloys and Compounds. 2020;**834**:155018. DOI: 10.1016/j. jallcom.2020.155018

[76] Li H, Bruning R, Liang J, Jiang W, Robichaud J, Liu J, et al. Alkaline steamingassisted conversion: A new strategy for the synthesis of pure $NiFe_2O_4$ and $CoFe_2O_4$ spinel ferrite nanoparticles. New Journal of Chemistry. 2022;**46**:20561-20571. DOI: 10.1039/ d2nj02281h

[77] Tuo Y, Liu G, Dong B, et al. Microbial synthesis of Pd/Fe_3O_4, Au/Fe_3O_4 and PdAu/Fe_3O_4 nanocomposites for catalytic reduction of nitroaromatic compounds. Scientific Reports. 2015;**5**(1):13515. DOI: 10.1038/srep13515

[78] Fomenko A, Kondranova A, Kazantsev S, Lozkomoev A, Pervikov A. Antimicrobial activity of $CuFe_2O_4$ nanoparticles obtained by electric explosion of Fe and Cu wires. In: AIP Conference Proceedings. Melville, NY, USA: AIP Publishing LLC; 2019. p. 2167(1). DOI: 10.1063/1.5131973

[79] Šepelák V, Baabe D, Litterst FJ, et al. Structural disorder in the high-energy milled magnesium ferrite. Journal of Applied Physics. 2000;**88**(10):5884-5893. DOI: 10.1063/1.1316048

[80] Haghniaz R, Rabbani A, Vajhadin F, et al. Anti-bacterial and wound healing-promoting effects of zinc ferrite nanoparticles. Journal of Nanobiotechnology. 2021;**19**:1-15. DOI: 10.1186/s12951-021-00776-w

[81] Novoselova JP, Safronov AP, Samatov OM, Beketov IV, Khurshid H, Nemati Z, et al. Laser target evaporation Fe_2O_3 nanoparticles for water-based ferrofluids for biomedical applications. IEEE Transactions on Magnetics. 2014;**50**(11):4600504. DOI: 10.1109/ TMAG.2014.2323354

[82] Semaltianos NG, Karczewski G. Laser synthesis of magnetic nanoparticles in liquids and application in the fabrication of polymer–nanoparticle composites. ACS Applied Nano Materials. 2021;**4**(7):6407-6440. DOI: 10.1021/acsanm.1c00715

[83] Zamani Kouhpanji MR, Stadler BJH. A guideline for effectively synthesizing and characterizing magnetic nanoparticles for advancing nanobiotechnology: A review. Sensors. 2020;**20**(9):2554. DOI: 10.3390/ s20092554

[84] Salih SJ, Mahmood WM. Review on magnetic spinel ferrite (MFe_2O_4) nanoparticles: From synthesis to application. Heliyon. 2023;**9**(6):e16601. DOI: 10.1016/j.heliyon.2023.e16601

[85] Qin H, He Y, Piao X, Huang D, Wang Z, Wang H, et al. Spinel ferrites (MFe_2O_4): Synthesis, improvement and catalytic application in environment and energy field. Advances in Colloid and Interface Science. 2021;**294**:102486. DOI: 10.1016/j.cis.2021.102486

[86] Fan G, Gu Z, Yang L, Li F. Nanocrystalline zinc ferrite photocatalysts formed using the colloid mill and hydrothermal technique. Chemical Engineering Journal. 2009;**155**:534-541. DOI: 10.1016/j. cej.2009.08.008

[87] Stergiou C. Magnetic, dielectric and microwave absorption properties of rare earth doped Ni–Co and Ni–Co–Zn spinel ferrite. Journal of Magnetism and Magnetic Materials. 2017;**426**:629-635. DOI: 10.1016/j.jmmm.2016.11.001

[88] Valenzuela R, Gaudisson T, Ammar S. Severe reduction of Ni–Zn ferrites during consolidation by Spark Plasma Sintering (SPS). Journal of Magnetism and Magnetic Materials. 2016;**400**:311-314. DOI: 10.1016/j.jmmm.2015.07.044

[89] Kim HJ, Choi H. Effect of plasma treatment on magnetic properties and heating efficiency of Ni-Zn nanoparticles. Journal of Magnetism and Magnetic Materials. 2019;**484**:14-20. DOI: 10.1016/j.jmmm.2019.03.102

[90] Gangaswamy DRS, Bharadwaj S, Varma MC, et al. Unusual increase in permeability in cobalt substituted Ni-Zn-Mg ferrite. Journal of Magnetism and Magnetic Materials. 2018;**468**:73-78. DOI: 10.1016/j.jmmm.2018.07.075

[91] Rao BP, Dhanalakshmi B, Ramesh S, et al. Cation distribution of Ni-Zn-Mn ferrite nanoparticles. Journal of Magnetism and Magnetic Materials. 2018;**456**:444-450. DOI: 10.1016/j.jmmm.2018.02.086

[92] Abdeen AM. Dielectric behaviour in Ni–Zn ferrites. Journal of Magnetism and Magnetic Materials. 1999;**192**(1):121-129. DOI: 10.1016/S0304-8853(98)00324-2

[93] Stankic S, Suman S, Haque F, et al. Pure and multi metal oxide nanoparticles: Synthesis, antibacterial and cytotoxic properties. Journal of Nanobiotechnology. 2016;**14**:1-20. DOI: 10.1186/s12951-016-0225-6

[94] Prabhakaran T, Mangalaraja RV, Denardin JC, et al. The effect of calcination temperature on the structural and magnetic properties of co-precipitated $CoFe_2O_4$ nanoparticles. Journal of Alloys and Compounds. 2017;**716**:171-183. DOI: 10.1016/J.JALLCOM.2017.05.048

[95] Liu S, Yu B, Wang S, et al. Preparation, surface functionalization and application of Fe_3O_4 magnetic nanoparticles. Advances in Colloid and Interface Science. 2020;**281**:102165. DOI: 10.1016/j.cis.2020.102165

[96] Rana G, Dhiman P, Kumar A, et al. Recent advances on nickel nano-ferrite: A review on processing techniques, properties and diverse applications. Chemical Engineering Research and Design. 2021;**175**:182-208. DOI: 10.1016/j.cherd.2021.08.040

[97] Melo RS, Banerjee P, Franco A. Hydrothermal synthesis of nickel doped cobalt ferrite nanoparticles: Optical and magnetic properties. Journal of Materials Science: Materials in Electronics. 2018;**29**:14657-14667. DOI: 10.1007/s10854-018-9602-2

[98] Pham TN, Huy TQ, Le AT. Spinel ferrite (AFe_2O_4)-based heterostructured designs for lithium-ion battery, environmental monitoring, and biomedical applications. RSC Advances. 2020;**10**(52):31622-31661. DOI: 10.1039/D0RA05133K

[99] Sun H, Yang X, Zhao L, Xu T, Lian J. One-pot hydrothermal synthesis of octahedral $CoFe/CoFe_2O_4$ submicron composite as heterogeneous catalysts with enhanced peroxymonosulfate activity. Journal of Materials Chemistry A. 2016;**4**:9455-9465. DOI: 10.1039/c6ta02126c

[100] Safi R, Ghasemi A, Shoja-Razavi R, et al. Rietveld structure refinement, cations distribution and magnetic

features of $CoFe_2O_4$ nanoparticles synthesized by co-precipitation, hydrothermal, and combustion methods. Ceramics International. 2016;**42**(5):6375-6382. DOI: 10.1016/j.ceramint.2016.01.032

[101] Dhiman M, Sharma R, Kumar V, et al. Morphology controlled hydrothermal synthesis and photocatalytic properties of $ZnFe_2O_4$ nanostructures. Ceramics International. 2016;**42**(11):12594-12605. DOI: 10.1016/j.ceramint.2016.04.115

[102] Sanpo N, Tharajak J, Li Y, et al. Biocompatibility of transition metal-substituted cobalt ferrite nanoparticles. Journal of Nanoparticle Research. 2014;**16**:1-13. DOI: 10.1007/s11051-014-2510-3

[103] Somvanshi SB, Khedkar MV, Kharat PB, et al. Influential diamagnetic magnesium (Mg^{2+}) ion substitution in nano-spinel zinc ferrite ($ZnFe_2O_4$): Thermal, structural, spectral, optical and physisorption analysis. Ceramics International. 2020;**46**(7):8640-8650. DOI: 10.1016/j.ceramint.2019.12.097

[104] Varma A, Mukasyan AS, Rogachev AS, et al. Solution combustion synthesis of nanoscale materials. Chemical Reviews. 2016;**116**(23):14493-14586. DOI: 10.1021/acs.chemrev.6b00279

[105] Sutka A, Mezinskis G. Sol-gel auto-combustion synthesis of spinel-type ferrite nanomaterials. Frontiers of Materials Science. 2012;**6**:128-141. DOI: 10.1007/s11706-012-0167-3

[106] Mukasyan AS, Epstein P, Dinka P. Solution combustion synthesis of nanomaterials. Proceedings of the Combustion Institute. 2007;**31**(2):1789-1795. DOI: 10.1016/j.proci.2006.07.052

[107] Ramakrishna A, Murali N, Margarette SJ, Mammo TW, Krishna JN, Sailaja B, et al. Studies on structural, magnetic, and DC electrical resistivity properties of $Co_{0.5}M_{0.37}Cu_{0.13}Fe_2O_4$ (M = Ni, Zn and Mg) ferrite nanoparticle systems. Advanced Powder Technology. 2018;**29**(11):2601-2607. DOI: 10.1016/j.apt.2018.07.005

[108] Gul S, Khan SB, Rehman IU, et al. A comprehensive review of magnetic nanomaterials modern day theranostics. Frontiers in Materials. 2019;**6**:179. DOI: 10.3389/fmats.2019.00179

[109] Rivero M, del Campo A, Mayoral Á, et al. Synthesis and structural characterization of Znx $Fe_{3-x}O_4$ ferrite nanoparticles obtained by an electrochemical method. RSC Advances. 2016;**6**(46):40067-40076. DOI: 10.1039/C6RA04145K

[110] Sayeed MA, OMullane, A. P. Electrodeposition at highly negative potentials of an iron-cobalt oxide catalyst for use in electrochemical water splitting. ChemPhysChem. 2019;**20**:3112-3119. DOI: 10.1002/cphc.201900498

[111] Tsuji M. Microwave-assisted synthesis of metallic nanomaterials in liquid phase. ChemistrySelect. 2017;**2**(2):805-819. DOI: 10.1002/slct.201700011

[112] Zhu YJ, Chen F. Microwave-assisted preparation of inorganic nanostructures in liquid phase. Chemical Reviews. 2014;**114**(12):6462-6555. DOI: 10.1021/cr400366s

[113] Shukla A, Bhardwaj AK, Singh S, Uttam K, Gautam N, Himanshu A, et al. Microwave assisted scalable synthesis of titanium ferrite nanomaterials. Journal of Applied Physics. 2018;**123**(16):161411. DOI: 10.1063/1.50087

[114] Diodati S, Walton RI, Mascotto S, Gross S. Low-temperature wet chemistry

synthetic approaches towards ferrites. Inorganic Chemistry Frontiers. 2020;7:3282-3314. DOI: 10.1039/D0QI00294A

[115] Kombaiah K, Vijaya JJ, Kennedy LJ, et al. Studies on the microwave assisted and conventional combustion synthesis of Hibiscus rosa-sinensis plant extract based $ZnFe_2O_4$ nanoparticles and their optical and magnetic properties. Ceramics International. 2016;**42**(2):2741-2749. DOI: 10.1016/j.ceramint.2015.11.003

[116] Chamani S, Khatamian M, Peighambardoust NS, Aydemir U. Microwave-assisted auto- combustion synthesis of binary/ternary Co_xNi_{1-x}Ferrite for electrochemical hydrogen and oxygen evolution. ACS Omega. 2021;**6**:33024-33032. DOI: 10.1021/acsomega.1c05183

[117] Yew YP, Shameli K, Miyake M, et al. Green biosynthesis of superparamagnetic magnetite Fe_3O_4 nanoparticles and biomedical applications in targeted anticancer drug delivery system: A review. Arabian Journal of Chemistry. 2020;**13**(1):2287-2308. DOI: 10.1016/J.ARABJC.2018.04.013

[118] Naik MM, Naik HSB, Nagaraju G, et al. Green synthesis of zinc ferrite nanoparticles in Limonia acidissima juice: Characterization and their application as photocatalytic and antibacterial activities. Microchemical Journal. 2019;**146**:1227-1235. DOI: 10.1016/j.microc.2019.02.059

[119] Chutia R, Chetia B. Biogenic $CuFe_2O_4$ magnetic nanoparticles as a green, reusable and excellent nanocatalyst for acetylation reactions under solvent-free conditions. New Journal of Chemistry. 2018;**42**(18): 15200-15206. DOI: 10.1039/C8NJ02685H

[120] Amiri M, Pardakhti A, Ahmadi-Zeidabadi M, et al. Magnetic nickel ferrite nanoparticles: Green synthesis by Urtica and therapeutic effect of frequency magnetic field on creating cytotoxic response in neural cell lines. Colloids and Surfaces B: Biointerfaces. 2018;**172**:244-253. DOI: 10.1016/j.colsurfb.2018.08.049

[121] Perez-Gonzalez T, Jimenez-Lopez C, Neal AL, et al. Magnetite biomineralization induced by Shewanella oneidensis. Geochimica et Cosmochimica Acta. 2010;**74**(3):967-979. DOI: 10.1016/j.gca.2009.10.035

[122] Amemiya Y, Arakaki A, Staniland SS, et al. Controlled formation of magnetite crystal by partial oxidation of ferrous hydroxide in the presence of recombinant magnetotactic bacterial protein Mms6. Biomaterials. 2007;**28**(35):5381-5389. DOI: 10.1016/j.biomaterials.2007.07.051

[123] Zhou W, He W, Zhang X, et al. Biosynthesis of iron phosphate nanopowders. Powder Technology. 2009;**194**(1-2):106-108. DOI: 10.1016/j.powtec.2009.03.034

[124] Bose S, Hochella MF Jr, Gorby YA, et al. Bioreduction of hematite nanoparticles by the dissimilatory iron reducing bacterium Shewanella oneidensis MR-1. Geochimica et Cosmochimica Acta. 2009;**73**(4):962-976. DOI: 10.1016/j.gca.2008.11.031

[125] Atrak K, Ramazani A, Taghavi FS. Green synthesis of $Zn0.5Ni0.5AlFeO_4$ magnetic nanoparticles and investigation of their photocatalytic activity for degradation of reactive blue 21 dye. Environmental Technology. 2020;**21**(41):2760-2770. DOI: 10.1080/09593330.2019.1581841

[126] Gayathri Manju B, Raji P. Green synthesis, characterization, and

antibacterial activity of lime-juice-mediated copper–nickel mixed ferrite nanoparticles. Applied Physics A. 2020;**126**(3):156. DOI: 10.1007/s00339-020-3313-2

[127] Tiwari P, Verma R, Kane SN, Tatarchuk T, Mazaleyrat F. Effect of Zn addition on structural, magnetic properties and anti-structural modeling of magnesium-nickel nano ferrites. Materials Chemistry and Physics. 2019;**229**:78-86. DOI: 10.1016/j.matchemphys.2019.02.030

[128] Naik MM, Naik H, Kottam N, Vinuth M, Nagaraju G, Prabhakara M. Multifunctional properties of microwave-assisted bioengineered nickel doped cobalt ferrite nanoparticles. Journal of Sol-Gel Science and Technology. 2019;**91**(3):578-595. DOI: 10.1007/s10971-019-05048-6

[129] Hujjatul Islam M, Paul MTY, Burheim OS, Pollet BG. Recent developments in the Sonoelectrochemical synthesis of nanomaterials. Ultrasonics Sonochemistry. 2019;**59**:104711. DOI: 10.1016/ j.ultsonch.2019.104711

[130] Wu K, Liu D, Lu W, Zhang K. One-pot sonochemical synthesis of magnetite@ reduced graphene oxide nanocomposite for high performance Li ion storage. Ultrasonics Sonochemistry. 2018;**45**:167-172. DOI: 10.1016/j.ultsonch.2018.03.015

[131] Almessiere M, Slimani Y, Guner S, Sertkol M, Korkmaz AD, Shirsath SE, et al. Sonochemical synthesis and physical properties of $Co_{0.3}Ni_{0.5}Mn0.2Eu_xFe_{2-x}O_4$ nano-spinel ferrites. Ultrasonics Sonochemistry. 2019;**58**:104654. DOI: 10.1016/j.ultsonch.2019.104654

[132] Slimani Y, Almessiere M, Korkmaz AD, Guner S, Güngünes H, Sertkol M, et al. $Ni_{0.4}Cu_{0.2}Zn_{0.4}TbxFe_{2-x}O_4$ nanospinel ferrites: ultrasonic synthesis and physical properties. Ultrasonics Sonochemistry. 2019;**59**:104757. DOI: 10.1016/j.ultsonch.2019.104757

[133] Ilosvai AM, Dojcsak D, Váradi C, et al. Sonochemical combined synthesis of nickel ferrite and cobalt ferrite magnetic nanoparticles and their application in glycan analysis. International Journal of Molecular Sciences. 2022;**23**(9):5081. DOI: 10.3390/ijms23095081

[134] Bang JH, Suslick KS. Applications of ultrasound to the synthesis of nanostructured materials. Advanced Materials. 2010;**22**(10):1039-1059. DOI: 10.1002/adma.200904093

[135] Savun-Hekimoğlu B. A review on sonochemistry and its environmental applications. Acoustics. 2020;**2**(4):766-775. DOI: 10.3390/acoustics2040042

[136] Pokhrel N, Vabbina PK, Pala N. Sonochemistry: Science and engineering. Ultrasonics Sonochemistry. 2016;**29**:104-128. DOI: 10.1016/j.ultsonch.2015.07.023

[137] Low SS, Yew M, Lim CN, Chai WS, Low LE, Manickam S, et al. Sonoproduction of nanobiomaterials – A critical review. Ultrasonics Sonochemistry. 2022;**82**:105887. DOI: 10.1016/j.ultsonch.2021.105887

[138] Amulya MAS, Nagaswarupa HP, Kumar MRA, Ravikumar CR, Prashantha SC, Kusuma KB. Sonochemical synthesis of $NiFe_2O_4$ nanoparticles: Characterization and their photocatalytic and electrochemical applications. Applied Surface Science Advances. 2020;**1**:100023. DOI: 10.1016/j.apsadv.2020.100023

[139] Yadav RS, Kuřitka I, Vilcakova J, Jamatia T, Machovsky M, Skoda D, et al.

Impact of sonochemical synthesis condition on the structural and physical properties of $MnFe_2O_4$ spinel ferrite nanoparticles. Ultrasonics Sonochemistry. 2020;**61**:104839. DOI: 101016/j.ultsonch.2019.104839

[140] Fuentes-García JA, Carvalho Alavarse A, Moreno Maldonado AC, et al. Simple sonochemical method to optimize the heating efficiency of magnetic nanoparticles for magnetic fluid hyperthermia. ACS Omega. 2020;**5**(41):26357-26364. DOI: 10.1021/acsomega.0c02212

[141] Yang S, Qiu X, Jin P, Dzakpasu M, Wang XC, Zhang Q, et al. MOF-templated synthesis of $CoFe_2O_4$ nanocrystals and its coupling with peroxymonosulfate for degradation of Bisphenol A. Chemical Engineering Journal. 2018;**353**:329-339. DOI: 10.1016/j.cej. 2018.07.105

[142] Huang Y, Zhang SL, Lu XF, Wu ZP, Luan D, Lou XW. Trimetallic spinel $NiCo_{2-x}Fe_xO_4$ nanoboxes for highly efficient electrocatalytic oxygen evolution. Angewandte Chemie (International Ed. in English). 2021;**60**:11841-11846. DOI: 10.1002/anie.202103058

[143] Patil PD, Shingte SR, Karade VC, Kim JH, Dongale TD, Mujawar SH, et al. Effect of annealing temperature on morphologies of metal organic framework derived $NiFe_2O_4$ for supercapacitor application. Journal of Energy Storage. 2021;**40**:102821. DOI: 10.1016/j.est.2021. 102821

[144] Pardavi-Horvath M. Microwave applications of soft ferrites. Journal of Magnetism and Magnetic Materials. 2000;**215**:171-183. DOI: 10.1016/S0304-8853(00)00106-2

[145] Rodrigues APG, Gomes DKS, Araújo JH, et al. Nanoferrites of nickel doped with cobalt: Influence of Co2+ on the structural and magnetic properties. Journal of Magnetism and Magnetic Materials. 2015;**374**:748-754. DOI: 10.1016/j.jmmm.2014.09.045

[146] Ünal B, Ünver İS, Güngüneş H, et al. Microwave, dielectric and magnetic properties of Mg-Ti substituted Ni-Zn ferrite nanoparticles. Ceramics International. 2016;**42**(15):17317-17331. DOI: 10.1016/j.ceramint.2016.08.028

[147] Valenzuela R. Magnetic Ceramics. Cambridge: Cambridge University Press; 2005

[148] Safari A, Gheisari K, Farbod M. Characterization of Ni ferrites powders prepared by plasma arc discharge process. Journal of Magnetism and Magnetic Materials. 2017;**421**:44-51. DOI: 10.1016/j.jmmm.2016.07.024

[149] Alarifi A, Deraz NM, Shaban S. Structural, morphological and magnetic properties of $NiFe_2O_4$ nano-particles. Journal of Alloys and Compounds. 2009;**486**(1-2):501-506. DOI: 10.1016/j.jallcom.2009.06.192

[150] Ditta A, Khan MA, Junaid M, et al. Structural, magnetic and spectral properties of Gd and Dy co-doped dielectrically modified Co-Ni ($Ni_{0.4}Co_{0.6}Fe_2O_4$) ferrites. Physica B: Condensed Matter. 2017;**507**:27-34. DOI: 10.1016/j.physb.2016.11.030

[151] Wahba AM, Mohamed MB. Structural and magnetic characterization and cation distribution of nanocrystalline $Co_xFe_{3-x}O_4$ ferrites. Journal of Magnetism and Magnetic Materials. 2015;**378**:246-252. DOI: 10.1016/j.physb.2016.11.030

[152] Coey JMD. Noncollinear spin structures. Canadian Journal of Physics. 1987;**65**(10):1210-1232. DOI: 10.1139/p87-197

[153] Pankhurst QA, Pollard RJ. Origin of the spin-canting anomaly in small ferrimagnetic particles. Physical Review Letters. 1991;**67**(2):248. DOI: 10.1103/PhysRevLett.67.248

[154] Meng F, Shang X, Zhang W, et al. Effect of In^{3+} substitution on structural and magnetic properties of Ni ferrite nanoparticles. Physica B: Condensed Matter. 2017;**504**:69-73. DOI: 10.1016/j.physb.2016.09.031

[155] Saravanakumar B, Rani BJ, Ravi G, et al. Reducing agent ($NaBH_4$) dependent structure, morphology and magnetic properties of nickel ferrite ($NiFe_2O_4$) nanorods. Journal of Magnetism and Magnetic Materials. 2017;**428**:78-85. DOI: 10.1016/j.jmmm.2016.12.017

[156] Van der Zaag PJ, Van der Valk PJ, Rekveldt MT. A domain size effect in the magnetic hysteresis of NiZn-ferrites. Applied Physics Letters. 1996;**69**(19):2927-2929. DOI: 10.1063/1.117326

[157] Sözeri H, Genç F, Ünal B, et al. Magnetic, electrical and microwave properties of Mn–Co substituted $Ni_xZn_{0.8x}Fe_2O_4$ nanoparticles. Journal of Alloys and Compounds. 2016;**660**:324-335. DOI: 10.1016/j.jallcom.2015.11.123

[158] Stergiou C. Microstructure and electromagnetic properties of Ni-Zn-Co ferrite up to 20 GHz. Advances in Materials Science and Engineering. 2016;**2016**(1):1934783. DOI: 10.1155/2016/1934783

[159] Jahan N, Chowdhury FUZ, Zakaria AKM, et al. Manipulation of magnetic properties of Cr-substituted Ni ferrite synthesized by conventional ceramic technique. Journal of Superconductivity and Novel Magnetism. 2017;**30**:261-268. DOI: 10.1007/s10948-016-3713-x

[160] Wu X, Chen W, Wu W, Li H, Lin C. Structural and magnetic properties evolution of Li-substituted $Co_{0.5}Ni_{0.5}Fe_2O_4$ ferrite. Journal of Electronic Materials. 2017;**46**:199-207. DOI: 10.1007/s11664-016-4877-7

[161] Khan K. Microwave absorption properties of radar absorbing nanosized cobalt ferrites for high frequency applications. Journal of Superconductivity and Novel Magnetism. 2014;**27**:453-461. DOI: 10.1007/s10948-013-2283-4

[162] Ravinder D, Latha K. Electrical conductivity of Mn–Zn ferrites. Journal of Applied Physics. 1994;**75**(10):6118-6120. DOI: 10.1063/1.355479

[163] Zaki HM. The influence of Zn ions substitution on the transport properties of Mg–ferrite. Physica B: Condensed Matter. 2009;**404**(20):3356-3362. DOI: 10.1016/j.physb.2009.05.012

[164] Lee SH, Yoon SJ, Lee GJ, et al. Electrical and magnetic properties of $NiCr_xFe_{2-x}O_4$ spinel ($0 \leq x \leq 0.6$). Materials Chemistry and Physics. 1999;**61**(2):147-152. DOI: 10.1016/S0254-0584(99)00136-4

[165] Gaudon M, Pailhé N, Wattiaux A, et al. Structural defects in AFe_2O_4 (a= Zn, Mg) spinels. Materials Research Bulletin. 2009;**44**(3):479-484. DOI: 10.1016/j.materresbull.2008.12.005

[166] Marinca TF, Chicinaş I, Isnard O. Structural and magnetic properties of the copper ferrite obtained by reactive milling and heat treatment. Ceramics International. 2013;**39**(4):4179-4186. DOI: 10.1016/j.ceramint.2012.10.274

[167] Kolahalam LA, Viswanath IVK, Diwakar BS, et al. Review on nanomaterials: Synthesis and applications. Materials Today:

Proceedings. 2019;**18**:2182-2190. DOI: 10.1016/j.matpr.2019.07.371

[168] Shakil M, Inayat U, Arshad MI, et al. Influence of zinc and cadmium co-doping on optical and magnetic properties of cobalt ferrites. Ceramics International. 2020;**46**(6):7767-7773. DOI: 10.1016/j.ceramint.2019.11.280

[169] Sinha AK, Singh MN, Achary SN, et al. Crystal field splitting and spin states of Co ions in cobalt ferrite with composition $Co_{1.5}Fe_{1.5}O_4$ using magnetization and X-ray absorption spectroscopy measurements. Journal of Magnetism and Magnetic Materials. 2017;**435**:87-95. DOI: 10.1016/j.jmmm.2017.03.070

[170] Chakrabarty S, Dutta A, Pal M. Enhanced magnetic properties of doped cobalt ferrite nanoparticles by virtue of cation distribution. Journal of Alloys and Compounds. 2015;**625**:216-223. DOI: 10.1016/j.jallcom.2014.10.179

[171] Maaz K, Karim S, Mashiatullah A, et al. Structural analysis of nickel doped cobalt ferrite nanoparticles prepared by coprecipitation route. Physica B: Condensed Matter. 2009;**404**(21):3947-3951. DOI: 10.1016/j.physb.2009.07.134

[172] Sharifianjazi F, Moradi M, Parvin N, et al. Magnetic $CoFe_2O_4$ nanoparticles doped with metal ions: A review. Ceramics International. 2020;**46**(11):18391-18412. DOI: 10.1016/j.ceramint.2020.04.202

[173] World Meteorological Organization, Guide to meteorological instruments and methods of observation. In: Secretariat of the World Meteorological Organization. Geneva, Switzerland: WMO; 1983

[174] Rezlescu N, Doroftei C, Rezlescu E, et al. Structure and humidity sensitive electrical properties of the Sn^{4+} and/or Mo^{6+} substituted Mg ferrite. Sensors and Actuators B: Chemical. 2006;**115**(2):589-595. DOI: 10.1016/j.snb.2005.10.028

[175] Seiyama T, Kato A, Fujiishi K, et al. A new detector for gaseous components using semiconductive thin films. Analytical Chemistry. 1962;**34**(11):1502-1503. DOI: 10.1021/ac60191a001

[176] Chen H, Chen H, Chen J, et al. Gas sensors based on semiconductor metal oxides fabricated by electrospinning: A review. Sensors. 2024;**24**(10):2962. DOI: 10.3390/s24102962

[177] Šutka A, Gross KA. Spinel ferrite oxide semiconductor gas sensors. Sensors and Actuators B: Chemical. 2016;**222**:95-105. DOI: 10.1016/j.snb.2015.08.027

[178] Brito VLO, Cunha SA, Araújo FF, et al. Processing and characterization of a Ni-Co ferrite for sensor applications. Cerâmica. 2015;**61**:341-349. DOI: 10.1590/0366-69132015613591893

[179] Dabbebi T, Hcini S, Alzahrani B, et al. Investigations of microstructural and impedance spectroscopic properties of $Mg_{0.5}Co_{0.5}Fe_{1.6}Al_{0.4}O_4$ ferrite prepared using sol–gel method. Journal of Materials Science: Materials in Electronics. 2021;**32**:12521-12534. DOI: 10.1007/s10854-021-05887-6

[180] Subramani AK, Kondo K, Tada M, et al. Spinel ferrite films by a novel solution process for high frequency applications. Materials Chemistry and Physics. 2010;**123**(1):16-19. DOI: 10.1016/j.matchemphys.2010.02.031

[181] Mohapatra PP, Singh HK, Kiran MSRN, et al. Co substituted Ni–Zn ferrites with tunable dielectric and magnetic response for high-frequency applications. Ceramics International. 2022;**48**(19):29217-29228. DOI: 10.1016/j.ceramint.2022.05.197

[182] Shakeel V, Gul IH, John P, et al. Biocompatible gelatin-coated ferrite nanoparticles: A magnetic approach to advanced drug delivery. Saudi Pharmaceutical Journal. 2024;**32**(6):102066. DOI: 10.1016/j.jsps.2024.102066

[183] Ahmed M, Zheng Y, Wang G, et al. Nanostructured spinel ferrites: A comprehensive study on the synthesis, characterization, and magnetic properties of Zn and Mn substituted magnetite nanoparticles produced through the co-precipitation method. Journal of Materials Research and Technology. 2024;**33**:5987-6010. DOI: 10.1016/j.jmrt.2024.10.215

[184] Mohammed Y, Hafeez HY, Mohammed J, et al. Hydrogen production via photocatalytic water splitting using spinel ferrite-based photocatalysts: Recent and future perspectives. Next Energy. 2024;**4**:100145. DOI: 10.1016/j.nxener.2024.100145

[185] Casbeer E, Sharma VK, Li XZ. Synthesis and photocatalytic activity of ferrites under visible light: A review. Separation and Purification Technology. 2012;**87**:1-14. DOI: 10.1016/j.seppur.2011.11.034

[186] Kefeni KK, Mamba BB, Msagati TAM. Application of spinel ferrite nanoparticles in water and wastewater treatment: A review. Separation and Purification Technology. 2017;**188**:399-422. DOI: 10.1016/j.seppur.2017.07.015

[187] Revathi J, Abel MJ, Archana V, et al. Synthesis and characterization of $CoFe_2O_4$ and Ni-doped $CoFe_2O_4$ nanoparticles by chemical Co-precipitation technique for photo-degradation of organic dyestuffs under direct sunlight. Physica B: Condensed Matter. 2020;**587**:412136. DOI: 10.1016/j.physb.2020.412136

[188] Anucha CB, Altin I, Bacaksiz E, et al. Titanium dioxide (TiO_2)-based photocatalyst materials activity enhancement for contaminants of emerging concern (CECs) degradation: In the light of modification strategies. Chemical Engineering Journal Advances. 2022;**10**:100262. DOI: 10.1016/j.ceja.2022.100262

[189] Boumaza S, Boudjemaa A, Bouguelia A, et al. Visible light induced hydrogen evolution on new hetero-system Zn Fe_2O_4/$SrTiO_3$. Applied Energy. 2010;**87**(7):2230-2236. DOI: 10.1016/j.apenergy.2009.12.016

[190] Dom R, Subasri R, Radha K, et al. Synthesis of solar active nanocrystalline ferrite, MFe_2O_4 (M: Ca, Zn, Mg) photocatalyst by microwave irradiation. Solid State Communications. 2011;**151**(6):470-473. DOI: 10.1016/j.ssc.2010.12.034

[191] Ao Y, Xu J, Fu D, et al. A novel magnetically separable composite photocatalyst: Titania-coated magnetic activated carbon. Separation and Purification Technology. 2008;**61**(3):436-441. DOI: 10.1016/j.seppur.2007.12.007

[192] Brame J, Li Q, Alvarez PJJ. Nanotechnology-enabled water treatment and reuse: Emerging opportunities and challenges for developing countries. Trends in Food Science & Technology. 2011;**22**(11):618-624. DOI: 10.1016/j.tifs.2011.01.004

[193] Rafie SF, Abu-Zahra N, Sillanpaa M. A comprehensive review of spinel ferrites and their magnetic composites as highly efficient adsorbents of rare earth elements. Emergy Contaminants. 2025;**11**:100429. DOI: 10.1016/j.emcon.2024.100429

[194] Reddy DHK, Yun YS. Spinel ferrite magnetic adsorbents: alternative future materials for water purification? Coordination Chemistry Reviews. 2016;**315**:90-111. DOI: 10.1016/j.ccr.2016.01.012

[195] Phouthavong V, Yan R, Nijpanich S, Hagio T, Ichino R, Kong L, et al. Magnetic adsorbents for wastewater treatment: Advancements in their synthesis methods. Materials. 2022;**15**(3):1053. DOI: 10.3390/ma15031053

[196] Ding HL, Zhang YX, Wang S, et al. Fe_3O_4@SiO_2 core/shell nanoparticles: The silica coating regulations with a single core for different core sizes and shell thicknesses. Chemistry of Materials. 2012;**24**(23):4572-4580. DOI: 10.1021/cm302828d

[197] Dong T, Yang L, Zhu M, et al. Removal of cadmium (II) from wastewater with gas-assisted magnetic separation. Chemical Engineering Journal. 2015;**280**:426-432. DOI: 10.1016/j.cej.2015.05.066

[198] Jiang C, Pang S, Ouyang F, et al. A new insight into Fenton and Fenton-like processes for water treatment. Journal of Hazardous Materials. 2010;**174**(1-3):813-817. DOI: 10.1016/j.jhazmat.2009.09.125

[199] Meyerstein D. Re-examining Fenton and Fenton-like reactions. Nature Reviews Chemistry. 2021;**5**(9):595-597. DOI: 10.1038/s41570-021-00310-4

[200] Sundararajan M, Sailaja V, Kennedy LJ, et al. Photocatalytic degradation of rhodamine B under visible light using nanostructured zinc doped cobalt ferrite: Kinetics and mechanism. Ceramics International. 2017;**43**(1):540-548. DOI: 10.1016/j.ceramint.2016.09.191

[201] Srivastava V, Kohout T, Sillanpää M. Potential of cobalt ferrite nanoparticles ($CoFe_2O_4$) for remediation of hexavalent chromium from synthetic and printing press wastewater. Journal of Environmental Chemical Engineering. 2016;**4**(3):2922-2932. DOI: 10.1016/j.jece.2016.06.002

[202] Kefeni KK, Mamba BB. Photocatalytic application of spinel ferrite nanoparticles and nanocomposites in wastewater treatment. Sustainable Materials and Technologies. 2020;**23**:e00140. DOI: 10.1016/j.susmat.2019.e00140

[203] Wang Y, Miao Y, Li G, et al. Engineering ferrite nanoparticles with enhanced magnetic response for advanced biomedical applications. Materials Today Advances. 2020;**8**:100119. DOI: 10.1016/j.mtadv.2020.100119

[204] Shobana MK. Nanoferrites in biosensors–a review. Materials Science and Engineering: B. 2021;**272**:115344. DOI: 10.1016/j.mseb.2021.115344

[205] Abdel Maksoud MIA, Ghobashy MM, Kodous AS, et al. Insights on magnetic spinel ferrites for targeted drug delivery and hyperthermia applications. Nanotechnology Reviews. 2022;**11**(1):372-413. DOI: 10.1515/ntrev-2022-0027

[206] Kefeni KK, Msagati TAM, Nkambule TTI, et al. Spinel ferrite nanoparticles and nanocomposites for biomedical applications and their toxicity. Materials Science and Engineering: C. 2020;**107**:110314. DOI: 10.1016/j.msec.2019.110314

[207] Joshi HM, Lin YP, Aslam M, et al. Effects of shape and size of cobalt ferrite nanostructures on their MRI contrast and thermal activation. The Journal of Physical Chemistry C. 2009;**113**(41):17761-17767. DOI: 10.1021/jp905776g

[208] Amiri S, Shokrollahi H. The role of cobalt ferrite magnetic nanoparticles in medical science. Materials Science and Engineering: C. 2013;**33**(1):1-8. DOI: 10.1016/j.msec.2012.09.003

[209] Montha W, Maneeprakorn W, Buatong N, et al. Synthesis of doxorubicin-PLGA loaded chitosan stabilized (Mn, Zn)Fe_2O_4 nanoparticles: Biological activity and pH-responsive drug release. Materials Science and Engineering: C. 2016;**59**:235-240. DOI: 10.1016/j.msec.2015.09.098

[210] Bahrami B, Hojjat-Farsangi M, Mohammadi H, et al. Nanoparticles and targeted drug delivery in cancer therapy. Immunology Letters. 2017;**190**:64-83. DOI: 10.1016/j.imlet.2017.07.015

[211] Mody A, Cox S, Shah A, Singh W, Bevins H. Parihar, magnetic nanoparticle drug delivery systems for targeting tumor. Applied Nanoscience. 2014;**4**:385-392. DOI: 10.1007/s13204-013-0216-y

[212] Faraji M, Yamini Y, Rezaee M. Magnetic nanoparticles: Synthesis, stabilization, functionalization, characterization, and applications. Journal of the Iranian Chemical Society. 2010;**7**:1-37. DOI: 10.1007/BF03245856

[213] Obaidat IM, Issa B, Haik Y. Magnetic properties of magnetic nanoparticles for efficient hyperthermia. Nanomaterials. 2015;**5**(1):63-89. DOI: 10.3390/nano5010063

[214] Nurjanah N, Adzkia QAA, Rustana R, et al. A review: Nanoparticles Ni Fe_2O_4 synthesis and its application as hyperthermia agents in biomedicine. Indonesian Journal of Chemical Science. 2022;**11**(2):103-113. DOI: 10.15294/ijcs.v11i2.52590

[215] Safi M, Sarrouj H, Sandre O, et al. Interactions between sub-10-nm iron and cerium oxide nanoparticles and 3T3 fibroblasts: The role of the coating and aggregation state. Nanotechnology. 2010;**21**(14):145103. DOI: 10.1088/0957-4484/21/14/145103

[216] Paramo L, Jiménez-Chávez A, Medina-Ramirez IE, et al. Biocompatibility evaluation of TiO_2, Fe_3O_4, and TiO_2/Fe_3O_4 nanomaterials: Insights into potential toxic effects in erythrocytes and HepG2 cells. Nanomaterials. 2023;**13**(21):2824. DOI: 10.3390/nano13212824

[217] Ehdaie B. Application of nanotechnology in cancer research: Review of progress in the National Cancer Institute's alliance for nanotechnology. International Journal of Biological Sciences. 2007;**3**(2):108. DOI: 10.7150/ijbs.3.108

[218] Syu WJ, Huang CC, Hsiao JK, et al. Co-precipitation synthesis of near-infrared iron oxide nanocrystals on magnetically targeted imaging and photothermal cancer therapy via photoablative protein denature. Nano. 2019;**3**(3):236. DOI: 10.7150/ntno.24124

[219] Armijo LM, Brandt YI, Mathew D, et al. Iron oxide nanocrystals for magnetic hyperthermia applications. Nanomaterials. 2012;**2**(2):134-146. DOI: 10.3390/nano2020134

[220] Silvestri N, Gavilán H, Guardia P, et al. Di- and tri-component spinel ferrite nanocubes: Synthesis and their comparative characterization for theranostic applications. Nanoscale. 2021;**13**(32):13665-13680. DOI: 10.1039/D1NR01044A

[221] Montazerabadi A, Beik J, Irajirad R, et al. Folate-modified and curcumin-loaded dendritic magnetite nanocarriers for the targeted

thermo-chemotherapy of cancer cells. Artificial Cells, Nanomedicine, and Biotechnology. 2019;**47**(1):330-340. DOI: 10.1080/21691401.2018.1557670

[222] Oh Y, Moorthy MS, Manivasagan P, et al. Magnetic hyperthermia and pH-responsive effective drug delivery to the sub-cellular level of human breast cancer cells by modified $CoFe_2O_4$ nanoparticles. Biochimie. 2017;**133**:7-19. DOI: 10.1016/j.biochi.2016.11.012

[223] Christophi C, Winkworth A, Muralihdaran V, et al. The treatment of malignancy by hyperthermia. Surgical Oncology. 1998;**7**(1-2):83-90. DOI: 10.1016/S0960-7404(99)00007-9

[224] Amiri M, Gholami T, Amiri O, et al. The magnetic inorganic-organic nanocomposite based on Zn Fe_2O_4-Imatinib-liposome for biomedical applications, in vivo and in vitro study. Journal of Alloys and Compounds. 2020;**849**:156604. DOI: 10.1016/j.jallcom.2020.156604

[225] Chauhan VP, Popović Z, Chen O, et al. Fluorescent nanorods and nanospheres for real-time in vivo probing of nanoparticle shape-dependent tumor penetration. Angewandte Chemie International Edition. 2011;**50**(48):11417-11420. DOI: 10.1002/anie.201104449

[226] Nie S. Understanding and overcoming major barriers in cancer nanomedicine. Nanomedicine. 2010;**5**(4):523-528. DOI: 10.2217/nnm.10.23

[227] Augusto PA, Castelo-Grande T, Vargas D, Pascual A, Hernandez L, Estevez AM, et al. Upscale design, process development, and economic analysis of industrial plants for nanomagnetic particle production for environmental and biomedical use. Materials. 2020;**13**(11):2477. DOI: 10.3390/ma13112477

[228] Abdullah M, Hasany S, Qureshi MA, Hussain S. Cost-effective synthesis of cobalt ferrite nanoparticles by sol-gel technique. Materials Science Forum. 2022;**1067**:213-219. DOI: 10.4028/p-jdlq11

Chapter 9

Transition Metal Chalcogenides for Wastewater Treatment

Sawaira Moeen, Iram Shahzadi, Ali Haider, Anum Shahzadi, Muhammad Ikram, Mohammed Benali Kanoun and Souraya Goumri-Said

Abstract

Rapid population expansion, industrialization, accelerated urbanization, and ineffective governmental policies in certain areas have heightened the expenditure of resources and diverse environmental practices contributing to water contamination. Several materials, including quantum dots, metal oxides, carbonaceous substances, and metal sulfides, have effectively eliminated benign contaminants from polluted water. Among these, transition metal chalcogenides (TMCs) have garnered attention owing to their appropriate bandgap energy, earth abundance, optoelectronic characteristics, and catalyst stability. In this chapter, the causes of water pollution and techniques for water purification, including chemical (precipitation, adsorption, and disinfection), physical (flocculation, sedimentation, filtration, and flotation), and biological (aerobic, anaerobic, and anoxic) methods, are thoroughly discussed. This chapter covers the conventional and advanced photocatalysis mechanisms of transition metal chalcogenides for wastewater purification. Lastly, strategies to improve the photocatalytic activity of transition metal chalcogenides are examined. In particular, the advantages of TMC-based photocatalysis over conventional treatment methods are elucidated. Critical considerations regarding the environmental impact of TMCs—such as the potential toxicity of cadmium-containing chalcogenides and the energy footprint of their synthesis—are evaluated alongside proposed solutions like safer catalyst designs and green synthesis techniques. The influence of environmental factors (pH, temperature, and contaminant levels) on the photocatalytic efficacy and stability of TMCs in realistic wastewater conditions is also assessed. Additionally, recent progress in applying artificial intelligence (AI) and machine learning (ML) to TMC development is highlighted, demonstrating how data-driven approaches can improve catalyst design, scalability, and long-term performance in photocatalytic wastewater treatment.

Keywords: transition metal chalcogenides, water purification, photocatalysis, doping, defect engineering, machine learning

1. Introduction

The sufficient supply of fresh and drinkable water has become a global concern, mainly owing to fast industrialization, high population growth rates, rising energy consumption, and significant lifestyle changes [1–3]. Essential sources of water pollution include human waste, toxic metal ions, and various cleaning chemicals and colors that adversely affect humans and aquatic life [4–6]. Saline seawater constitutes 97.5% of the world's water resource, while the remaining 2.5% is freshwater suitable for use by humans. Due to industrialization and various human activities, a growing array of contaminants, including dyes, pharmaceuticals, heavy metals, pesticides, detergents, and insecticides, contaminate water sources [7–11]. Specifically, previously unacknowledged public health hazards linked to the discharge of hormones, personal care items, medicines, and other chemical compounds into the water system are emerging, resulting in a heightened need for more stringent laws [12]. Numerous chemical and microbiological agents exhibit exceptional stability and long-term transport characteristics, rendering these pollutants nondegradable. The principal sources of these contaminants are commercial and agricultural waste, and industrial smoke, which may rapidly disseminate through water and air throughout different regions. **Figure 1** illustrates the schematic distribution of pollutants from sources to consumers. Emerging pollutants may also originate from drug waste and the

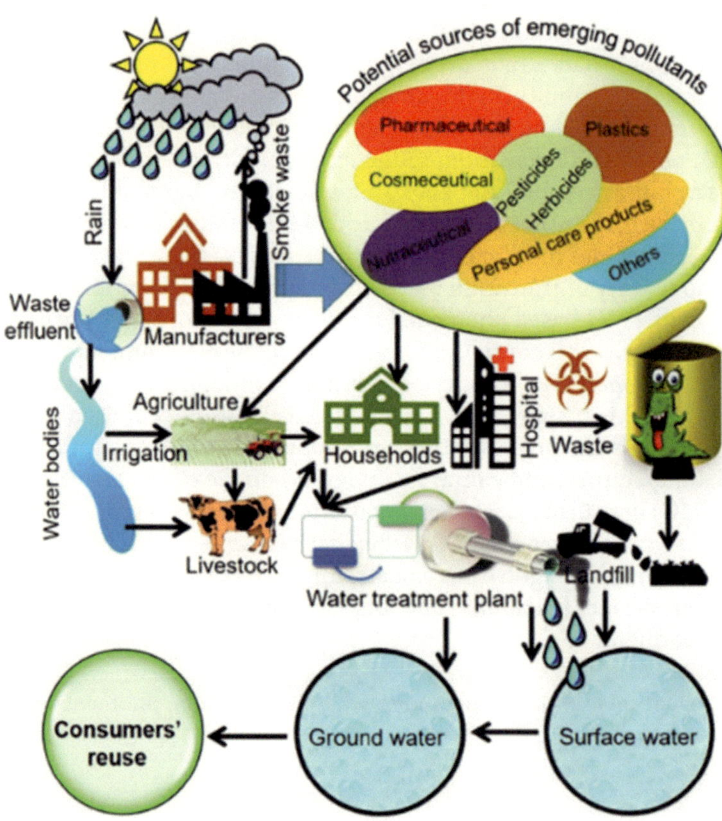

Figure 1.
Source-to-receptor transportation of contaminants [13].

breakdown of organic compounds in neutral areas [13]. Biological micropollutants may also be present in drinking water. The escalation of leaching processes resulting from the over-application of fertilizers in agricultural practices contributes to a rise in environmental pollutants in contaminated water [14]. A significant quantity of water contaminants remains to be addressed. Novel chemical compounds are constantly being formulated and launched into the market due to rapid industrialization. These novel compounds will ultimately infiltrate aquatic ecosystems. Moreover, pathogenic microorganisms transmitted by water are prevalent throughout. These potentially hazardous microorganisms enter streams via untreated storm drains, septic tanks, sewage, agricultural runoff, and other enterprises, including tanning and meatpacking facilities. Recently discovered poisons in the water may adversely affect the ecology of aquatic habitats and human health. Access to clean water to consume, free from hazardous contaminants and infectious viruses, is essential for human health [15–20]. Moreover, water supply contamination becomes a significant issue after a natural disaster. A diverse array of natural disasters, such as tsunamis, earthquakes, storms, floods, and volcanic eruptions, profoundly affect water quality. Therefore, it is necessary to obtain freshwater for human and aquatic life survival [11].

In response to environmental issues arising from water contamination, multiple scholars have committed their time to developing innovative water purification methods [21–23]. Numerous solutions have been devised to reduce wastewater effluents and alleviate the risks posed by contaminants, including chemical precipitation, adsorption, physical flocculation, sedimentation, filtration, and flotation, biological aerobic, anaerobic, and anoxic [24–26]. These techniques generally face substantial expenses due to the necessity of efficiently eliminating impurities to render the water fresh and recyclable. The typical treatment technologies that are accessible in the modern day do not eliminate the contaminants; rather, the pollutants are either accumulated or converted into another phase [27–29]. Among several strategies, photocatalytic activity (PCA) is regarded as one of the most effective ways for water purification due to certain features that are not as effective in other comparable approaches [30]. By contrast, photocatalytic oxidation has emerged as one of the most effective modern approaches for water purification [30] because it can achieve the complete mineralization of organic and inorganic pollutants into innocuous end-products using light energy. Unlike adsorption or precipitation methods that simply concentrate contaminants into sludge, photocatalysis can destroy pollutants in situ with minimal chemical additives, often harnessing solar irradiation as a driving force. Numerous researchers have focused their work on finding photocatalytic material possessing enormous active sites, cost-effective, low bandgap energy (E_g), and environmentally friendly [31, 32]. Recently, TMCs have garnered significant study attention owing to their adaptable chemical structure, adjustable bandgaps, abundance in the Earth's crust, absorbance, and catalytic stability. The term chalcogens (group XVI elements) derives from the fact that the majority of copper ores (Greek chalkos) are oxides or sulfides, which often comprise concentrations of Se and Te [33–36]. Senasu et al. reported the synthesis of a CdS catalyst for effective photocatalytic reduction (99%) of reactive red and Congo red dye [37]. Ayodhya et al. utilized CuS as the catalyst for photocatalytic degradation (90%) of methylene blue dye within 4 hours [38]. Kaviyarasu et al. reported a 70% photocatalytic reduction of rhodamine B dye using CuSe quantum dots within 6 hours [39]. This book chapter will critically assess the water purification techniques and conclude with the highly effective approach for contaminant removal using TMCs. Additionally, strategies to enhance the performance of TMCs were thoroughly discussed.

2. Techniques for water purification

Incremental developmental milestones are being achieved to establish new solutions for wastewater treatment and meet clean water standards [40]. Nonetheless, effectively treating discharged water containing contaminants with existing technologies has shown to be difficult [41]. A variety of strategies for the treatment of wastewater have been thoroughly documented in the literature. They generally encompass physical, chemical, and biological processes deemed sufficiently effective for water treatment in various manners, as seen in **Figure 2**. The unique characteristics of each treatment procedure may offer advantages in certain aspects while imposing limitations on others. Treatment methods that include substantial installation and operational expenses, prolonged processing durations, diminished output, and hazardous byproducts post-treatment are sometimes deemed less viable for industrial applications [23, 42]. Therefore, it is essential to identify an alternate treatment technique capable of fully degrading or eliminating pollutants [28].

2.1 Physical method

The mass transport strategy underpins the physical pollution removal techniques [43]. Its ease, flexibility, high effectiveness, and contaminant recyclability render it more likely to be utilized [44, 45]. The reduced chemical demand is another advantage of this approach. Physical treatment appears to be more dependable than alternative therapies as it is not reliant on live creatures. Recently, scientists have employed adsorption as a physical method owing to its great efficiency and cheap operational costs [23, 45, 46].

2.1.1 Adsorption

The adsorption technique is extensively employed in wastewater treatment owing to its outstanding efficacy, straightforward operation, and simplicity of regeneration [47].

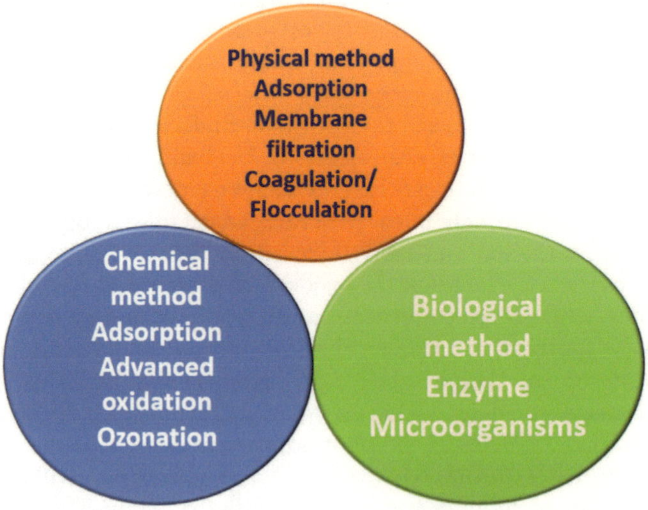

Figure 2.
Methods for wastewater treatment [23].

Adsorption can be used to eliminate heavy metals from wastewater, even at low concentrations. Consequently, adsorption is a pragmatic technique for application in wastewater treatment relative to alternative approaches. Adsorption procedures include several criteria to ascertain an efficient methodology, including pH, adsorbent dosage, temperature, and contact duration. The pH is crucial in the adsorption process; establishing the initial pH state is vital for enhancing pollutant removal efficacy. Because of the solution's pH, hydrogen ions in -COOH, -OH, and -NH functional groups interact with metal ions on the absorbent interface in different ways. The adsorbent dosage is a crucial component in the adsorption technique, as it influences the adsorbent capacity relative to the starting content of the adsorbate and is essential for establishing the optimal process [48, 49]. An increased adsorbent dosage may enhance removal capability owing to an expanded surface area and greater active adsorption sites [50]. The temperature is a characteristic that influences the method of adsorption by changing physicochemical processes. In endothermic processes, elevated temperatures enhance the process's reaction rate, but in exothermic reactions, elevated temperatures diminish the reaction rate. The adsorption process involves many sequential processes till its completion [51]. According to Anfar et al. [52], one of the phenomena that has received the most attention is adsorption by activated carbon. Numerous studies have demonstrated the efficacy of activated carbon and activated carbon composites as adsorbents for the removal of significant pollutants, such as dyes and heavy metals [52]. Because of their large specific surface area, decreased flocculant formation, and large number of accessible active sites for species adsorption, nanomaterials have recently attracted a lot of interest as adsorbents in wastewater remediation. These adsorbents are also very attractive and cost-effective due to their recyclable and reusable nature [53–56]. **Figure 3** illustrates the fundamental mechanism of pollutant adsorption onto the surfaces of TMCs as adsorbents.

2.1.2 Membrane filtration

The separation or elimination of dyes and organic substances using membrane filtration is one of the most efficient and cost-effective wastewater remediation methods across several sectors. Nonetheless, the advancement of membranes exhibiting adequate thermal durability and boosted efficacy remains a formidable challenge.

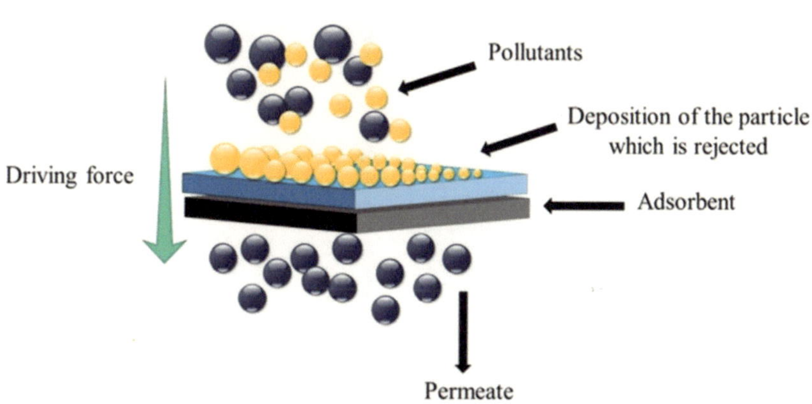

Figure 3.
Fundamental mechanism of pollutant adsorption on the surface of TMCs [35].

The hydrophilic ability and electrical conductivity of the membrane can influence its opposition, flexibility, and antifouling efficacy [57].

A membrane is a physical boundary that delineates two phases, creating a barrier to the transit of matter. Membranes possess small apertures that permit the passage of water molecules while obstructing bigger substances. Membrane filtration can be conducted as either dead-end or cross-flow filtration, as revealed in **Figure 4(a, b)**. In the first type of filtration, the water stream runs orthogonally to the membrane; all particles accumulate on the membrane interface during filtering and are subsequently detached by backwashing. The buildup of particles in dead-end filtration frequently leads to a reduced flux relative to the cross-flow type. In the second type of filtration, the influent water runs align to the membrane interface. The flow velocity over the membrane interface generates a shear stress that diminishes filter cake development. As most solids are retained with the retentate rather than accumulating on the outermost part of the membrane, the system may operate at elevated flux levels [58, 59].

Membrane processes may be categorized into two classes: (i) membrane filtration, which includes ultrafiltration and microfiltration, and (ii) reverse osmosis, which encompasses nanofiltration. Microfiltration membranes include substantial pore dimensions that exclude big particles and bacteria. Ultrafiltration membranes possess smaller holes than microfiltration membranes and can reject microorganisms and soluble macromolecules. Reverse osmosis membranes are predominantly nonporous and effectively reject particles and various low molecular weight species, including salt ions and tiny organic molecules. Nanofiltration membranes are regarded as "loose" reverse osmosis membranes, demonstrating performance characteristics that lie among ultrafiltration and reverse osmosis membranes [59–63].

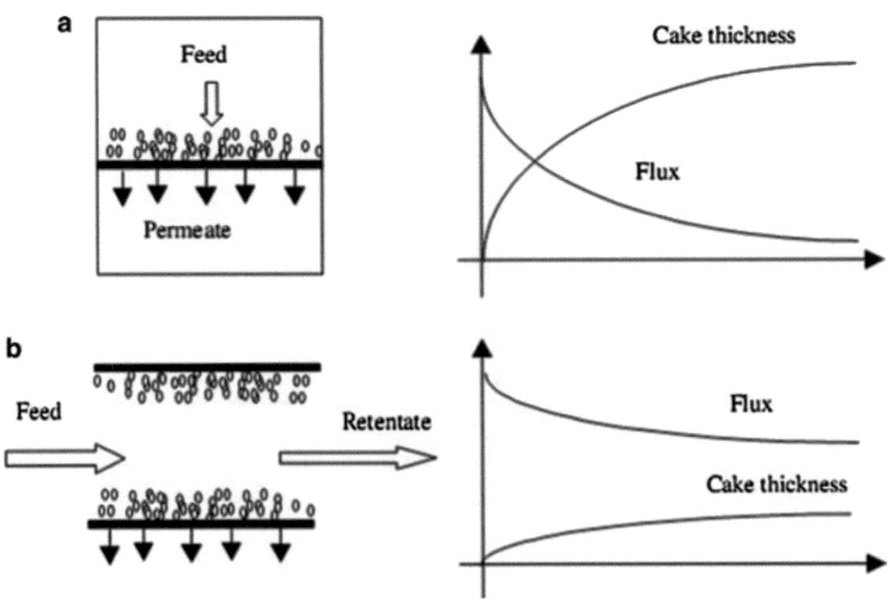

Figure 4.
Graphical representation of (a) dead-end and (b) cross-flow filtration.

2.1.3 Coagulation and flocculation

Coagulation and flocculation in contaminated water treatment entail the introduction of chemicals to modify the physical characteristics of absorbed and suspended particles, hence enhancing their removal by eventual sedimentation [64]. Many scholarly works utilize the terms coagulation and flocculation alternately and inconspicuously. Coagulation is often characterized as the instability of a suspension, resulting in aggregation. Conversely, flocculation refers to the process whereby destabilized particles are prompted to interact, creating bigger aggregates. The electrokinetic charge significantly affects the actions of colloidal fragments in water [65, 66]. Every particle possesses a similar charge, and most colloidal contaminants in water are negatively charged. In hydrophilic colloids, the principal charges predominantly arise from polar groups, including carboxylic and amine groups. Due to the dimensions of colloidal particles (about 0.01 to 1 μm), the attractive interactions among particles are less than the repulsive forces generated by their electrical charges. Consequently, these particles often remain distinct and distributed within the solution. Removing colloidal particles by settling is feasible only if they settle swiftly, about several hundred millimeters per hour [67, 68]. **Figure 5** illustrates the process, encompassing the sedimentation that transpires. Coagulation and flocculation employ coagulants such as $Ca(OH)_2$, $Al_2(SO_4)_3{\cdot}18H_2O$, $FeCl_3{\cdot}7H_2O$, and $Fe_2(SO_4)_3{\cdot}7H_2O$ to interact with contaminants and dispersed dyes, facilitating their removal via sorption, electrostatic forces, and bridging mechanisms. The sorption and bridging facilitate the coagulation and flocculation of colors and contaminants from wastewater, attributed to the hydrogenated amine groups and the elevated molecular weight polymer, respectively. Coagulation-flocculation diminishes dissolved compounds, coloring agents, non-settleable particles, and suspended matter in wastewater effluent [68–70].

2.2 Chemical methods

The elimination of persistent organic pollutants, such as pesticides and drugs, in contaminated water can be achieved via advanced oxidation processes (AOPs) or ozonation. These strategies eradicate such contaminants via mineralization or transform

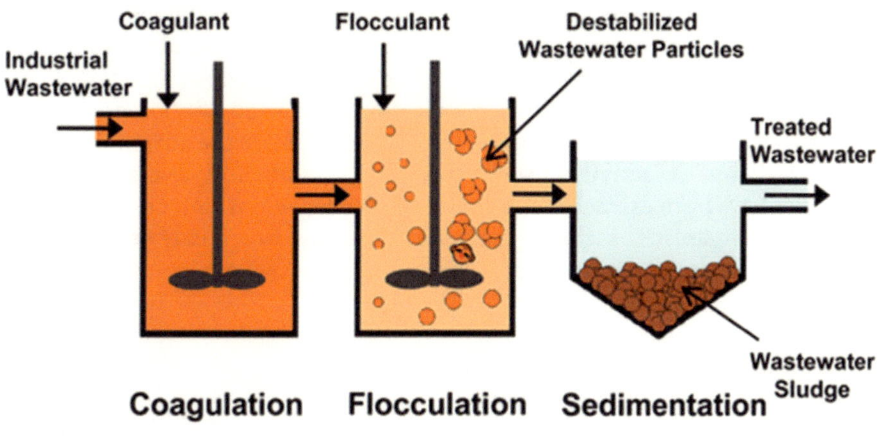

Figure 5.
Diagram illustrating the coagulation-flocculation treatment process [68].

them into less detrimental compounds to human beings and aquatic ecosystems. The aforementioned approach can serve as a preliminary for polluted water before biological processes, including activated sludge treatment. This technique may be ineffective in situations where organic and inorganic compounds predominate, as frequently observed with municipal polluted water, due to the extremely high oxidant requirements necessary for sufficient degradation of target contaminants at low concentrations. Consequently, it is probably more practical to use these methods as a tertiary approach following secondary processing. These are appropriate for treating potable water to eliminate potentially dangerous organic contaminants in source waters. During these processes, organic pollutants undergo a series of oxidation and naturally occurring reactions. In other terms, initial substances of degradation are frequently subjected to further deterioration. Effective treatment is not always indicated by the absence of parent substances because the degradation products may be just as toxic or have biological activity as the parent compounds. In order to guarantee the stability of the treated contaminated water, it is crucial to evaluate the residual toxicity or estrogenicity after treatment. These techniques produce breakdown byproducts that can be categorised based on their chemical composition, toxicity, and biodegradability [71–73].

2.3 Biological methods

In the second stage of treatment, microorganisms are used to break down the organic matter in the contaminated water, resulting in inorganic byproducts. The microorganisms are subsequently eliminated by the sedimentation process. The digestion of wastewater, both aerobically and anaerobically, is a part of the biological treatment. First, there is the activated growth process, containing a contact bed, trickling bed intermittent filter, and rotating biological contactor.

This method involves affixing biomass to an inert substrate, which can include porcelain, rock, slag, or plastic. Secondly, suspended growth methods involve aerated lagoons, sludge processes, and activated oxidation ponds. These techniques include converting organic waste by mechanically or naturally suspending bacteria. The water is sufficiently homogeneous for chemical treatment after the biological process removes the last of the organic and particulate particles from the physical method [74, 75]. When compared to physical or oxidation methods, the biological techniques for the complete reduction of textile wastewater have the following benefits: (a) reduced sludge generation; (b) cost-effectiveness; (c) reduced environmental sustainability; (d) the production of non-toxic metabolites or total mineralisation; and (e) reduced water usage (higher concentration or reduced dilution necessity). The efficacy of biological decomposition procedures relies on the adaptability of the chosen bacteria and the activity of their enzymes. Consequently, several microbes and enzymes have been extracted and tested for the decomposition of various colors. Different microorganisms, including bacteria, algae, and fungi, may decompose various dyes found in textile effluent [71].

3. Photocatalysis

The word photocatalyst is derived from two components: "photo," about photons, and "catalyst," which refers to a chemical that modifies the reaction rate in existence. Photocatalysts (PCs) are substances that alter the pace of a chemical process when exposed to light. PCA encompasses reactions in the existence of light and a material called a semiconductor. A PC is a semiconducting substrate that captures light and facilitates chemical reactions. PCA is a phenomenon wherein an e^--h^+ pair is produced when a semiconducting PC is illuminated [76, 77]. The process started with the photon absorption with wavelengths in the ultraviolet, visible, or infrared spectral ranges [78, 79]. It is an innovative energy-efficient process that directly transforms light energy into chemical energy, applicable for the elimination of persistent organic compounds. PCA is superior to existing methods because of its complete reduction of contaminants, clarity, lack of harmful byproducts, economics, and reduction of contaminants to ppm and ppb quantities [80–84]. Furthermore, PCA obviates all traditional wastewater treatment procedures since it represents a sustainable solution in the domain of environmental remediation [85].

3.1 Various mechanisms for photocatalysis

3.1.1 Conventional mechanism

Currently, there is a trend to clarify the complex physicochemical mechanisms of PCA using the perspective of energy band theory [86]. At absolute zero, the VB, characterized by low energy, is fully filled by e^-, but the CB, associated with higher energy, remains unoccupied. During the PCA, when the absorbed energy (hv) of the SC meets or exceeds the E_g, photogenerated e^- are excited from the VB to the CB, generating h^+ in the VB and resulting in the occupancy of e^- in the CB and h^+ in the VB of the SC (**Figure 6**) [88, 89]. Subsequently, oxidative h^+ and reductive e^- engage in the redox process on the SC surface, correspondingly [90]. To eliminate target contaminants, h^+ can directly oxidize them or react with H_2O/OH^- to produce $^\bullet OH$. Electrons generally grab O_2 to produce the $^\bullet O_2^-$, which, together with the produced $^\bullet OH$, participates in the breakdown of contaminants. In energy conversion, CO_2 or H^+ can be reduced to hydrocarbons or H_2, whereas H_2O/OH^- can be oxidized to O_2

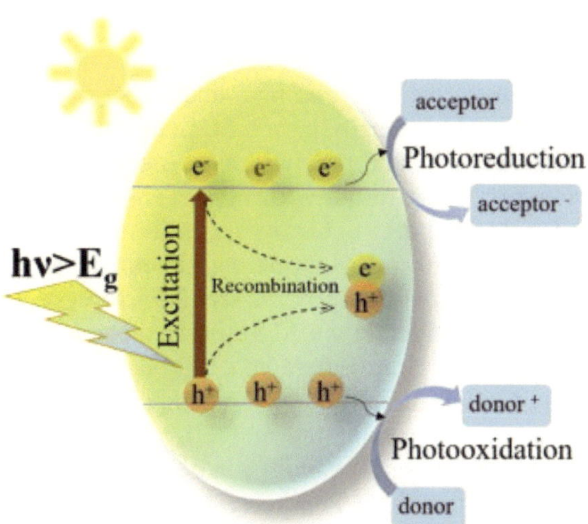

Figure 6.
Schematic representation of semiconductor photocatalytic mechanisms [87].

(Eqs. (1)–(5)). The critical process is the trapping of photogenerated e^- by O_2 to produce the $^{\bullet}O_2^-$, which not only prevents recombination of e^--h^+ but also extends the lifespan of the h^+ [91–94].

The production and recombination of e^--h^+ carriers in SC are the two principal variables that limit the PCA of PC. A single-component PC struggles to fulfill all requirements due to the propensity for recombination of e^--h^+ carriers in narrow band gap PC, while wide band gap PC faces challenges in generating sufficient photogenerated e^- and h^+, leading to suboptimal quantum efficiency and diminished PCA (Eqs. (1)–(5)) [93].

$$\text{Photocatalyst} + h\nu(\text{UV}) \rightarrow e^-_{cb} + h^+_{vb} \tag{1}$$

$$O_2 + e^-_{cb} \rightarrow {}^{\cdot}O_2^- \tag{2}$$

$$h^+_{vb} + H_2O \rightarrow {}^{\cdot}OH \tag{3}$$

$$\text{Dye} + h^+_{vb} \rightarrow \text{oxidation product} \tag{4}$$

$$\text{Dye} + e^-_{cb} \rightarrow \text{reduction product} \tag{5}$$

3.1.2 Advanced mechanism

The heterojunctions generated between two SC materials may be classified into three categories depending upon the alignment of their VB and CB: straddling-gap junctions or Type I, staggered-gap junctions or Type II, and broken-gap junctions or Type III, as seen in **Figure 7a–c**.

In Type I, e^- and h^+ accumulate in the SC with a smaller E_g, leading to a heightened possibility of interaction and consequently diminished PCA. In Type II, e^- and h^+ migrate to their corresponding smaller energy bands, with e^- congregating in photosystem II (PS II) and h^+ amassing in photosystem I (PS I). The PCA could be improved by promoting the physical separation of e^--h^+ carriers; however, the redox characteristics will be reduced when e^- and h^+ collect in bands with smaller energy potentials. In Type III, e^- and h^+ remain in their SC materials with no charge transfer, and thus, the two SCs work independently without synergy.

Furthermore, when two SC materials with Type II alignment (staggered band) are closely interfaced and transport of charges occurs among CB of e^- in PS II and the VB h^+ in PS I, as illustrated in **Figure 7d**, a direct Z-scheme is established. Despite the identical staggered band alignment of direct Z-scheme and Type II heterojunctions, e^- and h^+ migrate in opposing directions within these two configurations. Unlike Type II heterojunctions, where photogenerated e^- and h^+ migrate from regions of high potential to low potential, thereby compromising redox functions, a Z-scheme heterojunction facilitates the migration of e^- in PS II, which subsequently recombines with h^+ in PS I. This process retains high energetic h^+ at the VB of PS II and more energetic e^- at the CB of PS I for PCA. This heterojunction maintains the potent redox capacities of e^- and h^+ for involvement in PCA. Consequently, the PCA of direct Z-scheme photocatalysts generally exceeds that of Type II (staggered-gap) heterogeneous photocatalysts [94].

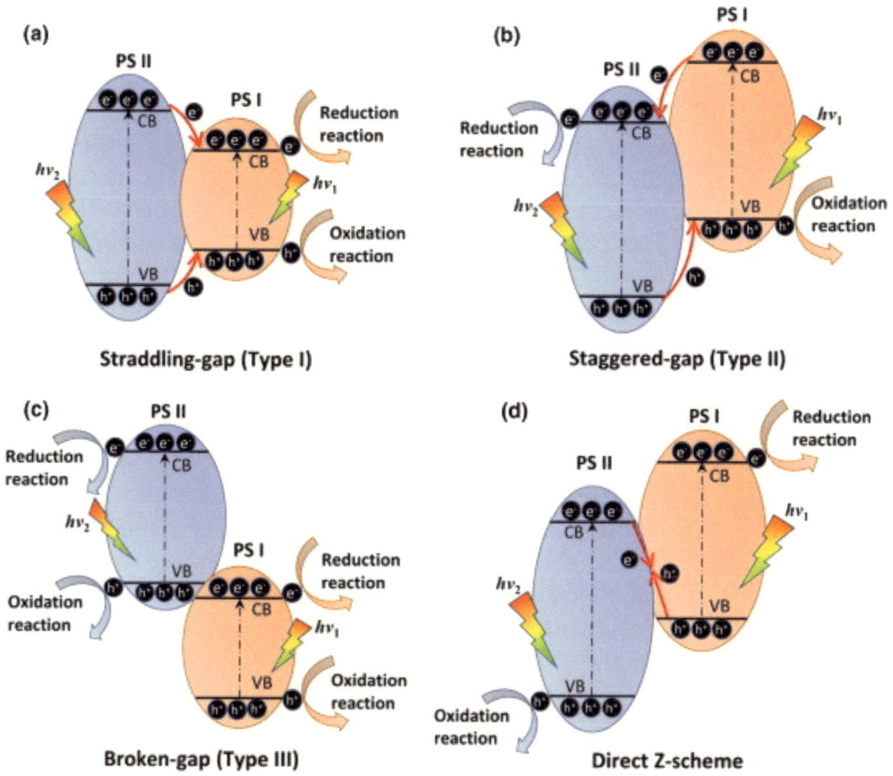

Figure 7.
Graphical representation of (a) Type I, (b) Type II, (c) Type III, and (d) direct Z-scheme [94].

4. Fundamental of transition metal chalcogenides

Chalcogenides belong to Group 16, which comprises O, S, Se, Te, and Po. Se and Te are classified as metalloid semiconductors, Po is classified as a metal, whereas oxygen and S are designated as nonmetals. TMCs encompass Te, Se, and S together referred to as M-tellurides, M-selenides, and M-sulfides (M = transition metals). TMCs have long been of interest, and with the advent of contemporary fabrication techniques and exfoliation procedures, it has been feasible to investigate and exploit the monolayers of these compounds [94]. Recent literature demonstrates the utilization of 2D-layered TMCs nanostructures owing to their superior physical and chemical characteristics. Each transition metal molecule is covalently bound to six chalcogenides (e.g., Te, S, and Se), whereas the individual layers are interconnected by van der Waals forces. Among various TMCs, 2D transition metal dichalcogenides (TMDs) provide an intriguing material resource. They present an appealing alternative to graphene (zero E_g) and possess the potential to surpass it in certain aspects, owing to their adjustable chemistry that encompasses true metals (e.g., NbS_2 VSe_2), semimetals (e.g., WTe_2, $TiSe_2$), semiconductors (e.g., WS_2, MoS_2), and insulators (e.g., HfS_2) [95, 96].

4.1 TMCs in photocatalytic reaction

Photocatalysts should have a large surface area, shape with larger active sites, anti-corrosion, adequate band potentials, and stability over time with narrow E_g [97].

The structure, arrangement, and wide range of metals, including d-orbitals in transition metals and the variety of TMCs, are crucial in the development of efficient metal chalcogenide photocatalysts [98]. Consequently, a diverse array of binary, ternary, and quaternary metal chalcogenide nanomaterials has demonstrated the majority of these attributes and, as a result, has been utilized in PCA [3]. Binary metal chalcogenides exhibit considerable diversity, typically including one metal and one chalcogenide in its formula. A diverse array of binary TMCs has been effectively produced and employed for photocatalysis. While chalcogenides dependent on alkali metals like NaX and KX (X = S, Se, and Te) show promise as photocatalysts, their substitutes, binary TMCs, have garnered a lot of interest. Consequently, TMCs (mostly S and Se) exhibit adjustable homogenous compositions, distinctive crystal structures, phase purity, and comparatively enhanced conductivity, which contribute to their appeal in PCA [31]. CdX chalcogenides have garnered significant interest as photocatalysts for hydrogen production under solar illumination, attributed to their notable E_g of 1.49 to 2.4 eV as bulk CdX [99]. Nevertheless, the E_g (**Figure 8**) has been documented to fluctuate with particle size, resulting in improved visible-light absorption. Consequently, QDs of diminutive dimensions and tunable E_g, namely CdX, have been shown to exhibit remarkable characteristics, including enhanced absorbance and effective separation of photogenerated e^- and h^+ for hydrogen evolution and CO_2 reduction, in addition to pollutant degradation. However, the disadvantage of utilizing both bulky and quantum dot CdX materials is their environmental toxicity if not

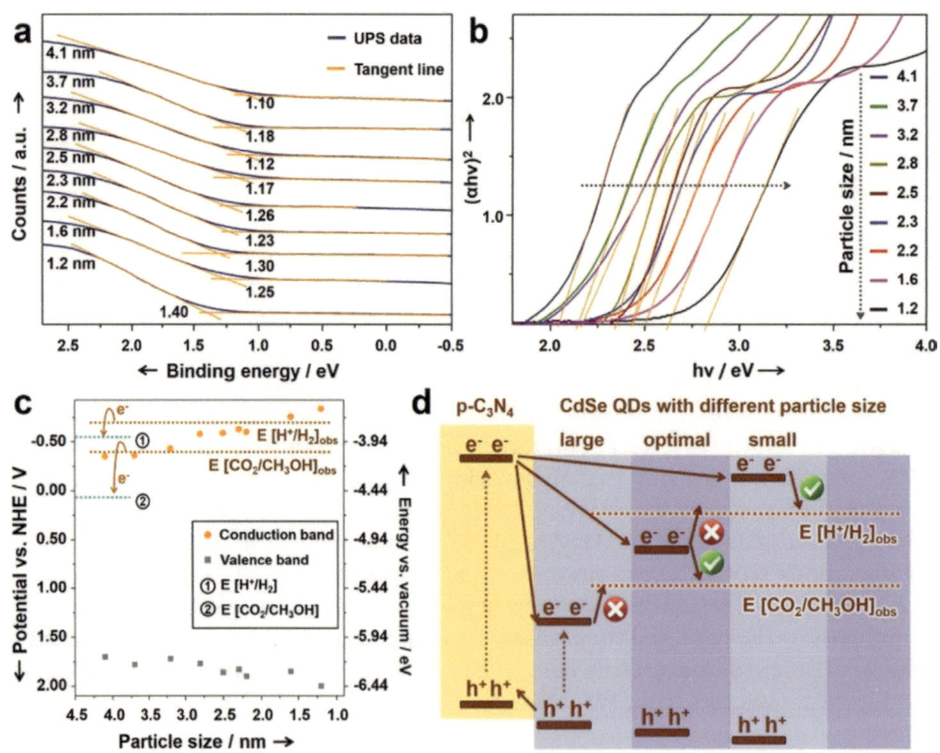

Figure 8.
(a) Ultraviolet photoelectron spectroscopy (UPS), (b) Tauc plots, (c) Photocatalytic CO_2 to CH_3OH and H/H_2O potential of CdSe at numerous particle sizes, and (d) PCA involved when p-C_3N_4 is combined with CdSe QDs [100].

well managed and recycled [31]. Consequently, with the use of C_3N_4, ZnS, and ZnSe, green-produced CdX QDs have been shown for hydrogen evolution, carbon dioxide reduction, and pollutant degradation.

MCSs are inorganic compounds of several component elements, with anion(s) belonging to the chalcogen group. Despite oxygen's classification under the chalcogen group, oxides are generally differentiated from chalcogenides owing to their distinct chemical properties. MCSc have garnered significant interest due to their markedly superior efficiency in photocatalytic processes relative to binary metal-based chalcogenides [101]. Consequently, a diverse array of MCSs has been synthesized as counterparts to binary metal-based chalcogenides, exhibiting enhanced photocatalytic capabilities attributed to synergistic interactions among the constituent components. The orbitals participating in the VB and CB of binary and MCSs exhibit distinct properties. The VB of quaternary Cu_2ZnSnS_4 comprises hybridized sulfide-p and copper-d orbitals, while the CB originates from the hybridization of sulfide-s/p and Sn-s orbitals, distinguishing it from binary TMCs like ZnS and CdS. Consequently, some MCSs, owing to their synergistic actions, impede the recombination of e^--h^+ pairs, enhance charge transfer potency, and mitigate photo corrosion under visible-light exposure [31]. However, the disadvantage of utilizing both bulky and quantum dot CdX materials is their environmental toxicity if not well managed and recycled. Consequently, with the use of C_3N_4, ZnS, and ZnSe, green-produced CdX QDs have been shown for hydrogen evolution, carbon dioxide reduction, and pollutant degradation. Ensuring that photocatalysts (especially in nanoparticle form) are recovered and recycled after use will be essential to avoid ecological harm.

4.2 Method to improve the photocatalytic activity of TMCS: Elemental doping

Elemental doping is a prevalent and effective technique for enhancing absorbance and redox capacity by modifying the surface properties of bulk TMCs. The potential enhanced mechanical flexibility in TMCs may lead to the incorporation of various size-dependent dopants inside the crystal relative to the bulk equivalent [102]. The possible increased mechanical flexibility in TMCs may provide improved incorporation of size-mismatched dopant ions within the crystal compared to the bulk counterpart. The dopant element can be either metal or nonmetal [103]. The nonmetal addition into TMCs often leads to a decrease in E_g, which is ascribed to overlapping among dopant orbitals and chalcogenide atoms [104]. Dopants induced the localized states within the E_g, resulting in a reduction of E_g [105]. In this scenario, the photogenerated e^- from the dopant gap state directly stimulates material CB, instead of the VB edge (**Figure 9a**). In metal doping, the dopant's ionic radius relative to the host metal cation determines occupied sites and electronic characteristics. They may influence the position of the Fermi level, surface properties, and the electrical conductivity inside the matrix. The nonmetallic element, including N, B, F, P, and C, was examined for doping TMCs. The doping process primarily involves the pre-treatment of precursors for synthesis, resulting in a low dopant concentration that typically induces little variations to the host crystal structure. Liu et al. reported the sol-gel synthesis of N-doped MoS_2 for photocatalytic reduction of RhB. The E_g of N-doped MoS_2 was lower than MoS_2, indicating improved light absorption for PCA (**Figure 9a**) [106]. In contrast, metal doping is a very rare method for the E_g engineering of TMCs. Li et al. reported the preparation of Mo-doped WO_3 nanosheets for RhB degradation. The ionic radius of Mo closely matched with W, allowing Mo ions to integrate into the host lattice without disturbing its crystal structure. A higher concentration of Mo doping

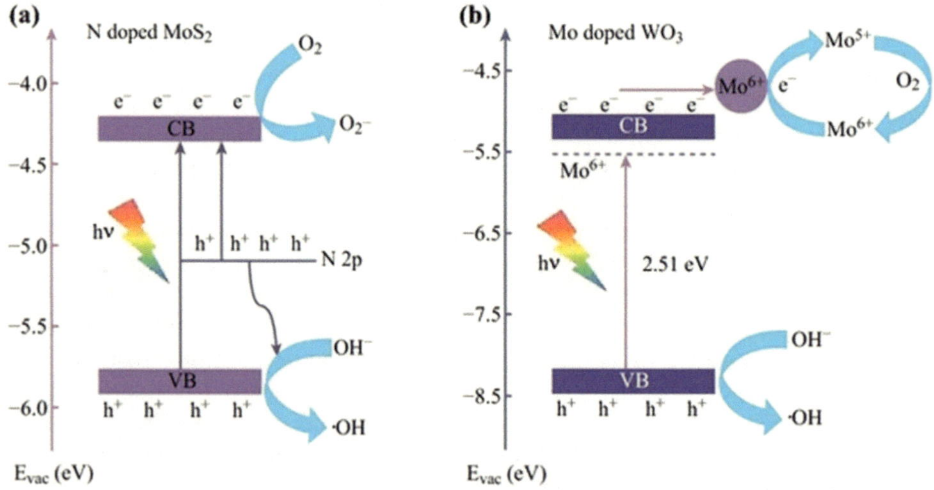

Figure 9.
Effect of elemental doping (a) N-doped MoS_2 and (b) Mo-doped WO_3 [106, 107].

established a donor level underneath the CB of WO_3, as well as decreased the E_g (2.56 to 2.36 eV), enhancing the absorbance of visible light for PCA (**Figure 9b**) [107].

5. Recent developments in transition metal chalcogenides for photocatalytic applications

Because of their strong visible-light absorption, earth abundance, high catalytic stability, and adjustable bandgap, transition metal chalcogenides (TMCs) have become increasingly popular in photocatalysis. Their photocatalytic performance has been further improved by recent developments in synthesis methods, structural alterations, and heterostructure engineering, increasing their viability for practical uses.

5.1 Nanostructured TMCs for enhanced photocatalysis

Nanostructuring has been very important in the last few years for making TMCs work better as photocatalysts. Two-dimensional (2D) transition metal dichalcogenides (TMCs), including MoS_2, WS_2 , and SnS_2, demonstrate enhanced photocatalytic activity attributable to their extensive surface area, distinctive electronic characteristics, and improved charge carrier mobility [108].

- Exfoliating bulk TMCs into monolayers or few-layered structures lowers charge recombination and raises the number of active sites. For example, monolayer MoS_2 has shown better photocatalytic hydrogen production than its bulk counterpart because it can break down excitons better and conduct electricity better [109].

- Quantum dot TMCs: Transition metal chalcogenide quantum dots (QDs) have come to light as good candidates for cleaning up wastewater. Because of quantum confinement effects, $MoSe_2$, CdSe, and WSe_2 quantum dots absorb more light in the visible range and separate charges better, which makes them better at breaking down pollutants through photocatalysis [110].

5.2 Defect engineering in 2D TMCs

Defect engineering, the deliberate introduction or manipulation of vacancies and other lattice defects, is another powerful approach to tune the bandgap and photocatalytic efficiency of 2D TMCs like $MoSe_2$ or WS_2. Creating chalcogenide vacancies (e.g., S or Se vacancies) or metal vacancies in the crystal lattice can introduce mid-gap electronic states that effectively narrow the bandgap or alter the band edge positions [110]. This bandgap modulation makes it possible for the material to absorb light over a wider range of wavelengths in the visible spectrum, which can help solar energy be used more effectively. For instance, the controlled creation of sulphur vacancies in monolayer MoS_2 has been demonstrated to redshift its photoluminescence peak (signifying a diminished effective bandgap), thus facilitating the absorption of lower-energy light. In $MoSe_2$, point defects like molybdenum vacancies were found to lower the bandgap by about 0.3 eV, which shows how defect states directly change the electronic structure.

Defects can also act as hot spots for catalysis. Vacancies frequently serve as adsorption sites, facilitating the stronger binding of pollutant molecules and enhancing their interaction with photogenerated charge carriers. Some defects can also trap electrons or holes in a way that keeps the charges from recombining right away, which makes it more likely that these charges will take part in surface reactions. For example, having the right amount of S-vacancies in MoS_2 or WS_2 can speed up photocatalytic hydrogen evolution and pollutant degradation by giving these processes active sites. However, having too many defects can be bad. For example, if there are too many intrinsic sulphur vacancies in 2D MoS_2, they can act as electron donors that create shallow trap states. This slows down the photoresponse and increases non-radiative recombination. pmc.ncbi.nlm.nih.gov. So, in defect engineering, the right balance must be found—there should be enough defects to make things happen, but not so many that the crystalline network becomes too disordered or the carriers get stuck.

There are many ways to add defects to a controlled manner:

- *Thermal annealing*: Mild annealing of TMC nanosheets in vacuum or inert atmospheres at temperatures below the material's decomposition point can cause chalcogen atoms to be lost, leaving vacancies. For instance, heating exfoliated MoS_2 to about 100°C (in a vacuum) creates sulphur vacancies, as shown by changes in the optical spectra [110]. You can change the vacancy concentration by changing the temperature and the amount of time.

- *Plasma and ion irradiation*: Ar plasma or focused ion beams can knock atoms out of the 2D lattice, creating vacancies in a way that can be controlled. For example, plasma-treated MoS_2 has a lot of S-vacancies and shows a change in bandgap and optical emission [110]. Ion irradiation enables the spatial patterning of defects and facilitates the engineering of defect profiles throughout a material.

- *Non-stoichiometric synthesis*: When chemicals are made (through hydrothermal, solvothermal, chemical vapour deposition, etc.), TMC crystals with built-in vacancies can be made by purposefully using a small amount of chalcogen precursor or adding reducing agents. Likewise, post-synthesis chemical etching (utilising suitable reactants that selectively eliminate chalcogen atoms) can produce defect-laden surfaces.

- *Electron beam irradiation*: A focused electron beam can be used to make holes or even patterns of defects in a TMC sheet with a transmission electron microscope. This method is very accurate, but it doesn't work very well for large amounts of data. It is, however, useful for basic research on defect properties.

Using these methods, scientists can deliberately add defects to 2D TMCs, which lets them change the band structure and surface chemistry of the materials. The main goal is to use defect sites to make photocatalytic performance better, either by increasing light absorption or by giving reactive centres, while minimising any negative effects on stability. Recent studies have utilised computational tools and AI-driven strategies to determine the optimal defect types and concentrations for enhancing photocatalytic activity in TMCs, illustrating the convergence of materials science and machine learning in this emerging field.

5.3 Heterostructure engineering and doping for improved photocatalytic performance

To further improve the photocatalytic performance of TMCs, researchers have focused on constructing heterostructures and introducing dopants to fine-tune their electronic properties.

- Type II and Z-scheme heterojunctions: The combination of TMCs with other semiconductors such as TiO_2, g-C_3N_4, and ZnO has led to significant improvements in charge separation efficiency and redox potential. For example, the MoS_2/g-C_3N_4 heterojunction has demonstrated enhanced visible-light-driven photocatalytic degradation of dyes due to efficient electron transfer and reduced recombination [111].

- Metal and nonmetal doping: Elemental doping with transition metals (e.g., Co, Fe, Ni) or nonmetals (e.g., N, P, S) can alter the band structure of TMCs, improving their optical absorption and carrier dynamics. Nitrogen-doped WS_2, for instance, has exhibited enhanced photocatalytic reduction of Cr(VI) due to the creation of mid-gap states that facilitate charge transfer [112].

5.4 Plasmonic enhancement and surface modifications

Surface plasmon resonance (SPR) effects have been explored to augment the light harvesting of TMC photocatalysts. By decorating TMCs with plasmonic metal nanoparticles (Ag, Au, etc.), one can exploit the intense local electromagnetic fields generated when the metal NPs resonate with incident light. For example, Ag-decorated MoS_2 has shown significantly improved photocatalytic hydrogen evolution and pollutant degradation rates, due to hot electron injection from the excited Ag into MoS_2 and the amplification of the local light intensity near the MoS_2 surface [113]. Plasmonic coupling extends the photoresponse of the system and provides additional electron-hole pairs via the metal's excitation.

Surface chemical modifications can also enhance TMC performance. Functionalizing TMC surfaces with organic molecules, polymers, or inorganic oxides can improve their dispersibility in water and stability against photocorrosion. For instance, coating MoS_2 with a thin layer of conductive polymer can protect it from oxidative degradation while facilitating charge transfer. Similarly, anchoring TMC nanostructures onto solid supports (like silica or alumina frameworks) helps in easy recovery of the catalyst and prevents aggregation. Such surface modifications are crucial for translating high laboratory efficiencies of TMCs into practical water treatment systems.

5.5 Machine learning and AI-driven material discovery for optimized photocatalysis

Artificial intelligence (AI) and machine learning (ML) have changed materials science by making it possible to quickly screen and find new photocatalysts. Using AI-driven methods to design and improve TMCs has sped up the creation of high-performance materials with specific properties. The bandgap of a material is one of the most important things that affects how well it works as a photocatalyst. Machine learning models trained on existing materials databases, like the Materials Project and the Open Quantum Materials Database (OQMD), can accurately predict the bandgap of new TMCs. Algorithms like Gaussian process regression and deep neural networks have been used to find new TMC compositions with the best bandgap values for photocatalysis with visible light [114]. High-throughput density functional theory (DFT) simulations, integrated with AI-driven workflows, have enabled the swift evaluation of thousands of TMC compositions. For example, a recent study used machine learning algorithms to look at more than 30,000 possible TMCs and found that $MoSe_2$-based heterostructures were very promising candidates for photocatalytic water splitting [115]. As the use of AI in materials development grows, researchers are also using these tools to solve real-world problems with photocatalysts, such as how to make them work on a larger scale and last a long time.

You can use data from durability tests and environmental simulations to train ML models to guess how different TMC compositions or nanostructures will hold up over time.

Making TMC photocatalysts in the lab can take a long time and cost a lot of money. AI-based methods like reinforcement learning and Bayesian optimisation have been used to fine-tune synthesis parameters like temperature, precursor concentration, and annealing time in order to get the best crystallinity, defect engineering, and phase control. For instance, an AI-optimized synthesis of WS_2 nanoparticles led to a 30% increase in photocatalytic efficiency compared to traditional synthesis methods [116–118].

5.6 Comparison with other materials

TMCs have shown great promise in photocatalysis, but other advanced materials like perovskites, metal-organic frameworks (MOFs), and hybrid nanocomposites are also very important for treating wastewater. **Table 1** below shows how these materials are better, worse, and more useful than TMCs:

TMCs are better at separating charges and staying stable, but they have problems with oxidation and toxicity. Perovskites are very efficient, but they are not very stable. MOFs, on the other hand, are very good at adsorbing but not very good at conducting electricity.

Hybrid nanocomposites, which combine different materials, offer the best of both worlds, but they need advanced synthesis methods.

Material type	Advantages	Limitations	Applications
TMCs	Tunable bandgap, high stability, excellent charge separation	Susceptibility to oxidation, potential toxicity of some compounds	Photocatalytic degradation of dyes, Cr(VI) reduction, hydrogen evolution
Perovskites	High light absorption, long charge carrier lifetime	Instability in aqueous environments, lead toxicity (Pb-based perovskites)	Solar-driven photocatalysis, water splitting
MOFs	High surface area, tunable pore structure	Low conductivity, stability issues under harsh conditions	Adsorption, photocatalytic degradation of organic pollutants
Hybrid nanocomposites	Synergistic properties from multiple materials, enhanced charge separation	Complex synthesis, potential agglomeration issues	Visible-light photocatalysis, enhanced pollutant degradation

Table 1.
Comparison of advanced materials for photocatalytic and environmental applications.

6. Conclusion and future perspective

Transition metal chalcogenides (TMCs) have shown great promise as materials for treating wastewater because of their amazing photoelectrochemical, structural, optical, and morphological properties. They are very good for catalytic and adsorption-based remediation techniques because they are not very toxic and have a narrow bandgap (E_g).

Photocatalytic activity (PCA) has shown a lot of promise as a treatment method because it is cheap, good for the environment, and easy to use. Improving the performance of TMCs by adding metals and nonmetals, making composites, and changing the surface has made them even better at breaking down organic pollutants and speeding up redox reactions. Even with these improvements, there are still some problems that need to be solved before these materials can be used on a large scale. To make sure that wastewater treatment is effective and long-lasting, it is important to deal with problems like long-term stability, charge carrier recombination, and material degradation. Also, thorough toxicity tests must be done to stop secondary contaminants from getting into the environment. For commercial success and widespread use, it will be important to create synthesis methods that can be used on a large scale and are not too expensive. Recent advances in nanostructuring, heterojunction engineering, plasmonic enhancement, and defect tuning have greatly enhanced the photocatalytic properties of TMCs. The combination of artificial intelligence (AI) and machine learning (ML) has also changed the way materials are discovered by making high-throughput screening, predictive modelling, and automated synthesis optimisation possible.

The combination of experimental research and computational methods is expected to speed up the creation of next-generation photocatalysts, which will lead to water purification technologies that are very efficient, scalable, and long-lasting.

Author details

Sawaira Moeen[1], Iram Shahzadi[2], Ali Haider[3], Anum Shahzadi[4], Muhammad Ikram[1], Mohammed Benali Kanoun[5] and Souraya Goumri-Said[6]*

1 Advance Nanomaterials Research Lab, Department of Physics, Government College University Lahore, Pakistan

2 School of Pharmacy, University of Management and Technology, Lahore, Pakistan

3 Department of Clinical Sciences, Faculty of Veterinary and Animal Sciences, Muhammad Nawaz Shareef, University of Agriculture, Multan, Punjab, Pakistan

4 Department of Pharmacy, COMSATS University Islamabad, Lahore, Pakistan

5 Department of Mathematics and Sciences, College of Humanities and Sciences, Prince Sultan University, Riyadh, Saudi Arabia

6 Department of Physics, College of Science and General Studies, Alfaisal University, Riyadh, Saudi Arabia

*Address all correspondence to: sosaid@alfaisal.edu

IntechOpen

© 2025 The Author(s). Licensee IntechOpen. This chapter is distributed under the terms of the Creative Commons Attribution License (http://creativecommons.org/licenses/by/4.0), which permits unrestricted use, distribution, and reproduction in any medium, provided the original work is properly cited.

References

[1] Kato S, Kansha Y. Comprehensive review of industrial wastewater treatment techniques. Environmental Science and Pollution Research. 2024;**31**(39):51064-51097

[2] Meena M et al. Wastewater treatment techniques: An introduction. In: Removal of Emerging Contaminants through Microbial Processes. Singapore: Springer; 2021. pp. 161-182

[3] Rezai B, Allahkarami E. Wastewater treatment processes—Techniques, technologies, challenges faced, and alternative solutions. In: Soft Computing Techniques in Solid Waste and Wastewater Management. Netherlands: Elsevier; 2021. pp. 35-53

[4] Ajmal A et al. Principles and mechanisms of photocatalytic dye degradation on TiO_2 based photocatalysts: A comparative overview. RSC Advances. 2014;**4**(70):37003-37026

[5] Meybeck M. Global analysis of river systems: From Earth system controls to Anthropocene syndromes. Philosophical Transactions of the Royal Society of London. Series B: Biological Sciences. 2003;**358**(1440):1935-1955

[6] Selvaraj V et al. An over review on recently developed techniques, mechanisms and intermediate involved in the advanced azo dye degradation for industrial applications. Journal of Molecular Structure. 2021;**1224**:129195

[7] Mishra RK. Fresh water availability and its global challenge. British Journal of Multidisciplinary and Advanced Studies: Sustainability. 2023;**4**(3):1-78. DOI: 10.37745/bjmas.2022.0208

[8] Routoula E, Patwardhan SV. Degradation of anthraquinone dyes from effluents: A review focusing on enzymatic dye degradation with industrial potential. Environmental Science & Technology. 2020;**54**(2):647-664

[9] Wang H et al. Comparison of dye degradation efficiency using ZnO powders with various size scales. Journal of Hazardous Materials. 2007;**141**(3):645-652

[10] Lorimer J et al. Degradation of dye effluent. Pure and Applied Chemistry. 2001;**73**(12):1957-1968

[11] Rafiq A et al. Photocatalytic degradation of dyes using semiconductor photocatalysts to clean industrial water pollution. Journal of Industrial and Engineering Chemistry. 2021;**97**:111-128

[12] Schwarzenbach RP et al. The challenge of micropollutants in aquatic systems. Science. 2006;**313**(5790):1072-1077

[13] Rasheed T et al. Environmentally-related contaminants of high concern: Potential sources and analytical modalities for detection, quantification, and treatment. Environment International. 2019;**122**:52-66

[14] Piai L et al. Biodegradation and adsorption of micropollutants by biological activated carbon from a drinking water production plant. Journal of Hazardous Materials. 2020;**388**:122028

[15] Cooper C. Biological effects of agriculturally derived surface water pollutants on aquatic systems—A review. Journal of Environmental Quality. 1993;**22**(3):402-408

[16] Malik D et al. A review on impact of water pollution on freshwater fish

species and their aquatic environment. In: Advances in Environmental Pollution Management: Wastewater Impacts and Treatment Technologies. Vol. 1. Haridwar (Uttarakhand), India: Agro Environ Media, Publication Cell Agriculture and Environmental Science Academy; 2020. pp. 10-28

[17] Qadri R, Faiq MA. Freshwater pollution: Effects on aquatic life and human health. In: Fresh Water Pollution Dynamics and Remediation. Springer; 2020. pp. 15-26

[18] Halder JN, Islam MN. Water pollution and its impact on the human health. Journal of Environment and Human. 2015;**2**(1):36-46

[19] Münzel T, Hahad O, Lelieveld J, Aschner M, Nieuwenhuijsen MJ, Landrigan PJ, et al. Soil and water pollution and cardiovascular disease. Nature Reviews Cardiology. 2025;**22**(2):71-89

[20] Wang Q, Yang Z. Industrial water pollution, water environment treatment, and health risks in China. Environmental Pollution. 2016;**218**:358-365

[21] Nassar MY, Ali EI, Zakaria ES. Tunable auto-combustion preparation of TiO_2 nanostructures as efficient adsorbents for the removal of an anionic textile dye. RSC Advances. 2017;**7**(13):8034-8050

[22] Crini G, Lichtfouse E. Advantages and disadvantages of techniques used for wastewater treatment. Environmental Chemistry Letters. 2019;**17**:145-155

[23] Rashid R et al. A state-of-the-art review on wastewater treatment techniques: The effectiveness of adsorption method. Environmental Science and Pollution Research. 2021;**28**:9050-9066

[24] Xu Z et al. Development of a novel mixed titanium, silver oxide polyacrylonitrile nanofiber as a superior adsorbent and its application for MB removal in wastewater treatment. Journal of the Brazilian Chemical Society. 2018;**29**(3):560-571

[25] Kamarudin NS et al. Potential nanomaterials application in wastewater treatment: Physical, chemical and biological approaches. Materials Today: Proceedings. 2021;**42**:107-114

[26] Shukla H, Gautam R, Kumari N. A comprehensive review: Photodegradation of dyes with rare earth doped metal oxide nanoparticles for wastewater treatment. Journal of Physics and Chemistry of Solids. 2025:112593

[27] Khan AH et al. Application of advanced oxidation processes followed by different treatment technologies for hospital wastewater treatment. Journal of Cleaner Production. 2020;**269**:122411

[28] Yahya N et al. A review of integrated photocatalyst adsorbents for wastewater treatment. Journal of Environmental Chemical Engineering. 2018;**6**(6):7411-7425

[29] Slatni I et al. Mesoporous silica synthesized from natural local kaolin as an effective adsorbent for removing of Acid Red 337 and its application in the treatment of real industrial textile effluent. Environmental Science and Pollution Research. 2020;**27**:38422-38433

[30] Al-Mamun M et al. Photocatalytic activity improvement and application of UV-TiO_2 photocatalysis in textile wastewater treatment: A review. Journal of Environmental Chemical Engineering. 2019;**7**(5):103248

[31] Munonde TS, Nomngongo PN. Review on metal chalcogenides and metal

chalcogenide-based nanocomposites in photocatalytic applications. Chemistry Africa. 2023;**6**(3):1127-1143

[32] Singh P et al. Nanomaterials photocatalytic activities for waste water treatment: A review. Environmental Science and Pollution Research. 2022;**29**(46):69294-69326

[33] Ejeromedoghene O et al. Environmental decontamination using transition metal dichalcogenides based materials: A review. Journal of Materials & Environmental Sustainability Research. 2022;**2**(1):1-18

[34] Priya A, Sagadevan S. Chalcogenides-based nanomaterials for contaminant removal in wastewater treatment. In: Structural and Chemical Features of Chalcogenides. IntechOpen; 2024

[35] Yadav S, Yashas SR, Shivaraju HP. Transitional metal chalcogenide nanostructures for remediation and energy: A review. Environmental Chemistry Letters. 2021;**19**(5):3683-3700

[36] Brammer L. Halogen bonding, chalcogen bonding, pnictogen bonding, tetrel bonding: Origins, current status and discussion. Faraday Discussions. 2017;**203**:485-507

[37] Senasu T, Hemavibool K, Nanan S. Hydrothermally grown CdS nanoparticles for photodegradation of anionic azo dyes under UV-visible light irradiation. RSC Advances. 2018;**8**(40):22592-22605

[38] Ayodhya D et al. Photocatalytic degradation of dye pollutants under solar, visible and UV lights using green synthesised CuS nanoparticles. Journal of Experimental Nanoscience. 2016;**11**(6):418-432

[39] Kaviyarasu K et al. Solution processing of CuSe quantum dots: Photocatalytic activity under RhB for UV and visible-light solar irradiation. Materials Science and Engineering B. 2016;**210**:1-9

[40] Ahmed SN, Haider W. Heterogeneous photocatalysis and its potential applications in water and wastewater treatment: A review. Nanotechnology. 2018;**29**(34):342001

[41] Nguyen CH, Juang R-S. Efficient removal of cationic dyes from water by a combined adsorption-photocatalysis process using platinum-doped titanate nanomaterials. Journal of the Taiwan Institute of Chemical Engineers. 2019;**99**:166-179

[42] Wawrzkiewicz M, Polska-Adach E, Hubicki Z. Application of titania based adsorbent for removal of acid, reactive and direct dyes from textile effluents. Adsorption. 2019;**25**:621-630

[43] Samsami S et al. Recent advances in the treatment of dye-containing wastewater from textile industries: Overview and perspectives. Process Safety and Environmental Protection. 2020;**143**:138-163

[44] Foroutan R et al. Characteristics and performance of Cd, Ni, and Pb bio-adsorption using Callinectes sapidus biomass: Real wastewater treatment. Environmental Science and Pollution Research. 2019;**26**:6336-6347

[45] Chai WS et al. A review on conventional and novel materials towards heavy metal adsorption in wastewater treatment application. Journal of Cleaner Production. 2021;**296**:126589

[46] Wong JKH et al. Potential and challenges of enzyme incorporated nanotechnology in dye wastewater

treatment: A review. Journal of Environmental Chemical Engineering. 2019;**7**(4):103261

[47] Anfar Z et al. Microwave assisted green synthesis of Fe_2O_3/biochar for ultrasonic removal of nonsteroidal anti-inflammatory pharmaceuticals. RSC Advances. 2020;**10**(19):11371-11380

[48] Huang W, Liu Z-m. Biosorption of Cd (II)/Pb (II) from aqueous solution by biosurfactant-producing bacteria: Isotherm kinetic characteristic and mechanism studies. Colloids and Surfaces B: Biointerfaces. 2013;**105**:113-119

[49] Afroze S, Sen TK, Ang HM. Adsorption removal of zinc (II) from aqueous phase by raw and base modified Eucalyptus sheathiana bark: Kinetics, mechanism and equilibrium study. Process Safety and Environmental Protection. 2016;**102**:336-352

[50] Kumar A et al. Intensive insight into the enhanced utilization of rice husk ash: Abatement of rice mill wastewater and recovery of silica as a value added product. Ecological Engineering. 2016;**91**:270-281

[51] Xu J et al. A review of functionalized carbon nanotubes and graphene for heavy metal adsorption from water: Preparation, application, and mechanism. Chemosphere. 2018;**195**:351-364

[52] Anfar Z et al. Recent trends on numerical investigations of response surface methodology for pollutants adsorption onto activated carbon materials: A review. Critical Reviews in Environmental Science and Technology. 2020;**50**(10):1043-1084

[53] Yang C et al. Fabrication and characterization of a high performance polyimide ultrafiltration membrane for dye removal. Journal of Colloid and Interface Science. 2020;**562**:589-597

[54] Sadegh H et al. The role of nanomaterials as effective adsorbents and their applications in wastewater treatment. Journal of Nanostructure in Chemistry. 2017;**7**:1-14

[55] Tan KB et al. Adsorption of dyes by nanomaterials: Recent developments and adsorption mechanisms. Separation and Purification Technology. 2015;**150**:229-242

[56] Sahoo TR, Prelot B. Adsorption processes for the removal of contaminants from wastewater: The perspective role of nanomaterials and nanotechnology. In: Nanomaterials for the Detection and Removal of Wastewater Pollutants. Netherlands: Elsevier; 2020. pp. 161-222

[57] Wiesner MR, Chellam S. Mass transport considerations for pressure-driven membrane processes. Journal-American Water Works Association. 1992;**84**(1):88-95

[58] Li K. Ceramic Membranes for Separation and Reaction. United States: John Wiley & Sons; 2007

[59] Liang R et al. Fundamentals on adsorption, membrane filtration, and advanced oxidation processes for water treatment. In: Nanotechnology for Water Treatment and Purification. Springer; 2014. pp. 1-45

[60] Hube S et al. Direct membrane filtration for wastewater treatment and resource recovery: A review. Science of the Total Environment. 2020;**710**:136375

[61] Yuan K, Chen R, Zeng Y. Fabrication of composite membrane by constructing helical carbon nanotubes in ceramic

support channels for efficient emulsion separation. Membranes. 2025;**15**(5):150

[62] Abdel-Fatah MA. Nanofiltration systems and applications in wastewater treatment. Ain Shams Engineering Journal. 2018;**9**(4):3077-3092

[63] Bartels CR et al. Design considerations for wastewater treatment by reverse osmosis. Water Science and Technology. 2005;**51**(6-7):473-482

[64] Alexander JT, Hai FI, Al-Aboud TM. Chemical coagulation-based processes for trace organic contaminant removal: Current state and future potential. Journal of Environmental Management. 2012;**111**:195-207

[65] Gregory J. Particles in Water: Properties and Processes. United States: CRC Press; 2005

[66] Bratby J. Coagulation and Flocculation in Water and Wastewater Treatment. United Kingdom: IWA Publishing; 2016

[67] Press C. Physical–Chemical Treatment of Water and Wastewater. CRC Press; 2000

[68] Teh CY et al. Recent advancement of coagulation–flocculation and its application in wastewater treatment. Industrial & Engineering Chemistry Research. 2016;**55**(16):4363-4389

[69] Zhao C et al. Application of coagulation/flocculation in oily wastewater treatment: A review. Science of the Total Environment. 2021;**765**:142795

[70] Abujazar MSS et al. Recent advancement in the application of hybrid coagulants in coagulation-flocculation of wastewater: A review. Journal of Cleaner Production. 2022;**345**:131133

[71] Holkar CR et al. A critical review on textile wastewater treatments: Possible approaches. Journal of Environmental Management. 2016;**182**:351-366

[72] Ikehata K, Gamal El-Din M, Snyder SA. Ozonation and advanced oxidation treatment of emerging organic pollutants in water and wastewater. Ozone: Science & Engineering. 2008;**30**(1):21-26

[73] Rekhate CV, Srivastava JK. Recent advances in ozone-based advanced oxidation processes for treatment of wastewater—A review. Chemical Engineering Journal Advances. 2020;**3**:100031

[74] Raju S et al. Improved methodology to determine the fate and transport of microplastics in a secondary wastewater treatment plant. Water Research. 2020;**173**:115549

[75] Kumar R, Sarmah AK, Padhye LP. Fate of pharmaceuticals and personal care products in a wastewater treatment plant with parallel secondary wastewater treatment train. Journal of Environmental Management. 2019;**233**:649-659

[76] Ameta R et al. Chapter 6—Photocatalysis. In: Ameta SC, Ameta R, editors. Advanced Oxidation Processes for Waste Water Treatment. Netherlands: Elsevier; 2018. pp. 135-175

[77] Nagajyothi PC et al. Green synthesis: Photocatalytic degradation of textile dyes using metal and metal oxide nanoparticles-latest trends and advancements. Critical Reviews in Environmental Science and Technology. 2020;**50**(24):2617-2723

[78] Yousif E, Haddad R. Photodegradation and photostabilization of polymers, especially polystyrene: Review. Springerplus. 2013;**2**(1):398

[79] Ali S, Ismail PM, Khan M, Dang A, Ali S, Zada A, et al. Charge transfer in TiO 2-based photocatalysis: Fundamental mechanisms to material strategies. Nanoscale. 2024;**16**(9):4352-4377

[80] Burrows HD et al. Reaction pathways and mechanisms of photodegradation of pesticides. Journal of Photochemistry and Photobiology B: Biology. 2002;**67**(2):71-108

[81] Sindelar FW et al. Treatment of effluent from the agate dyeing industry using Photodegradation and electrodialysis processes. Separation Science and Technology. 2015;**50**(1):142-147

[82] Ayodhya D, Veerabhadram G. A review on recent advances in photodegradation of dyes using doped and heterojunction based semiconductor metal sulfide nanostructures for environmental protection. Materials Today Energy. 2018;**9**:83-113

[83] Pouretedal HR, Kiyani M. Photodegradation of 2-nitrophenol catalyzed by CoO, CoS and CoO/CoS nanoparticles. Journal of the Iranian Chemical Society. 2014;**11**(1):271-277

[84] Zheng Y et al. Microscale flower-like magnesium oxide for highly efficient photocatalytic degradation of organic dyes in aqueous solution. RSC Advances. 2019;**9**(13):7338-7348

[85] Ahmed S et al. Advances in heterogeneous photocatalytic degradation of phenols and dyes in wastewater: A review. Water, Air, & Soil Pollution. 2011;**215**(1):3-29

[86] Fox MA, Dulay MT. Heterogeneous photocatalysis. Chemical Reviews. 1993;**93**(1):341-357

[87] Huang D et al. Artificial Z-scheme photocatalytic system: What have been done and where to go? Coordination Chemistry Reviews. 2019;**385**:44-80

[88] Ida S et al. Potential gradient and photocatalytic activity of an ultrathin p–n junction surface prepared with two-dimensional semiconducting nanocrystals. Journal of the American Chemical Society. 2014;**136**(5):1872-1878

[89] Xue W et al. Performance and toxicity assessment of nanoscale zero valent iron particles in the remediation of contaminated soil: A review. Chemosphere. 2018;**210**:1145-1156

[90] Abe R et al. Photocatalytic overall water splitting under visible light by TaON and WO_3 with an IO_3^-/I^- shuttle redox mediator. Chemical Communications. 2005;**30**:3829-3831

[91] Sasaki Y et al. Solar water splitting using powdered photocatalysts driven by Z-schematic interparticle electron transfer without an electron mediator. The Journal of Physical Chemistry C. 2009;**113**(40):17536-17542

[92] Zeng Z et al. Insight into the charge transport correlation in Au_x clusters and graphene quantum dots deposited on TiO_2 nanotubes for photoelectrochemical oxygen evolution. Journal of Materials Chemistry A. 2018;**6**(24):11154-11162

[93] Zhou C et al. Highly porous carbon nitride by supramolecular preassembly of monomers for photocatalytic removal of sulfamethazine under visible light driven. Applied Catalysis B: Environmental. 2018;**220**:202-210

[94] Li X et al. A review of material aspects in developing direct Z-scheme photocatalysts. Materials Today. 2021;**47**:75-107

[95] Saini M et al. 2D metal chalcogenides for photodegradation and disinfection

of wastewater pollutants. In: Advanced Two-Dimensional Nanomaterials for Environmental and Sensing Applications. United States: CRC Press; 2024. pp. 183-204

[96] Boeurn O et al. Assessment of electrochemical removal mechanisms of diverse transition metal dichalcogenides for caesium ion removal in wastewater treatment. 2D Materials. 2024;**11**(4):045020

[97] Ifires M et al. One-step electrodeposition of superhydrophobic NiO-Co$(OH)_2$ urchin-like structures on Si nanowires as photocatalyst for RhB degradation under visible light. Journal of Alloys and Compounds. 2019;**774**:908-917

[98] Han C. Synthesis of Nanostructured Metal Chalcogenides Used for Energy Conversion and Storage [PhD dissertation]. University of Wollongong; 2015

[99] Xiang X et al. Cadmium chalcogenide (CdS, CdSe, CdTe) quantum dots for solar-to-fuel conversion. Advanced Photonics Research. 2022;**3**(11):2200065

[100] Li A et al. Adjusting the reduction potential of electrons by quantum confinement for selective photoreduction of CO_2 to methanol. Angewandte Chemie International Edition. 2019;**58**(12):3804-3808

[101] Khan MD, Revaprasadu N. Metal–Organic Precursors for Ternary and Quaternary Metal Chalcogenide Nanoparticles and Thin Films. RSC; 2020

[102] Zhang H. Ultrathin two-dimensional nanomaterials. ACS Nano. 2015;**9**(10):9451-9469

[103] Zhang H, Chen G, Bahnemann DW. Photoelectrocatalytic materials for environmental applications. Journal of Materials Chemistry. 2009;**19**(29):5089-5121

[104] Asahi R et al. Visible-light photocatalysis in nitrogen-doped titanium oxides. Science. 2001;**293**(5528):269-271

[105] Bilmes S et al. Surface and electronic structure of titanium dioxide photocatalysts. The Journal of Physical Chemistry B. 2000;**104**(42):9851-9858

[106] Li N et al. Synthesis of Mo-doped WO_3 nanosheets with enhanced visible-light-driven photocatalytic properties. RSC Advances. 2015;**5**(115):95394-95400

[107] Jiang T et al. Optical and photocatalytic properties of Mn-doped CuO nanosheets prepared by hydrothermal method. Crystal Research and Technology. 2016;**51**(1):58-64

[108] Chhowalla M, Shin HS, Eda G, Li LJ, Loh KP, Zhang H. The chemistry of two-dimensional layered transition metal dichalcogenide nanosheets. Nature Chemistry. 2013;**5**(4):263-275. DOI: 10.1038/nchem.1589

[109] Lv R, Robinson JA, Schaak RE, Sun D, Sun Y, Mallouk TE, et al. Transition metal dichalcogenides and beyond: Synthesis, properties, and applications of single-and few-layer nanosheets. Accounts of Chemical Research. 2015;**48**(1):56-64

[110] Zhu S-C, Xiao F-X. Transition metal chalcogenides quantum dots: Emerging building blocks toward solar-to-hydrogen conversion. ACS Catalysis. 2023;**13**(11):7269-7309

[111] Wang C, Yang G, Shi W, Matras-Postolek K, Yang P. Construction of 2D/2D MoS2/g-C3N4 heterostructures

for photoreduction of Cr (VI). Langmuir. 2021;**37**(20):6337-6346

[112] Arunpandian M, Tae Hwan O. Nitrogen-doped hollow carbon spheres-decorated Co2SnO4/WS2 heterostructures with improved visible-light photocatalytic degradation of organic dye. Molecules. 2025;**30**(9):2081

[113] Hattori Y, Gutiérrez Álvarez S, Meng J, Zheng K, Sá J. Role of the metal oxide electron acceptor on gold–plasmon hot-carrier dynamics and its implication to photocatalysis and photovoltaics. ACS Applied Nano Materials. 2021;**4**(2):2052-2060. DOI: 10.1021/acsanm.0c03358

[114] Butler KT, Davies DW, Cartwright H, Isayev O, Walsh A. Machine learning for molecular and materials science. Nature. 2018;**559**(7715):547-555. DOI: 10.1038/s41586-018-0337-2. Epub 2018 Jul 25

[115] Pollice R, dos Passos Gomes G, Aldeghi M, Hickman RJ, Krenn M, Lavigne C, et al. Data-driven strategies for accelerated materials design. Accounts of Chemical Research. 2021;**54**(4):849-860. DOI: 10.1021/acs.accounts.0c00785

[116] Belkhode PN, Awatade SM, Prakash C, et al. An integrated AI-driven framework for maximizing the efficiency of heterostructured nanomaterials in photocatalytic hydrogen production. Scientific Reports. 2025;**15**:24936. DOI: 10.1038/s41598-025-10785-6

[117] Zhang Y, Kuwahara Y, Mori K, Yamashita H. Defect engineering of MoS2 and its impacts on electrocatalytic and photocatalytic behavior in hydrogen evolution reactions. Chemistry: An Asian Journal. 2019;**14**:278. DOI: 10.1002/asia.201801594

[118] Lu JM, Wang HF, Guo QH, et al. Roboticized AI-assisted microfluidic photocatalytic synthesis and screening up to 10,000 reactions per day. Nature Communications. 2024;**15**:8826. DOI: 10.1038/s41467-024-53204-6

Chapter 10

Palladium-Based Photonic Microresonator Hydrogen Sensors

Kenan Cicek

Abstract

This chapter explores palladium-based photonic microresonator sensors, leveraging palladium's unique hydrogen absorption properties to enhance sensor performance. These sensors exhibit high sensitivity and rapid response due to changes in optical properties induced by hydrogen-palladium interactions, including refractive index shifts and volumetric expansion. Various photonic architectures, such as microring resonators and plasmonic nanostructures, are analyzed for their efficacy in hydrogen detection. Emphasis is placed on the integration of palladium with PICs to develop compact, efficient, and highly sensitive hydrogen sensors. The chapter concludes with an evaluation of performance metrics such as sensitivity, Q-factor, and FOM, highlighting the potential of these advanced sensing platforms to revolutionize hydrogen monitoring across diverse applications.

Keywords: hydrogen gas sensor, microresonators, ring resonators, palladium, photonic sensor

1. Introduction

Hydrogen has emerged as an environmentally friendly alternative in the energy industry. Recognized as a clean energy source, it holds great potential for reducing greenhouse gas emissions and decreasing reliance on fossil fuels. However, alongside its benefits, hydrogen also poses significant risks due to its flammable nature and small molecular structure, which increases the likelihood of leaks.

Hydrogen interacts with atmospheric oxygen, making it a highly explosive element. The lower explosive limit (LEL) of 4% represents the minimum concentration of hydrogen in the air that can ignite, while the upper explosive limit (UEL) of 75% indicates the maximum concentration at which hydrogen remains flammable [1]. This wide flammability range distinguishes hydrogen from many other combustible gases, emphasizing the critical need for robust safety measures.

Due to its small molecular size and lightweight structure, hydrogen is particularly tent to leakage. This necessitates constant monitoring and detection of leaks in hydrogen usage and transportation systems. Consequently, developing effective detection methods is essential to ensure its safe use.

Modern technologies for hydrogen detection range from electrochemical sensors to acoustic and semiconductor-based sensors [2–4]. However, these existing methods

 IntechOpen

often face limitations such as susceptibility to environmental factors (e.g., temperature, electromagnetic interference), unsuitability for harsh conditions, high costs, and complex structures. This creates a demand for more practical and efficient detection techniques.

In recent years, optical-based technologies have gained significant attraction in sensor development, paralleling advancements in other fields. Optical methods offer several advantages over traditional detection techniques, including corrosion resistance, immunity to electromagnetic interference, spark-free operation, simple design, low cost, and rapid response times. Literature highlights various optical sensor configurations—such as waveguide-based sensors [5], fiber-optic sensors [6], photonic crystal sensors [7], and microdisk or microring sensors [8]—as highly effective for detecting hydrogen. These sensors demonstrate precise detection capabilities and fast response times, ensuring safer monitoring solutions.

2. Palladium: Hydrogen detection *via* hydrides

Palladium, symbolized as Pd, is a chemical element classified within the platinum group metals, which share similar chemical properties with platinum, rhodium, ruthenium, iridium, and osmium. Palladium is widely used to convert harmful gases into nontoxic substances. Additionally, it is one of the metals, along with yttrium (Y), Titanium (Ti), and magnesium (Mg), that exhibit a high capacity for hydrogen absorption at room temperature. These metals absorb hydrogen to form metal hydrides. The incorporation of hydrogen atoms into the metal lattice results in lattice expansion [9], which causes changes in the metal's electrical conductivity. Specifically, the reduction in the volumetric density of free electrons following hydride formation leads to variations in electrical conductivity. These changes can be explained by the Drude model [10].

$$\epsilon(\omega) = 1 - \frac{Ne^2}{[\varepsilon_0 m(\omega^2 + i\gamma\omega)]} \tag{1}$$

Here, N and γ represent the volumetric density of free electrons and the damping frequency, respectively, while m symbolizes the electron mass and ω denotes the optical frequency of light. In this context, the electrical conductivity of palladium, influenced by hydrogen absorption, is formulated with Eq. (2) [11].

$$\epsilon_{Pd}(c\%) = h(c\%) \times \epsilon_{Pd}(0\%) \tag{2}$$

Here, $\epsilon_{Pd}(c\%)$ represents the electrical permittivity of palladium depending on the hydrogen concentration in the atmosphere, $\epsilon_{Pd}(0\%)$ denotes the electrical permittivity of palladium in the absence of hydrogen, and h(c%) is a nonlinear coefficient that varies with hydrogen density. For instance, h(0) = 1 at 0% hydrogen concentration, and h(4) = 0.8 at 4% hydrogen concentration [12]. Thus, this change in electrical permittivity affects reflection, transmission intensity, or wavelength shift depending on the sensor configuration, enabling hydrogen concentration to be monitored through optical signals.

The ease with which hydrogen bonds to palladium atoms and the reversibility of this reaction have made palladium an indispensable element in hydrogen sensing

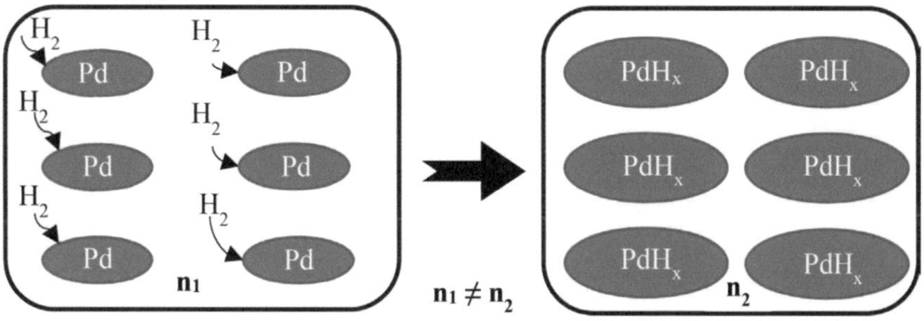

Figure 1.
The palladium element undergoes a volume expansion upon hydrogen absorption, forming palladium hydride, were x = [Pd]/[H].

applications. As shown in **Figure 1**, the compound formed when palladium absorbs hydrogen is called palladium hydride, denoted as PdH_x, where:

$$x = \frac{[H]}{[Pd]} \tag{3}$$

Here, x represents the ratio of absorbed hydrogen to palladium, typically less than 1. When $x < 0.017$, the palladium hydride exists in the α-phase, while $x > 0.58$ corresponds to the β-phase. In intermediate values, a mixture of the α and β phases occurs. In the α-phase, hydrogen atoms are accommodated within the palladium lattice without forming direct bonds with palladium atoms. In contrast, in the β-phase, hydrogen atoms bond with palladium, creating a more ordered structure [13].

As also shown in **Figure 1**, hydrogen absorption by palladium causes volumetric expansion. Hydrogen atoms embedded within palladium expand its crystal structure, increasing its volume. This expansion occurs because the hydrogen atoms fill voids in the palladium lattice and increase the interatomic distances. In the literature, it is stated that this expansion of the Pd lattice is 0.087% under 1% hydrogen gas concentration [14]. The volumetric change is observable during hydrogen absorption and desorption processes. Apart from volumetric expansion, hydrogen absorption alters both the density and mass of palladium, leading to changes in the refractive index of both palladium and its surrounding environment, as depicted in **Figure 1**. This response of palladium to hydrogen absorption is widely utilized in optical hydrogen sensing applications.

Palladium-based optical sensors detect changes in optical properties induced by the interaction between palladium and hydrogen. In such sensors, palladium is typically deposited as a thin film or nanoparticles onto a photonic structure. When hydrogen comes into contact with palladium, two significant changes occur:

1. Refractive index change: Palladium's refractive index shifts, directly affecting the sensor's optical properties, such as light transmission or reflection.

2. Volume change: The volumetric expansion modifies the interaction between the sensor and the evanescent field in the surrounding medium, causing changes in the wavelength, phase, or amplitude of the light.

These optical changes in photonic structures are used to detect hydrogen presence and concentration. For highly sensitive and rapid hydrogen detection, multiple palladium-based optical sensor designs have been developed. These sensors are primarily categorized into interferometric, fiber-optic, and evanescent-field (Microresonator).

3. Palladium: Based optical sensors

3.1 Interferometric optical sensors

Optical interferometers are devices that measure the interference patterns created by combining two or more light beams, using one as a reference as sketched in **Figure 2**. When factors such as the refractive index, temperature, strain, or other environmental conditions change along the other path (reference path is stable), a phase difference is introduced between the paths. Upon recombination of these paths, an interference pattern is generated at the output, providing a sensing mechanism based on the output amplitude. The most common types of interferometers include the Michelson, Mach-Zehnder, Fabry-Pérot, and Sagnac interferometers. Each type has its own design and operational features, but they all exploit the same fundamental principle of interference to achieve high-precision measurements. Due to their capability to measure changes in optical path length with exceptional precision, numerous interferometer-based optical sensor studies have been presented in the literature, operating on the same principle despite having different geometries [15–18].

Optical interferometers are widely used sensor structures in hydrogen detection applications. Most interferometric hydrogen sensors employ a specialized design incorporating a palladium metal layer. They typically operate within the wavelength range of 600 nm to 1700 nm [19, 20], depending on the application and material configuration. In most cases, interferometers are formed using optical fibers, with the operating wavelength typically around 1550 nm range, as it offers low optical losses in fibers, making it ideal for high-precision applications.

In the Mach-Zehnder interferometer shown in **Figure 2**, the palladium-coated arm experiences a change in refractive index in a hydrogen environment, which affects the optical path of the light passing through it. This change creates a phase difference between the arms of the interferometer, leading to variations in the output amplitude, thereby providing information about the hydrogen concentration. An interferometric optical hydrogen sensor typically has an overall device length on the scale of a few millimeters.

Interferometric sensors generally fail to achieve relatively high sensitivity and performance levels. This type of hydrogen sensor demonstrates sensitivity at the level of picometers per percent of hydrogen [21]. Additionally, the large amount of palladium required for hydrogen detection extends the response time. An interferometric

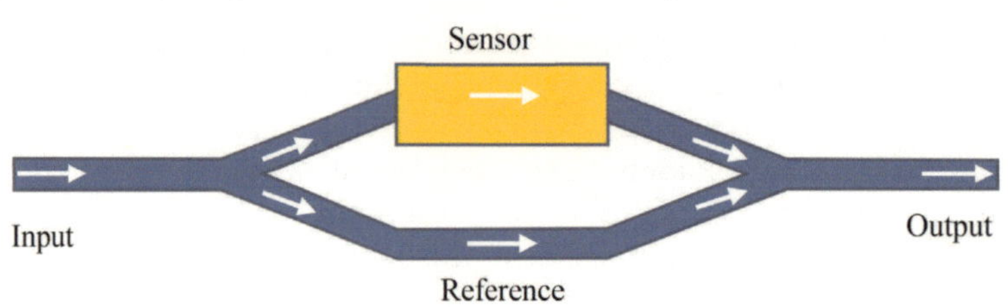

Figure 2.
Basic interferometer sensor geometry.

sensor typically responds to hydrogen changes within seconds but requires several minutes for full recovery. For these reasons, optical interferometers are not frequently preferred as hydrogen sensors.

3.2 Fiber optic sensors

Fiber optic sensors offer an alternative method for hydrogen detection using Pd. These sensors operate based on the principle of guiding light through an optical fiber and utilizing a section of the fiber coated with Pd as shown in **Figure 3**. When Pd interacts with hydrogen, its optical properties change, which alters the characteristics of light within the fiber, enabling the detection of hydrogen.

The Pd used in fiber optic hydrogen sensors can adopt various structures depending on the fiber geometries, as reflected in the fundamental sensor configurations shown in **Figure 3**. For instance, **Figure 3(a)** depicts a sensor design where Pd is coated on sections of a fiber etched on both sides. When hydrogen is detected, the optical properties and physical structure of Pd change, leading to a variation in the effective refractive index of the mode within the fiber. This results in changes in phase, wavelength, or amplitude at the output, providing information about the detected hydrogen concentration.

The sensor structures [22] shown in **Figure 3(b–c)** operate based on a similar principle. There are differences in sensitivity and performance coefficients among these three designs. The structure with a higher amount of Pd metal has a lower performance coefficient due to the losses caused by the absorption of the metal. However, the increased Pd content enhances the measurement sensitivity.

Finally, in **Figure 3(d)**, a thin Pd film placed at the tip of the fiber changes the amount of reflected light at the fiber's end when hydrogen is detected, providing information about the hydrogen density [23]. This structure offers an effective alternative for hydrogen detection.

Fiber optic sensors typically operate within a wavelength range of 400 to 1700 nm, with the 1550 nm region being the most commonly used due to the low attenuation of

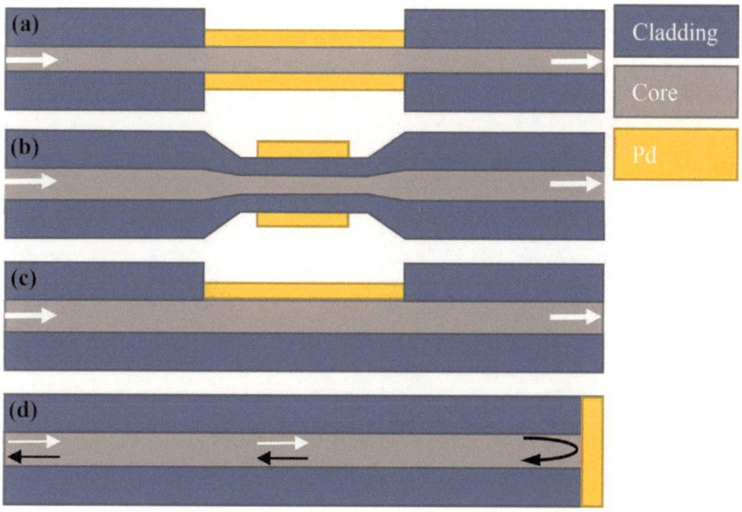

Figure 3.
Different geometries of fiber optic hydrogen sensors.

silica fibers, as also highlighted in interferometric sensor applications. These sensors are effective over a wide range of hydrogen concentrations, from 0.005 to 20% H_2, depending on the design and coating of the sensor [24]. The sensitivity of these sensors depends on the sensor type and its coating material, typically reaching the level of picometers per % of hydrogen concentration.

While fiber optic hydrogen sensors are used at significant densities, their practicality is limited due to compatibility issues with Photonic Integrated Circuit (PIC) technology. PIC technology allows the integration of photonic and electronic components on the same chip, enabling the development of smaller, faster, and energy-efficient sensors. However, the integration of fiber optic sensors with this technology is challenging, restricting their use in large-scale applications. Similar to optical interferometric sensors, the response time of fiber optic sensors is in the order of seconds, and they also require several minutes to recover before the next measurement.

3.3 Evanescent: Field sensors (microresonators)

Evanescent field-based optical sensors gather information about the surrounding environment by utilizing the evanescent wave that arise during the interaction of light with a surface. These sensors operate based on the principle that a portion of the light guided through waveguides spreads into a region known as the evanescent field, as shown in **Figure 4(a–b)**. The evanescent wave field forms on the surface of the waveguide and extends a certain distance, interacting with molecules or particles in the surrounding medium. This interaction alters the properties of the light in the waveguide (such as phase, wavelength, and intensity). When these changes are detected, significant information about the environmental parameters is obtained.

Although fiber optic sensors based on this technique exist, microresonators are the most commonly used structure. Optical microresonators are versatile devices with a wide range of applications across various sensing fields, including biosensing [25], chemical sensing [26], environmental monitoring, temperature sensing [27], strain and pressure sensing [28], and refractive index sensing [29]. The primary reasons for the popularity of microresonators include advantages such as high quality-factor (Q-factor), sensitivity, selectivity, compact size, and most importantly, compatibility with PIC technology.

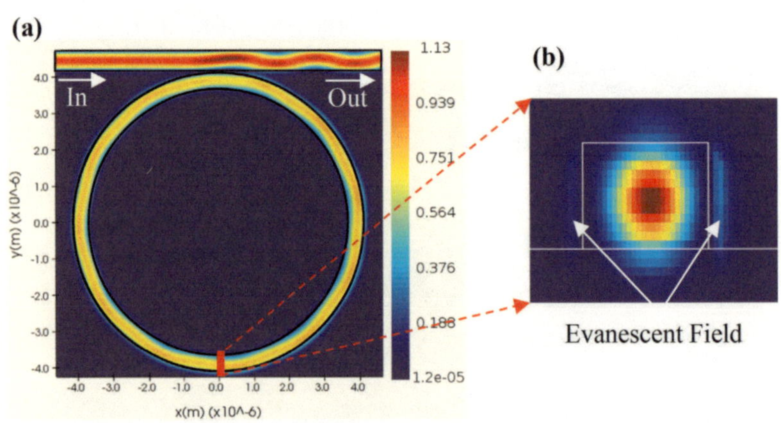

Figure 4.
(a) Electric field distribution in the ring-resonator. (b) Electric field distribution and the evanescent wave in the cross-section of the ring-resonator.

3.3.1 Microresonator hydrogen sensors

Microresonators are micrometer-scaled wavelength selective optical elements that can be used for a wide range of purposes. Whispering gallery modes (WGM) are the specific eigen-modes of electromagnetic waves inside the microresonator.

The light waves that are circling around the cavity is confined *via* continuous total internal reflection off the cavity surface. When the circulating light returns to the point where it started to travel inside the cavity, the waves interfere constructively with themselves, forming standing waves (**Figure 5(a)**). In case of constructive interference, two or more waves of the same frequency and phase interfere in such a way that results in mutual reinforcement generating single amplitude that is equal to the sum of the amplitudes of the individual waves. Thus, a light beam propagating within a microresonator loops back on itself multiple times and causes constructive interference. As such, the resonance occurs when the optical path length of the resonator is exactly the multiples of the wavelength. Therefore, as shown in **Figure 5(b)**, the ring-resonator supports multiple resonances with different free spectral ranges (FSR) depending on the resonator radius. In the absence of losses, in such a system of resonator, the intensity of the light transmitted from the resonator will be equal to the intensity of the light fed into the system. Optical microresonators come in various shapes and configurations, each optimized for different applications. The most common types of optical microresonators are characterized by their geometry, mode confinement, and resonance properties. These common resonators are microring, microdisk, micro toroid, microsphere, and microwire, as shown in the **Figure 5(c)**.

Regardless of the shape, the conventional resonators, which are mostly used in microscopic detection applications, usually include perfectly smooth resonator in order to sense any imperfections within the medium [30–32]. Among all micro resonators, microring resonators (MRRs) tend to be widely used in optical sensing

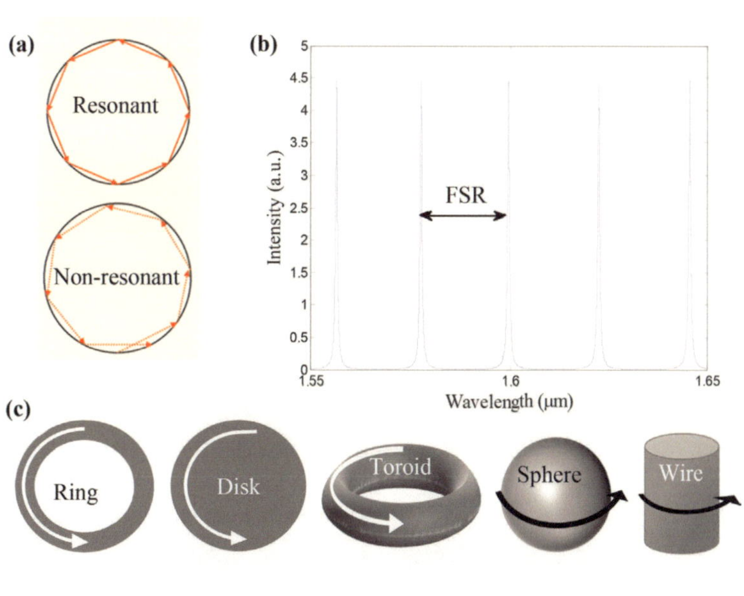

Figure 5.
The evolution of (a) resonant and non-resonant modes, (b) Resonance wavelengths spectrum of microresonator, and (c) Most common type of microresonator geometries.

applications due to their excellent sensitivity, high Q – factor, and compactness, combined with ease of fabrication and integration into integrated photonic systems.

An MRR system that has two ports with a single input provides WGMs in one particular direction, **Figure 6**. Each WGM's resonant wavelength λ in the cavity is given by:

$$\lambda = \frac{2Ln_{eff}}{m} \tag{4}$$

Where L, n_{eff}, and m are optical length—circumference of MRR, effective refractive index, and mode number, respectively. Any change in the medium of the MRR, sensing medium, affects the effective refractive index n_{eff} of WGMs that can be expressed as:

$$\Delta n_{eff} = \frac{\partial n_{eff}}{\partial n_{med}} \Delta n_{med} \tag{5}$$

Here, Δn_{med} is the refractive index change of sensing medium. Due to the change in effective refractive index of the WGMs, a resonance wavelength shift of $\Delta\lambda$ is introduced that is expressed as:

$$\Delta\lambda = \frac{2L}{m} \Delta n_{eff} \tag{6}$$

Thus, the shift in resonance wavelength $\Delta\lambda$ due to a change in the refractive index of the resonators' sensing medium is:

$$\Delta\lambda = \frac{2L}{m} \frac{\partial n_{eff}}{\partial n_{med}} \Delta n_{med} \tag{7}$$

Eq. (7) demonstrates that the wavelength shift is directly proportional to the refractive index change in the sensing medium of the MRR.

On the other hand, performance of a sensor in detecting changes in the refractive index Δn is measured by its sensitivity value. The sensitivity, S, is mathematically expressed as follows:

Figure 6.
The mode in smooth optical microring resonator.

$$S = \frac{\Delta\lambda}{\Delta n} \tag{8}$$

For hydrogen measurement performance, however, Eq. (8) needs to be modified because concentration becomes a factor. The change in concentration affects the refractive index, which in turn causes a shift in the resonance wavelength. Therefore, taking the hydrogen concentration directly into account, the sensitivity formula based on concentration change ΔC can be expressed as:

$$S = \frac{\Delta\lambda}{\Delta C} \tag{9}$$

Another key parameter for evaluating the performance of an optical sensor is the Q-factor, which quantifies the relationship between the energy-storing capacity and the energy loss of the resonator. A high Q-factor indicates that the resonator can store energy efficiently with minimal losses, which is essential for achieving high sensitivity and precision in optical sensing applications. Mathematically, Q-factor is formulated as:

$$Q - \text{factor} = \frac{\lambda_{res}}{\Delta\lambda} \tag{10}$$

λ_{res} is resonance wavelength. To compare different optical sensors or design improvements, the Figure of Merit (FOM) is a crucial performance metric. A high FOM value represents high sensitivity and low losses in optical sensors, but this interpretation can vary depending on the type of optical sensor and its application. The sensitivity-based FOM is:

$$\text{FOM} = \frac{S}{FWHM} \tag{11}$$

FWHM (Full Width at Half Maximum) is a statistical measure that describes the width of a resonant peak. It should be noted that in evaluating the performance of microresonator sensors; Q-factor, FOM, and sensitivity are significant metrics. However, the importance of each these metrics varies depending on the application requirements.

3.3.2 Microring hydrogen sensors

Microring resonators are being used for hydrogen detection applications by leveraging palladium's hydrogen absorbing characteristics. The sensor structures used for hydrogen detection typically consist of MRRs with radii of a few micrometers. These resonators are operated with a palladium component as the active sensing material, and they are in contact with evanescent field.

In the basic hydrogen detection system, a TE polarized incident beam with a Gaussian spectrum is introduced into the system through an access waveguide, as illustrated in **Figure 7**. The excited resonant WGMs within the microring resonator interact with the sensor medium *via* evanescent field, which extends outward from the resonator's ring cross-section, as shown in **Figure 7(a)**. When hydrogen gas is introduced into the sensing environment, as illustrated in **Figure 7(b)**, the Pd component inside the resonator undergoes expansion. This expansion occurs because Pd

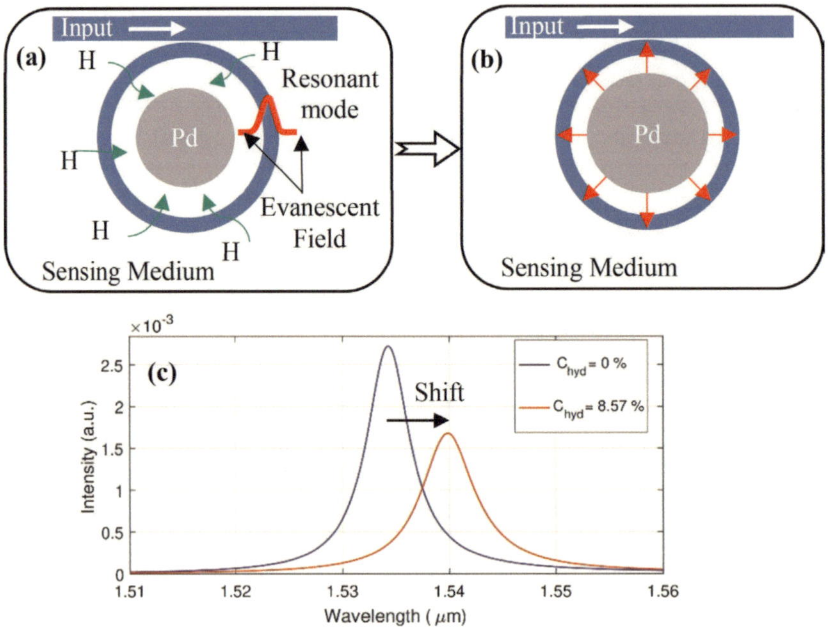

Figure 7.
The volumetric expansion of the palladium element causes a change in the effective refractive index of the WGMs within the waveguide in the sensing environment.

disk absorbs hydrogen, causing it to swell. As previously mentioned, the WGMs of the MRRs are highly sensitive to changes in the surrounding environment, thanks to the resonator's high Q-factor. The Q-factor ensures that even small alterations in the physical properties of the sensor, such as changes in the material size, will significantly affect the resonant modes. Consequently, any variation in the size or shape of the Pd material—due to the absorption of hydrogen—induces a shift in the resonance wavelength of the modes, as shown in **Figure 7(c)**. This shift is a direct result of changes in the effective refractive index of the resonator, which is influenced by the expansion of the Pd disk. By monitoring this shift, it is possible to quantify the concentration of hydrogen present in the sensor's vicinity. The shift in resonance wavelength provides a precise, real-time measurement of hydrogen levels, making the system highly effective for detecting even trace amounts of hydrogen gas. This mechanism allows for highly sensitive and accurate gas detection, driven by the direct interaction between the Pd material and the resonant WGMs of the microring resonators.

A study [8] proposes a novel four-port MRR sensor geometry designed on SOI technology for hydrogen sensing purpose. In this configuration also shown in **Figure 8 (a)**, a Pd disk with strategically positioned air holes on is surrounded by concentric MRR. The Pd disk is assumed to be in contact with the substrate only at its center. To achieve this, the nanoscale air holes are used to etch away the excess substrate beneath the Pd disk, allowing for greater freedom of expansion and improving the sensor's sensitivity.

The MRR resonator used here is designed with an outer radius R_{out} and an inner radius R_{in}, resulting in a waveguide width of $W = R_{out} - R_{in}$. The height of the structure is the typical height of a classical photonic waveguide, which is 220 nm.

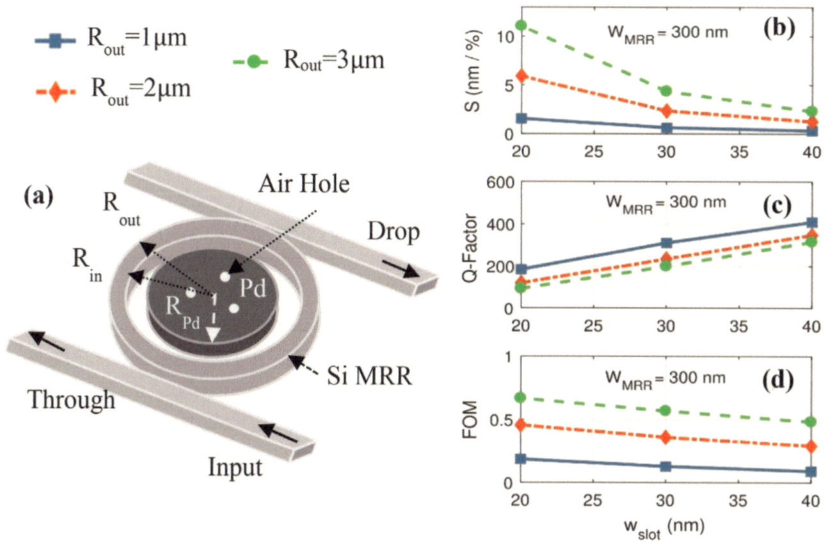

Figure 8.
The volumetric expansion of the palladium element causes a change in the effective refractive index of the WGMs within the waveguide in the sensing environment [8].

Inside the MRR, a coaxially placed Pd disk is separated by a narrow slot. Metals like Pd exhibit both Drude and Lorentz components in their dielectric function. Therefore, the optical properties of Pd are described using the combined Lorentz–Drude model. Upon absorbing hydrogen, the Pd expands, reducing the slot width and increasing the refractive index of the surrounding medium. These changes induce a redshift in the whispering gallery mode (WGM) wavelengths, which are tracked to detect hydrogen. The sensor geometry, including ring width, slot width, and radius, is optimized for maximum sensitivity and performance. Through Finite-Difference Time-Domain (FDTD) simulations, key geometrical parameters—such as the width of the microring (W_{MRR}), the slot width (W_{slot}), and the outer radius of the ring (R_{out})—are optimized to maximize the sensitivity.

As shown in **Figure 8(b)**, the sensitivity of the sensor increases as the W_{slot} decreases. This is because a smaller slot width enhances the interaction between the evanescent field and the Pd disk, leading to a larger wavelength shift ($\Delta\lambda$). However, a smaller slot width may also increase absorption losses, which could reduce the Q-factor of the resonator as can be seen in **Figure 8(c)**. On the other hand, FOM balances sensitivity and Q-factor, see in **Figure 8(d)**. The best FOM of 0.92 was achieved for R_{out} = 3.0 µm, W_{MRR} = 400 nm, and W_{slot} = 20 nm. Thus, with an optimal configuration, a sensitivity of 11.038 nm/% hydrogen that is 23 times higher sensitivity compared to earlier hydrogen sensors is achieved while maintaining practical functionality.

Pd-based microscale hydrogen sensors tend to respond slowly to changes in hydrogen concentration because the physical transformation of palladium requires time to complete. This characteristic presents a significant disadvantage for this detection technique. For the sensor to fully detect hydrogen, the physical transformation of the palladium must be entirely realized. Consequently, incorporating a large palladium component in the sensor design further slows the response time, delaying accurate readings. One approach to reducing response time is to use smaller Pd components, allowing the physical transformation to complete more quickly and enabling accurate readings in a shorter period. Plasmonic technology addresses this requirement by utilizing smaller Pd elements within a more compact sensor design. However, with current microscale technology, reducing the size of Pd decreases the sensor's sensitivity, necessitating the miniaturization of the entire sensor geometry. This approach, however, encounters a significant limitation due to the diffraction limit barrier. The surface plasmon polariton (SPP) technique offers an effective solution to overcome this limitation, enabling the miniaturization and enhanced performance of photonic devices. SPPs are electromagnetic waves that propagate along the interface between a metal and a dielectric or a metal and air, tightly bound to the surface. These waves typically occur at visible or infrared frequencies and are generated by the interaction between the electromagnetic field and the collective oscillations of free electrons on the metal's surface (surface plasmons). Although microscale ring resonators perform well in hydrogen sensing, in addition to the response time issue, with advancements in integrated photonics, the size of the devices must also be reduced to compete with CMOS technology. This situation has led to the introduction of various sensor structures in the literature, inspired by these ideas, to enable faster hydrogen sensing.

Figure 9 illustrates a nanoscale air ring resonator featuring an inner radius of 80 nm and an outer radius of 250 nm, coupled to a 40 nm wide access air waveguide on a silver (Ag) plasmonic platform [33]. In this sensor configuration, a Pd disk with a radius of 45 nm is positioned within the air ring resonator, allowing it to interact with the circulating optical mode. When the Pd disk interacts with hydrogen, the sensor's percentage of response η for readings can be theoretically determined using the following equation:

$$\eta = 1 - \frac{8}{\pi^2} exp\left(-D\frac{\pi^2 t}{4T^2}\right) - \frac{8}{9\pi^2} exp\left(-D\frac{9\pi^2 t}{4T^2}\right) \tag{12}$$

Here, D represents diffusion coefficient with a value of 10 nm^2/s and T denotes Pd thickness. The equation calculates the percentage of elapsed time required for complete stabilization (100%) of the Pd element within the sensor system. However, most

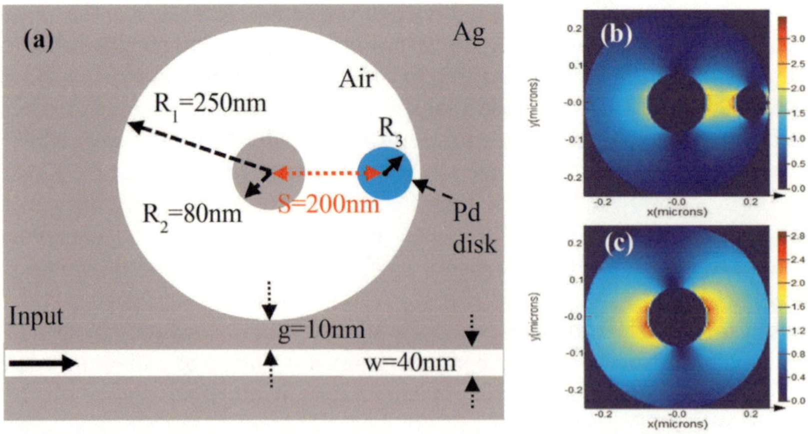

Figure 9.
(a) The geometry of an SPP sensor and the E-field distribution at the resonant mode for configurations (b) with and (c) without a Pd disk [33].

references consider 90% stabilization sufficient for a stable reading [33, 34]. Consequently, the response time for the sensor geometry shown in **Figure 9** can be expressed as:

$$RT = 0.85\frac{T^2}{D} \tag{13}$$

The numerical analysis of the proposed sensor geometry demonstrated good hydrogen gas detection performance with a sensitivity of 267 pm/(v/v–% hydrogen) and a rapid response time of 172.125 s. Furthermore, the sensor's compact footprint of 0.6×0.6 μm highlights the strong potential of this technique for integration with SOI technology, enabling a high-performance, rapid hydrogen detection platform.

As previously mentioned, when Pd absorbs hydrogen, it expands, causing a change in the refractive index around the resonator, which enables the detection process. It has been realized that refractive index-based optical sensors often encounter challenges related to selectivity. In hydrogen detection, the primary sensing mechanism relies on measuring changes in the refractive index of the sensing area. Unfortunately, these changes are not always exclusive to hydrogen; other gases or impurities can also affect the refractive index, potentially causing inaccurate readings. To address this issue and improve detection reliability, a self-referenced technique can be implemented to the sensor geometry, ensuring more accurate and dependable measurements.

In **Figure 10**, a self-referenced hydrogen nano sensor based on a SPP resonator system, designed to enhance hydrogen detection in complex environments [35]. As presented in **Figure 10(a)**, the system consists of a photonic integrated circuit platform with dual ring resonator design—one with a silver core and another with a palladium core—both coupled to an access waveguide with a g = 60 nm gap. By optimizing the design parameters—setting the silver-core air resonator with an outer

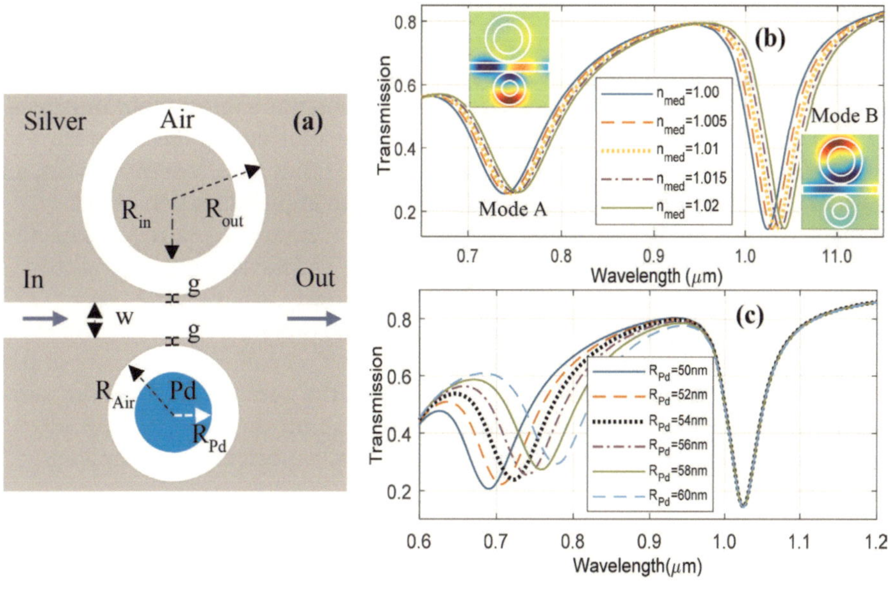

Figure 10.
Self referenced SPP based hydrogen sensor (a) geometry, (b) E-field distribution and spectrum under n_{med} change and (c) Pd size change [35].

radius (R_{out}) of 140 nm and an inner radius (R_{in}) of 90 nm, and the Pd-core resonator with a radius (R_{Pd}) of 50 nm and the air ring with a radius (R_{Air}) of 100 nm—a dual resonance spectrum is generated **Figure 10(b-c)**. This spectrum features resonance peaks at wavelengths of 0.69 μm (mode A) and 1.023 μm (mode B), corresponding to the Pd resonator and the silver-core (reference) resonator, respectively, as shown in **Figure 10(b)**. Insets in the figure depict the field distributions for Modes A and B, showing where the light is concentrated.

The figure demonstrates that both resonance peaks in the sensor's transmission spectrum shift in response to a label-free refractive index change (from 1.00 to 1.02) in the sensing medium. However, when the Pd core size is increased (from 50 to 60 nm) as if it changes due to hydrogen absorption, only the Mode A resonance peak associated with the Pd core resonator exhibits a shift. This behavior highlights the self-referenced approach of the sensor, where dual resonance modes are used to isolate hydrogen-induced effects from environmental interference, such as temperature fluctuations and other impurities. The palladium core's unique hydrogen absorption properties lead to measurable changes in refractive index, ensuring accurate detection while maintaining a stable resonance mode as a reliable reference. Numerical analysis conducted for the geometry in question reveals a sensitivity of 350 pm/% hydrogen, making this sensor compact, cost-effective, and reliable for multiparameter applications in a complex environment.

4. Conclusion

In conclusion, this chapter has highlighted the critical role of palladium in photonic microresonator-based hydrogen sensor applications, focusing on their design and operation principles. Palladium, with its high affinity for hydrogen and the resulting structural expansion and refractive index changes, has been shown to be an ideal material for hydrogen sensing. The integration of palladium with microresonator structures, particularly microrings, has enabled the development of highly sensitive and real-time hydrogen sensors. Notably, optimized microring designs have achieved sensitivities up to 11.038 nm/% hydrogen, marking a remarkable 23-fold improvement over previous technologies. Additionally, innovative configurations such as plasmonic nanoscale sensors provide faster response times, smaller sensor sizes, and outstanding reliability in complex environments. The implementation of self-referenced designs further enhances selectivity and accuracy in hydrogen detection, even in the presence of environmental interference. The advantages of photonic integration in creating compact and efficient sensor designs have also been highlighted. Ultimately, these advancements have the potential to revolutionize fields such as energy, environmental monitoring, and safety, signaling the promise of future developments and broader applications. Ultimately, the potential of these sensors to revolutionize fields such as energy, environmental monitoring, and safety underscores their significance, paving the way for future advancements and broader applications.

Conflict of interest

The authors declare no conflict of interest.

Author details

Kenan Cicek
Igdir University, Igdir, Turkey

*Address all correspondence to: eexkc@my.bristol.ac.uk

IntechOpen

© 2025 The Author(s). Licensee IntechOpen. This chapter is distributed under the terms of the Creative Commons Attribution License (http://creativecommons.org/licenses/by/4.0), which permits unrestricted use, distribution, and reproduction in any medium, provided the original work is properly cited.

References

[1] Utgikar VP, Thiesen T. Safety of compressed hydrogen fuel tanks: Leakage from stationary vehicles. Technology in Society. 2005;**27**(3): 315-320

[2] Lee E-B, Hwang I-S, Cha J-H, Lee H-J, Lee W-B, Pak JJ, et al. Micromachined catalytic combustible hydrogen gas sensor. Sensors and Actuators B: Chemical. 2011;**153**(2):392-397

[3] Kim S, Choi J, Jung M, Joo S, Kim S. Silicon carbide-based hydrogen gas sensors for high-temperature applications. Sensors. 2023;**13**(10): 13575-13583

[4] Dong S, Bai F, Li J, Viehland D. Sound-resonance hydrogen sensor. Applied Physics Letters. 2003;**82**: 4590-4592

[5] Fong NR, Berini P, Tait RN. Modeling and design of hydrogen gas sensors based on a membrane-supported surface plasmon waveguide. Sensors and Actuators B: Chemical. 2012;**161**(1): 285-291

[6] Wang J, Dai J, Hu W, Zhang F, Yang M. Improved performance of fiber-optic hydrogen sensor of porous Pt/WO3 based on ZIF-8. International Journal of Hydrogen Energy. 2024;**51**: 909-916

[7] Afsari A, Sarraf MJ. Design of a hydrogen sulfide gas sensor based on a photonic crystal cavity using graphene. Superlattices and Microstructures. 2020; **138**:106362

[8] Cicek K, Eryürek M, Kiraz A. Single-slot hybrid microring resonator hydrogen sensor. Journal of the Optical Society of America B. 2027;**34**(7): 1465-1470

[9] Peisl H. Lattice strains due to hydrogen in metals. In: Alefeld G, Völkl J, editors. Hydrogen in Metals I: Basic Properties. Berlin, Springer: Berlin Heidelberg; 1978. pp. 53-74

[10] Ashcroft NW, Mermin ND. Solid State Physics. 1st ed. New York: Brooks Cole; 1976

[11] Bévenot X, Trouillet A, Veillas C, Gagnaire H, Clément M. Hydrogen leak detection using an optical fibre sensor for aerospace applications. Sensors and Actuators B: Chemical. 2000;**67**(1):57-67

[12] Chen K, Yuan D, Zhao Y. Review of optical hydrogen sensors based on metal hydrides: Recent developments and challenges. Optics and Laser Technology. 2021;**137**:106808

[13] Suzana AF, Wu L, Assefa TA, Williams BP, Harder R, Cha W, et al. Structure of a seeded palladium nanoparticle and its dynamics during the hydride phase transformation. Communications Chemistry. 2021;**4**:64

[14] Noh J-S, Lee JM, Lee W. Low-dimensional palladium nanostructures for fast and reliable hydrogen gas detection. Sensors. 2011;**11**(1):825-851

[15] Butler MA. Optical fiber hydrogen sensor. Applied Physics Letters. 1984; **45**(10):1007-1009

[16] Zeakes J, Murphy K, Elshabini-Riad A, Claus R. Modified extrinsic Fabry-Perot interferometric hydrogen gas sensor. In: Proceedings of IEEE LEOS'94, Boston, MA, USA. 1994

[17] Farahi F, Leilabady PA, Jones JDC, Jackson DA. Interferometric fibre-optic hydrogen sensor. Journal of Physics E: Scientific Instruments. 1987;**20**(4): 432-434

[18] Wang G, Feng W. On-chip Mach-Zehnder interferometer sensor with a double-slot hybrid plasmonic waveguide for high-sensitivity hydrogen detection. Optics Express. 2023;**31**(24): 39500-39513

[19] Maciak E, Opilski Z. Hydrogen gas detection by means of a fiber optic interferometer sensor. Journal de Physique IV France. 2006;**137**:135-140

[20] Kim YH, Kim MJ, Rho BS, Kwack KH, Lee BH. Mach-Zehnder interferometric hydrogen sensor based on a single mode fiber having core structure modification at two sections. In: Sensors, 2010 IEEE. HI, USA: Waikoloa; 2010

[21] Luo J, Liu S, Chen P, Lu S, Zhang Q, Chen Y, et al. Fiber optic hydrogen sensor based on a Fabry–Perot interferometer with a fiber Bragg grating and a nanofilm. Lab on a Chip. 2021; **21**(9):1752-1758

[22] Villatoro J, Diez A, Cruz JL, Andres M. Highly sensitive optical hydrogen sensor using circular Pd-coated singlemode tapered fibre. Electronics Letters. 2001;**37**(16): 1011-1012

[23] Butler MA. Fiber optic sensor for hydrogen concentrations near the explosive limit. Journal of the Electrochemical Society. 1991;**138**(9): 46-47

[24] Shen C, Xie Z, Huang Z, Yan S, Sui W, Zhou J, et al. Review of the status and prospects of fiber optic hydrogen sensing technology. Chemosensors. 2023;**11**:473

[25] Malmir K, Habibiyan H, Ghafoorifard H. Ultrasensitive optical biosensors based on microresonators with bent waveguides. Optik. 2020;**216**: 164906

[26] Wang H, Yuan L, Kim C-W, Han Q, Wei T, Lan X, et al. Optical microresonator based on hollow sphere with porous wall for chemical sensing. Optics Letters. 2012;**37**:94-96

[27] Herter J, Wunderlich V, Janeczka C, Zamora V. Experimental demonstration of temperature sensing with packaged glass bottle microresonators. Sensors. 2018;**18**(12):4321

[28] Lei L, Tang J, Zhang T, Guo H, Li Y, Xie C, et al. Strain gauge using Si-based optical microring resonator. Applied Optics. 2014;**53**(36):8389-8394

[29] Chaudhary P, Mishra AK. Refractive index sensing near exceptional point of a system of triple microcavity. Sensors and Actuators A: Physical. 2023;**364**:114786

[30] Guider R, Gandolfi D, Chalyan T, Pasquardini L, Samusenko A, Pederzolli C, et al. Sensitivity and limit of detection of biosensors based on ring resonators. Sensing and Bio-Sensing Research. 2015;**6**:99-102

[31] Grimaldi IA, Bernini GTAR. Flow through ring resonator sensing platform. RSC Advances. 2015;**5**(86):70156-70162

[32] Witzens J, Hochberg M. Optical detection of target molecule induced aggregation of nanoparticles by means of high-Q resonators. Optics Express. 2011; **19**(8):7034-7061

[33] Cicek K. Fast hydrogen detection via plasmonic metal-insulator-metal technology. Applied Optics. 2018;**57**(34): 10032-10035

[34] Liu Y, Li Y, Huang P, Song H, Zhang G. Modeling of hydrogen atom diffusion and response behavior of

hydrogen sensors in Pd–Y alloy nanofilm. Scientific Reports. 2016;**6**: 37043

[35] Cicek K. Self-referenced hydrogen nanosensor based on SPP resonator system. Photonics and Nanostructures - Fundamentals and Applications. 2020; **39**:100786

Chapter 11

Metal-Metal Oxide, Metal-Organic Framework Nanocomposites for Energy Storage, Thermal Management, Radiation Shielding, and Magnetic Investigations with Quantum Computing

Bülend Ortaç, Saliha Mutlu, Ahmet Hakan Yilmaz, Nergis Arsu and Sevil Savaskan Yilmaz

Abstract

This chapter covers metal-organic framework (MOF) design, synthesis, multifunctional applications, energy storage, thermal regulation, phase change materials (PCMs), and radiation shielding using metal/metal oxide nanoparticles and their polymer nanocomposites. The study discusses their potential in energy storage, PCMs, radiation shielding, electromagnetic wave absorption, and quantum computing. Researchers are investigating the magnetism of MOFs for application in molecular spintronics and quantum information systems, perhaps as logic gates or qubits. The chapter claims radiation resistance in MOF geometries influences protective coatings, space missions, and nuclear shield applications. High-Z metal oxides and conductive polymers in MOFs produce unusual UV protection paths and temperature control. Next-generation multifunctional material developments are made possible by the multidisciplinary effort—material science, nanotechnology, and quantum physics. We also investigate MOFs' magnetic properties and possible applications in molecular spintronics and quantum computing. The work demonstrates how MOF spin properties might support logical operations and function as qubits in quantum technologies. Studying magnetical moment-lattice geometry interactions helps this work to open new quantum technologies.

Keywords: metal-organic frameworks, thermal management, phase change material, radiation shielding, quantum computing

1. Introduction

Using either vapor or liquid phase methods [1], one can make metal and metal oxide NPs by mixing precursors into porous MOF crystals that have already been

IntechOpen

formed. Heat [2], reducing chemicals [3], or radiation [4] can subsequently initiate particle production within the MOF. Alternatively, you can use a 'one-pot' methodology to crystallize the MOF alongside pre-formed functional NPs [5], thanks to the benign synthetic conditions typically used in MOF synthesis, that do not alter the chemical composition, structure, or morphology of most metal or metal oxide NPs.

1.1 Thermal management

MOF-water working pairs are anticipated to represent the next significant advancement in current sorption-based thermal energy storage and conversion systems. Currently, worldwide energy consumption has significantly escalated due to growing apprehensions regarding depleting fuel reserves and environmental deterioration. A lot of research has been done on thermal energy storage and conversion (TESC) systems because they could help reduce our reliance on fossil fuels and increase the use of renewable and waste heat energy, which would lower our CO_2 emissions [6]. MOFs, which are three-dimensional porous materials made up of metal centers connected by organic connectors, have recently become popular as materials that can store heat energy [7]. Compared to conventional porous adsorbents, MOFs provide benefits such as high crystallinity, well-defined porous structures, exceptionally large surface areas, and ease of chemical modifications, among others [8].

Zhou enhanced the thermal conductivity of MOF-based PCMs by utilizing PVP to link MIL-101(Cr)-NH_2 with rGO, resulting in a 472% increase [9]. Chen synthesized a dual-functional phase change material (PCM) exhibiting thermal storage and photoluminescence by concurrently doping salicylic acid (SA) and carbon CQD into the MIL-101(Cr)-NH_2 framework, demonstrating excellent SA adsorption capabilities and efficiently mitigating fluorescence quenching [10]. Their remarkable tunability and elevated interior surface areas primarily render MOFs suitable for a diverse array of applications. However, a feature that necessitates consideration in the practical applications of MOFs is their heat conductivity [11]. Currently, MOFs are generally shown to possess poor thermal conductivities ($< 2\ W\ m^{-1}\ K^{-1}$) [12]. It can be hard to find out experimentally what the intrinsic thermal conductivity of a MOF single crystal is compared to the system thermal conductivity of a compacted powder bed. Interfaces, grain boundaries, and inter-particle voids significantly influence measured thermal conductivity values, complicating the determination of intrinsic heat transfer mechanisms within the bulk crystal and the correlation of observed thermal transport behavior to the MOF structure or chemical composition [13].

MOFs are gradually advancing toward industrial applications, with a growing focus on critical thermal engineering concerns, including thermal stability and heat management. During charging, they must rapidly dissipate the heat of adsorption, as elevated temperatures significantly reduce maximum absorption. To prevent the system from overheating, the charging rate must be limited. The heat of desorption during discharging will cause a reduction in temperature, which will increase the amount of unusable adsorbates retained in the pores at the depletion pressure. The quantity of gas stored or delivered under realistic dynamic settings is consistently less than that under ideal isothermal conditions. Consequently, in applications utilizing pressure swing adsorption (PSA), minor temperature fluctuations adversely affect their performance [14–16].

Enthalpies constrain the use of conventional phase change materials (PCMs) in electronic thermal management. Chenxi Wang et al. utilized sorbents and their desorption mechanisms to achieve effective thermal management. Sorbents like MIL-101(Cr) that have high cyclic water loadings are much better than other PCMs because they absorb a lot of heat during the process of temperature-induced desorption. The thermal control of electronics is currently receiving increasing attention. The utilization of latent heat presents an appealing method for intermittent heat dissipation. However, the relatively low enthalpies of conventional PCMs constrain them. They illustrate a passive thermal management approach *via* the desorption of water within sorbents [17]. **Figure 1** shows the thermal management strategy with metal, metal oxide NPs, and MOF PCMs.

1.2 Phase change materials

PCMs are functional substances that harness latent heat through phase transitions to facilitate heat storage and utilization. To address the leakage issue of phase-change materials, certain researchers have utilized porous materials for the adsorption of phase-change substances, including graphene, expanded graphite, carbon nanotubes, bentonite, and MOFs [18]. MOFs are becoming more popular as adsorption carriers for PCMs because they have a large specific surface area, pores that can be made to fit different sizes, and strong capillary action [19–21]. By selecting suitable organic linkers and metal centers, they can systematically create three-dimensional MOFs that self-assemble *via* the coordination of organic linkers and metal nodes [22, 23]. Advanced cluster chemistry has sped up the progress of many different fields that are related to MOFs and have more potential uses [24, 25]. Recently, phase change thermal energy storage has utilized MOFs due to their adjustable pore size, ultrahigh surface area, substantial pore volume, and customizable chemical properties [26].

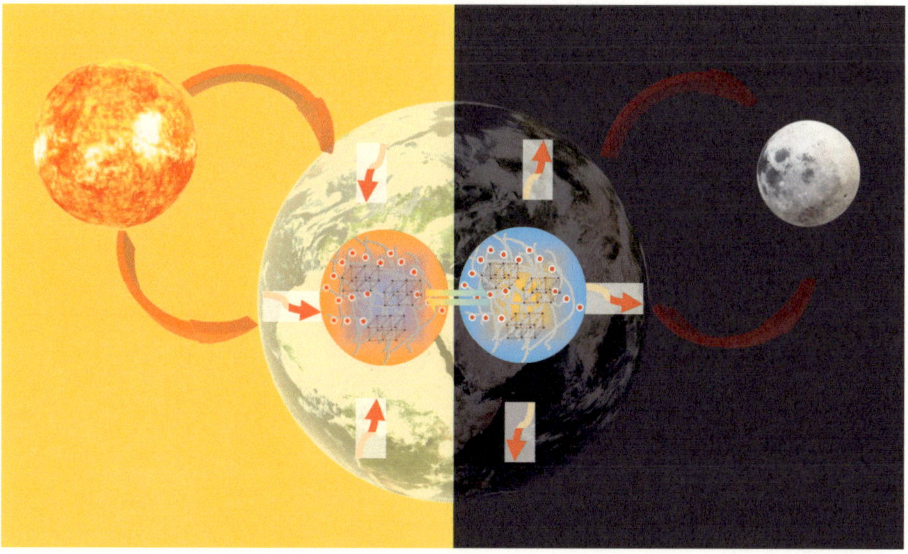

Figure 1.
Schematic representation of the thermal management strategy with metal, metal oxide NPs, and MOF PCMs.

Nozari et al. produced Cu-based MOF ionic liquid nanocomposite phase change materials, including ionic liquids (ILs). The researchers investigated phase transitions in an IL@MOF composite utilizing CuBTC (copper benzene-1,3,5-tricarboxylate) and [EMIM][TFSI] (1-ethyl-3-methylimidazolium bis(trifluoromethanesulfonyl)imide). The integration of apparent Cp data indicated that IL@MOF yielded a latent heat of 16.8 Jg^{-1} for the transition reaction, with 8.9 Jg^{-1} observed during the cooling scan [27].

1.3 Radiation shielding

Various domains extensively utilize MOFs due to their distinctive properties, which include high specific surface area, adjustable particle size, tunable pore dimensions, and ease of functionalization. In comparison with other inorganic fillers, MOFs' high specific surface area, three-dimensional porous structure, and ease of functionalization provide superior compatibility with epoxy resins [28], thereby inhibiting the formation of interfacial oxygen. Also, the secondary building units (SBU) in MOFs are better at resisting gamma rays, and MOFs have a lot of organic ligands that make it easier for them to get rid of free radicals. MOFs have gotten a lot of attention because they can be used for many things, such as luminescence detection, adsorption and separation of radionuclides, and capture of radioactive gases [29]. Nevertheless, numerous MOFs are unsuitable for applications involving strong radiation environments due to their inadequate thermal and chemical stability [30]. Furthermore, given the radiation environment within a nuclear reactor, the irradiated targets must exhibit superior radiation resistance. Certain researchers studied how stable different MOFs are to gamma radiation and found the most important properties of these materials that make them most stable to radiation, including their crystallinity, density, surface area, aromaticity, and linker connectivity [31, 32]. Gilson et al. [33] reported administering a maximal dose of 6 MGy to examine the radiation stability of a Np(V) Neptune MOF. Patrick L. Feng et al. looked into how the structure and properties of five MOF-based scintillators are related. They focused on the unique benefits of these crystalline and highly adaptable compounds as a starting point for creating new materials for specific radiation detection uses. Recently, two examples of MOF-based scintillators were reported. They both used the linker group 4,4′-trans-stilbenedicarboxylic acid (H_2SDC) [34]. **Figure 2** shows interactions between polymer nanocomposites-doped metal-metal oxide or MOF with gamma-rays.

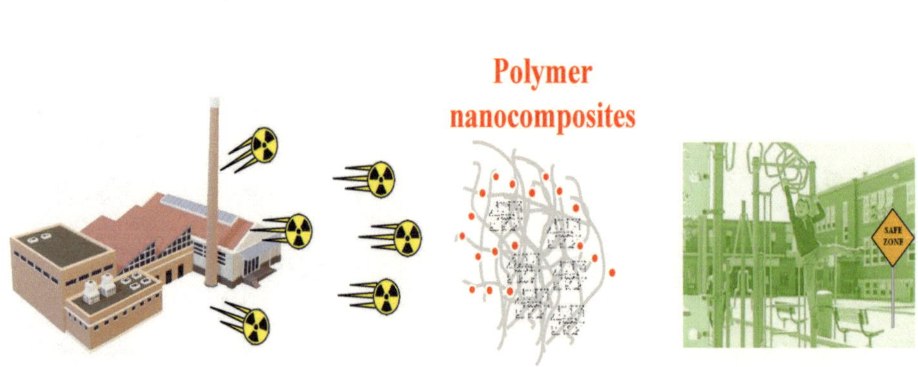

Figure 2.
Schematic illustration of interactions between polymer nanocomposites-doped metal-metal oxide or MOF with gamma-rays.

Chromium-51 (^{51}Cr) is a valuable radionuclide for diagnostic purposes, primarily utilized in the radiolabeling of red blood cells and platelets. However, to make ^{51}Cr that can be sold, it has to be activated by neutrons in nuclear reactors, which takes a long time and requires complicated separation methods. In this study, Chao Ma et al. synthesized five MOFs, namely MIL-100 (Cr), MIL-100 (Fe), MIL-100 (Al), MIL-101 (Cr), and aluminum fumarate MOF (FuAl), and investigated the effects of gamma ray irradiation at a high dosage rate, with a maximum dose of 6 MGy. The chromium-based MOFs, MIL-100 (Cr) and MIL-101 (Cr), were chosen as radiation targets to generate high specific activity (^{51}Cr) *via* the Szilard-Chalmers effect. Their aim was to examine their stability at elevated gamma radiation exposure. The selection of five distinct MOFs sought to explore the processes of radiation-induced damage and evaluate their resilience in intense radiation settings, such as those present in nuclear reactors [35]. Moreover, MIL-100 (Cr) and MIL-101 (Cr), characterized by their substantial surface area, extensive pore structure, and superior thermal, chemical, and radiation stability [36], were examined as irradiation target materials for the generation of high specific activity (^{51}Cr) through the Szilard Chalmers effect.

Metals such as Fe, Co, and Ni, along with their alloys or metal oxides, exhibit significant attraction due to their many loss mechanisms, including dielectric and magnetic losses [37]. MOFs represent a novel category of porous materials, attracting considerable attention for their use in electromagnetic wave absorbers. MOFs exhibit remarkable synergy between magnetic loss materials and dielectric materials due to their chemical composition, which includes both metal ions and organic ligands [38]. MOFs have been used to absorb microwaves because they have porous microstructures and precise host-guest coordination, which let them attenuate microwaves in a number of ways and give them adjustable dielectric and magnetic properties. Moreover, the components' dielectric and magnetic characteristics contribute to the regulation of electromagnetic parameters, enhancing polarization resonances and improving polarization at the interfaces [39]. The pyrolysis process frequently transforms MOFs into electromagnetic wave (EMW) absorbent materials.

In 2015, the researchers documented the first use of MOF-derived materials for EMW absorption. They utilized a hollow structure ZIF-67, composed of Co metal ions and an organic linker, and subjected it to high temperatures to produce Co/C composite nanomaterials with superior EMW attenuation efficacy. Nickel-based MOFs encompass Ni-ZIF and Ni-BTC, which were synthesized utilizing dimethylimidazole and trimesic acid as organic ligands, respectively. The pyrolysis process turned both MOFs into Ni@C-ZIF and Ni@C-BTC that were very good at absorbing electromagnetic waves [40]. Researchers have explored iron-based MOFs as an additional category of MOFs for electromagnetic wave absorption applications. Fe/C nanocubes with a structured geometry on an amorphous carbon framework have been synthesized from Prussian blue MOFs using pyrolysis in an inert atmosphere. The core-shell Fe@ graphitic carbon nanoparticles (CNP) were suitably distributed on the carbon framework [41]. The ideal pyrolysis temperature ranged from 600 to 700°C, while carbon species produced at lower temperatures were unable to absorb energy from incident electromagnetic waves due to their insufficient capacity to establish a continuous conductive network. The pyrolysis of Cu-based MOFs ($Cu_3(TMA)_2(\text{-}H_2O)_3$) has been utilized to synthesize the Cu_2O@nanoporous carbon (Cu_2O@NPC) composite. The concentrations of C and Cu_2O in Cu_2O@NPC were modifiable, facilitating the adjustment of permittivity and permeability, which led to enhanced ultrabroad-bandwidth absorptivity. The improved absorber demonstrated a very high RL min value of −31.1 dB at 5.6 GHz. The effective bandwidth, where RL is less than −10 dB, has been extended to 7.3 GHz. A CoZn-MOF was subjected to pyrolysis in a nitrogen environment to produce porous Co/ZnO/C microrods, which demonstrated superior microwave absorption compared to composites formed from individual metal-MOFs [42]. Huang and colleagues documented the synthesis of Co/C composites *via* the pyrolysis of CoZn-ZIF. During fabrication, one can adjust the composition of these materials to create efficient electromagnetic wave absorbers by evaporating zinc at 800°C [43]. Different morphologies of MOFs based on Ni/Fe metals have been synthesized by manipulating the molar ratio of ionic metals, Ni^{2+}: Fe^{3+}, in conjunction with various organic linkers [44]. The magnetic properties of these metals are crucial for regulating and adjusting the alignment of both magnetic and dielectric characteristics. Pyrolysis subsequently synthesized the NiFe/carbon porous (Ni/Fe@C) composite. With an RL min above −40 dB, these composites had great absorption properties. This meant that over 99.99% of electromagnetic wave energy was blocked. The synergy in both loss components elucidates the mechanism underlying the superior performance of high EMWs. Lu and his colleagues studied how NiCo-MOF breaks down at 600°C to make NiCo NPs/nanoporous carbon (NiCo NPs@NPC) composites that can be used as light electromagnetic wave absorbers [45]. Nanoflakes of NiO/$NiCo_2O_4$, resembling the geometry of lotus roots, were synthesized through the carbonization of NiCo-MOF [46]. The effective bandwidth demonstrated a reflection loss value of less than 10 dB, signifying the absorption of 90% of the electromagnetic energy. Zhang et al. made a group of conductive bimetallic organic frameworks called (Zn_xCu_{3-x})(hexahydroxytriphenylene)$_2$, or (ZnCu-HHTP), by changing the amounts of zinc and copper metal ions to change the exact spacing between the layers of the two-dimensional frameworks [46]. Zn_0Cu_1-HHTP has great dielectric properties and is very good at absorbing microwaves in the gigahertz range, with an absorber thickness of 2.95 mm and a minimum reflection loss (RL_{min}) of 81.62 dB at 7.28 GHz.

The degradation of the ozone layer, which serves as Earth's protective barrier, heightens everyday concerns regarding the escalation of ultraviolet radiation penetration. Experts have unequivocally established the adverse effects of these waves on human health, including pruritus and skin cancer [47]. Ultraviolet radiation can penetrate the silicate glass utilized in residences and museums, resulting in alterations in color, degradation of household items (plastic and wood), as well as damage to historical collections, including paintings and antiques [48].

Li et al. documented the synthesis of transparent and uniform PMMA-ZnO hybrid materials, altering the surfaces of ZnO NPs using MEA [49]. Xiong et al. made hybrid microspheres of ZnO@polymer by using MAA to change the surface of ZnO [50]. Recently, Matsuyama et al. demonstrated that TPM was a more effective coupling agent between ZnO and PMMA, enhancing the uniform dispersion of ZnO QDs inside the PMMA matrix. Somayeh Tarasi and his colleagues made clear films from nanocomposites based on MOF (ZnO/C) so that UV light could be absorbed at first [51]. The carbonization of the MOF under varying conditions substantially influences the composite composition, surface area, morphology, and ultimately the performance of the composites. Scientists led by Xia used MOF-5 as a starting material for the carbonization process at 800 and 1000°C. This made two products with different photocatalytic properties [52]. Reports have documented numerous MOF-derived composites for the absorption of electromagnetic microwave (low-frequency) radiation, but none for the absorption of UV light [52, 53] To reach this goal, TMU-21 MOF was used as a building block to make ZnO/C and ZnO/C-W nanocomposites at 800°C in argon and argon atmospheres full of water vapor, respectively. Subsequently, two hybrid films were made utilizing PVA and nanocomposites to examine the absorption characteristics of the nanocomposites. The films were initially produced by synthesizing ZnO/C and ZnO/C-W, followed by their dispersion in a PVA polymer layer. The transparent ZnO/C-W/P material demonstrates significant UV-shielding capabilities.

Researchers have thoroughly examined MOFs as adsorbents for nuclear waste remediation and radionuclide generation due to their elevated specific surface area, adjustable functional groups that facilitate great selectivity, and commendable chemical stability [54, 55]. A lot of MOFs have shown great results in separating radioactive gases (^{85}Kr, ^{129}I, ^{135}Xe, and ^{222}Rn), cleaning seawater, removing radionuclides from wastewater (^{59}Fe, ^{65}Zn, ^{137}Cs, and ^{235}U), and making new radionuclides [56]. Documentation of these MOFs' resistance to ionizing radiation is scarce, despite their exceptional chemical stability and significant promise in waste treatment applications. To fully harness the potential of MOFs in these domains, it is essential to assess their radiation resistance. Only a limited number of MOFs have undergone gamma radiation stability examination to date. Elsadi et al. [56], studied different SIFSIX^{-3} MOFs under beta and gamma radiation. They found that SIFSIX^{-3}-Cu was more resistant to gamma and beta radiation, withstanding doses of up to 50 kGy and 25 MGy, respectively. The creation of novel materials to adequately shield astronauts from detrimental radiation has been a significant concern since the inception of space exploration. The diverse forms of radiation present in space encompass galactic cosmic radiation (GCR) and solar energetic particles (SEPs). Spaceship fabrication extensively uses aluminum, especially for structural elements. Nonetheless, aluminum is not an efficient shield against GCR and SEPs. This requires the incorporation of additional shielding materials into the spacecraft design, leading to a heavier and larger vehicle. A diverse array of materials has been examined, encompassing hydrogenated graphite nanofibers, lithium hydride, polysulfone, polyetherimide, liquid methane, carbon fiber composites, epoxy, and polyethylene (PE). Recent investigations have integrated nanoscale particles composed of metals, metal oxides, and coated polymers into high-strength polymer matrices [57]. Attention toward radiation shielding has increasingly focused on polymer composites due to their potential for high protection efficiency and superior specific qualities, including specific strength and specific modulus [58]. Hydrogen-rich materials are known for their effectiveness in shielding against galactic cosmic rays (GCR) and solar energetic particles (SEP); therefore, composites with a hydrogen-rich matrix and ultra-high molecular weight polyethylene (UHMWPE) fiber reinforcement can provide both structural and protective benefits.

Exposure to high-energy radiation (such as gamma rays, X-rays, and UV radiation) induces many alterations in the physical properties of glasses, including modifications to their mechanical, chemical, optical, structural, electrical, and magnetic features. As ionizing radiation, gamma rays predominantly create absorption bands in the visible and ultraviolet parts of the spectrum. The irradiation process collects emitted electron-hole pairs, leading to the creation of defect centers. Numerous different lead (Pb) glass systems have demonstrated efficacy as effective shielding materials. However, the toxic properties of Pb generally discourage the use of Pb-based glass systems. Therefore, it is necessary to choose a suitable alternative.

Furthermore, due to the widespread application of radiation in diverse sectors, including medicine, agriculture, the nuclear power industry, and space exploration, there is a persistent impetus among researchers to develop advanced, next-generation, and environmentally friendly shielding materials. Various researchers have determined that glasses composed of heavy metals, such as PbO, Bi_2O_3, and MO_3, demonstrate shielding properties when subjected to repeated gamma irradiation [59].

The swift advancement of information and communication sectors has led to the detrimental effects of electromagnetic radiation on human health and gadget malfunction. The absorption and attenuation of electromagnetic waves have been accomplished in several materials. The distinctive changeable spatial configuration renders metal-organic MOFs advantageous for electromagnetic shielding and adsorption. With the progression of MOF research, numerous large-scale MOF-derived materials have been created. For example, the spatial structure of MOFs has evolved from two-dimensional to three-dimensional to accommodate more ligands. Advancements in synthetic methodologies for MOFs and their derivatives are progressing, emphasizing large-scale production and environmentally sustainable synthesis [60].

The two-dimensional network interlinks organic ligands and metal ions through coordination bonds inside each layer. These layers have a greater lateral dimension than 3D MOFs, facilitating material flow between layers and demonstrating low resistance and high flux properties. Miao et al. [61] developed a recyclable 2D MOF-CuHT (HT = 4-hydroxythiophenol) utilizing an ethanol suspension technique. The presence of the conductive honeycomb Cu_2S layer endows CuHT with exceptional microwave absorption properties, and its potential for large-scale production indicates significant industrial applicability. Liu et al. [62] developed one-dimensional sponge-like metallic Co was directly synthesized through significant magnetic loss, and Co/C composite materials were produced as nanosheets utilizing MOF derivative precursors. The SET value can attain 43.29 dB at 18 GHz, whereas the effective absorption frequency range of the C band is approximately 3.84 GHz. Liu et al. [63] utilized an electrostatic self-assembly technique in conjunction with a heat treatment process to construct 2D $Ti_3C_2T_x$ MXene/MOFs. MXene functioned as the substrate, while CNFs acted as a cross-linked network to improve the chemical stability of the composite material. The incorporation of magnetic CoNi particles into the composite material enhanced its magnetic characteristics and loss capacity, resulting in superior microwave absorption performance of 2D $Ti_3C_2T_x$ MXene/MOFs.

1.4 Magnetic molecules and MOFs quantum information in qubits quantum computing

The emergence of computers in the twentieth century was important to the advancement of computational chemistry. The numerical techniques employed to determine the static and dynamic characteristics of chemicals transformed chemistry as a field of study. The birth and evolution of quantum computing theory since the 1980s have demonstrated the potential for equally transformative advancements. Let us now examine the realm of quantum computers. Let us analyze the application of quantum computers in chemistry and MOFs. The bit—the fundamental unit of information in a computer, represented by the binary choices of 0 or 1—lays the foundation for the modern computerized universe. From the standpoint of contemporary silicon-based circuitry, this corresponds to a transistor existing in one of two states (i.e., being "on" or "off").

Unlike classical bits, qubits intrinsically demonstrate quantum mechanical phenomena, including superposition and entanglement. Superposition is a quantum mechanical phenomenon in which quantum systems, such as qubits, can simultaneously exist in a probabilistic linear combination of distinct states. The dimmable light switch analogy illustrates that a superposition state of a single qubit resembles the concept of light existing in a state between completely off and completely on. Entanglement is a quantum mechanical event that illustrates how several qubits display non-classical correlations, rendering the state of one qubit inseparable from the states of the others. The exponential increase in classical computational power needed to address quantum mechanical problems has led to the notion that employing a quantum mechanical computer may provide a more effective means of simulating quantum systems. **Figure 3** presents bloch sphere.

A quantum bit, or qubit, is frequently constructed from a two-level quantum system, such as the ground and excited states of an atom or the vertical and horizontal polarizations of a single photon. A qubit, denoted as $|A\rangle$, is the essential component of a quantum computer. It is capable of simultaneously or sequentially storing two states, either 0 or 1. An alternative perspective on a qubit is the superposition of these two states, which exists concurrently. Classical bits exist in one of two distinct states, identified as 1 and 0, whereas this is a linear combination of the binary states $|1\rangle$ and $|0\rangle$ ($\alpha|0\rangle + \beta|1\rangle$, where α and β denote the probability of occupying the $|0\rangle$ and $|1\rangle$ states, respectively). The new information units, known as quantum bits or qubits, operate and possess capabilities very different from classical static bits. Their foundation in quantum physics explains all of this. Quantum computers aim to tackle unsolved problems in classical computing. These computers run at rates that are truly astonishing and imitate human cognitive processes within the framework of neural networks, which include both deep learning and machine learning [64].

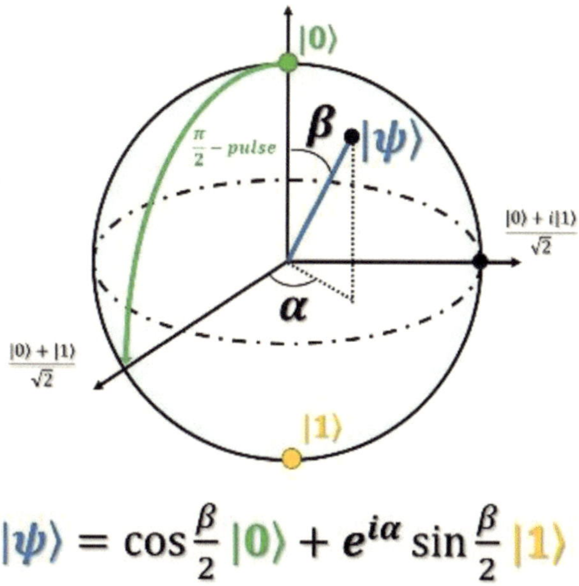

Figure 3.
Illustration of Bloch sphere.

2. Results and discussion

2.1 Thermal management

MOFs, recognized as adsorbent materials, have garnered heightened study attention in thermal energy storage and conversion. MOF-water working pairs are anticipated to represent a significant breakthrough in current sorption-based thermal energy storage and conversion systems. Tahmid Hasan Rupam et al. synthesized a promising metal-organic framework for high water sorption, MIL–100(Fe), along with its transition metal-doped derivatives (Ni^{2+} and Co^{2+}), using an environmentally friendly method. The cooling cycles indicated that the nickel-doped MIL–100(Fe) exhibits the highest specific cooling effect value of 450.79 kJ kg^{-1} [65].

Using a technique known as liquid-phase exfoliation—a means of thin-layer breakdown of bulk materials—Raza et al. produced nanosheets of molybdenum disulfide (MoS_2). They added several transition metals—more especially, silver (Ag), cobalt (Co), bismuth (Bi), and zirconium (Zr)—to raise the photocatalytic performance of MoS_2—that is, its capacity to accelerate chemical processes under light. These metals are p-type dopants as they increase the positive charge carriers (holes), therefore enhancing the electrical properties of the material. Using a hydrothermal synthesis technique—where materials are treated under high temperature and pressure in a water-based solution—the scientists embedded these metals into the MoS_2. The concentration of the transition metals was kept at 7.5% by weight. Doped MoS_2 nanosheets had significant antibacterial efficacy when tested against pathogens. *Escherichia coli* effectiveness ranged from 0% (in undoped samples) to 31.7–55.2% after doping. The activity of *Staphylococcus aureus* increased from 0% to between 34.2 and 69.23%. This indicates that doping MoS_2 enhances its ability to kill or inhibit bacteria. Theoretical studies confirmed that doping with Ag, Co, Bi, and Zr reduces the band gap of MoS_2. The band gap is the energy difference between the material's conduction band and valence band, and a smaller band gap allows the material to absorb light more efficiently. The conduction band (where electrons move freely) shifts to lower energies, while the valence band (where holes reside) shifts to higher energies. This band structure adjustment enhances the overall photocatalytic and antibacterial performance of the material [66].

2.2 Phase change materials

2.2.1 Preparation of the PCMs

PCMs are created from a range of materials, mostly including polymers, cross-linked structures, carbon-based porous materials, aerogels, and phase change fibers [67], and metal oxides/CNT/GO [68–70].

The methods of preparation for PCM composites are described in this part of the work. The simple physical immersion approach used produced the composite PCMs. An impregnant process [71] was used to synthesize octadecane@ACHT composite PCMs. A physical mixing and impregnation approach [72] efficiently produced SA@MIL-101(Cr)-NH_2@(Cr)-S-Si. The GO/PEG composite phase change material (PCM) was synthesized using ultrasound-assisted physical blending [73]. The SA was subjected to temperatures of 150°C in a thermostat, resulting in its liquefaction, after which the CNT sponge was immersed in the liquid SA for encapsulation purposes.

After several minutes, the composites were refloated, then filtered, and dried, resulting in the acquisition of the SA/CNT sponge phase change composite [74].

Multifunctional polyurethane-based PCMs, including ZnO and graphene aerogel (GA), were synthesized by a simple impregnation methods [75]. Composite PCMs based on polyethylene glycol (PEG) were synthesized using a vacuum impregnation technique [76]. PEG/SWCNs are synthesized using vacuum impregnation, whereas PEG/GNPs are produced *via* mixing and impregnating methods [77]. 3D interpenetrating network phase-change composites are produced using graphene-oxide (GO) aerogel (GA) infused with a minör quantity of carbon nanotubes (CNTs) and carbon spheres (CSs) as support materials and vacuum impregnation of molten PEG as the phase-change material [78]. Before compression into tablets, measured weights of PS-PEG block copolymers, metal oxide nanoparticles, and BN nanoparticles were mixed in an agar mortar and homogenized. Twenty minutes at 22°C under hydraulic pressure of 10 MPa produced the PCM nanocomposite tablets. A BTS [68–70] tool measured the tablet's thickness at 12,051 μm. Hydrothermal treatment in an autoclave [260] produced Cu-MOF/PANI.

2.2.2 Applications of PCMs

However, inadequate thermal conductivity and suboptimal shape stability continue to hinder phase change materials (PCMs), despite their significant latent heat and isothermal characteristics during phase transitions, which hold considerable potential in thermal management and energy storage/conversion. To address these bottlenecks, they proposed a simple strategy to fabricate PCMs with a core-shell distribution of binary fillers, a departure from traditional complex and energy-intensive methods. This approach employs straightforward physical mixing procedures, using a "core" of PEG nanocomposites integrated with graphene nanoplates (GNPs) and a "shell" filled with hexagonal boron nitride (h-BN). So, the composite that was made has a thermal conductivity of 3.63 $W{\cdot}m^{-1}\,K^{-1}$ when the GNPs and h-BN loadings are 0.625 and 15 vol%, respectively. This is about 1.5 times higher than the composite that had random filler distribution. Additionally, it exhibits significant latent heat retention and may preserve its original form at temperatures up to 90°C, hence enhancing its extensive applicability [79]. The inadequate photon capture ability of pristine MOF-based composite PCMs hinders solar energy usage. In order to reach this goal, Panpan Liu et al. [80] created advanced, high-performance photo-thermal composite PCMs by adding a photon absorber (polypyrrole, or PPy) and a thermal storage agent (1-octadecanol or ODA) to a MOF (Cr-MIL-101-NH_2). The coated PPy layer on the ODA@MOF surface functions as both a photon harvester and a phonon enhancer. The ODA@MOF/PPy composite PCMs show strong and broad-spectrum light absorption properties in the ultraviolet, visible, and NIR range. They also have better heat transfer capabilities than ODA@MOF. The photothermal conversion and storage efficiency of ODA@MOF/PPy-6% is 88.3%. Their engineered MOF-based photothermal composite PCMs also show long-lasting stability in stopping leaks, storing energy, and changing temperatures through photothermal conversion. The suggested coating approach and comprehensive understanding of the process are anticipated to advance the creation of high-efficiency MOF-based photothermal composite PCMs for solar energy applications.

The synthesis processes of some PCMs reveal that ODA@MOF/PPy, developed by Panpan et al. [80], was synthesized in the following manner. Initially, 96, 95, and 94 mg of ODA@MOF were submerged in 50 mL of 1 wt% ferric chloride solution. Subsequently, 4, 5, and 6 mL of pyrrole monomers was added into the aforementioned mixture to facilitate the chemical polymerization reaction while stirring at 25°C for 8 h, until the solution became black. Ultimately, ODA@MOF/PPy was obtained by centrifugation and subsequent drying of the mixed solution. The synthesized composite phase change materials were designated as ODA@MOF/PPy-4%, ODA@MOF/PPy-5%, and ODA@MOF/PPy-6%. Tang et al. [71] produced octadecane@ACHT composite phase change materials. Initially, ACHT was produced by coordinating chromium with alkylated bridging ligands through coordination bonds. The alkylated bridging ligands exhibiting structure-directing properties can create cylindrical assemblies that promote the aggregation of the alkylated ligands, subsequently allowing Cr^{3+} ions to couple with the carboxyl groups of the alkylated bridging ligands, resulting in the creation of hollow tubes. Secondly, octadecane@ACHT composite FSPCMs were synthesized using an impregnation technique. Sijia Liu et al. [72] synthesized SA@MIL-101(Cr)-NH_2@(Cr)-S-Si@SiO_2 composites using the co-condensation of dual precursors. About 0.5 g of SA@MIL-101(Cr)-NH_2@(Cr)-S-Si powder and TEOS was solubilized in 40 mL of formamide and subjected to magnetic stirring for 30 minutes. The pH was adjusted to 3.5 with a 0.1 mol/L HCl solution, and the sample was magnetically agitated and allowed to react for 7 days. The samples were thereafter allowed to stand and then filtered. Following a 12-hour drying period at 60°C in a drying oven, the SA@MIL-101(Cr)-NH_2@(Cr)-S-Si@SiO_2 composites, designated as PCM1, PCM2, and PCM3, were produced. To further investigate the influence of the silicon precursor on the thermal energy storage and impermeability of SA@MIL-101(Cr)-NH_2@(Cr)-S-Si@SiO_2, the quantity of TEOS was modified. SA@MIL-101(Fe)@(Fe)-S-Si@SiO_2, SA@UiO-68@(Zr)-S-Si@SiO_2, and SA@HKUST-1@(Cu)-S-Si@SiO_2 were synthesized using the identical approach as PCM2 to evaluate the method's universality. The subsequent section encompasses several studies, namely those related to phase transition samples, including cement/polymer/metal oxide and polymer/metal oxide compositions.

In their work, Tang et al. [71] created a MIL-101(Cr)-based hollow tube that was functionalized with alkyl groups by changing the ligand. The results showed that the MOF's mesoporous-macroporous structure had more latent heat than regular materials, which made it easier for PCMs to change states. However, MOFs with open pores rely only on weak capillary forces to adsorb phase change materials. This means that there is a big chance of leakage during many cycles of absorbing and releasing heat. Consequently, enhancing the impermeability of MOF-based PCMs is presently a significant area of research. Sijia Liu et al. [72] achieved exceptional impermeability and thermal storage performance in metal-organic framework (MOF)-based composite PCM by employing an encapsulation method that utilized unsaturated metal sites of MOF, resulting in the preparation of the PCM composite SA@MOFs@SiO_2. Recently, photothermal PCMs based on graphite, graphene, graphene derivatives, and carbon nanotubes have advanced significantly as leading contenders. Li and Wang [73] synthesized a graphene oxide (GO) @PEG composite PCM for photothermal conversion. Zhang and Liu [74] utilized a CNT sponge to fabricate a sebacic acid/CNT sponge composite PCM for photothermal conversion. Zhou et al. [75] predicted photosensitive hybrid materials with photoresponsive photo-/electro-thermal energy conversion efficiency. They put PEG and ZnO inside graphene aerogel at the same time to make a composite PCM. This PCM had an amazing photothermal conversion efficiency of 80.1%, which they attributed to ZnO's ability to absorb both visible (Vis) and ultraviolet (UV) light. In a similar manner, Yang et al.

[76] developed an arrayed hybrid scaffold of GO and BN for the encapsulation of PEG. The GO/BN hybrid scaffold provided the composite PCM with a remarkable photothermal conversion efficiency of 83.8%. Qian et al. [77] used a mix of graphene and CNT to surround PEG, creating a composite PCM that was 86.0% effective at converting light into heat. Bao et al. [78] constructed 3D interpenetrating network scaffolds by including GO aerogels, CNTs, and carbon spheres. The encapsulated PEG composite PCM demonstrated a photothermal conversion efficiency of 89.3%, which is 36.10% greater than that of GO aerogel@PEG. There are a lot of problems with making graphene, graphene derivatives, and carbon nanotubes because they tend to stick together and manage their pores, which makes them less useful for photothermal conversion and storage. Consequently, there is an urgent need for more promising nanoporous support materials that exhibit facile dispersibility and precise pore management to optimize thermophysical properties. PCMs that are thermally conductive were synthesized by combining BN/PbO(lead oxide) NPs with crosslinked poly (styrene-block-ethylene glycol di methyl methacrylate) (PS-PEG DM) copolymer. Yilmaz et al. demonstrate that PCM nanocomposites with latent energies ranging from 34.6 to 67.6 $J\ g^{-1}$ can reduce interior temperature fluctuations, improve thermal comfort, and conserve electrical energy. PbO nanoparticle doping improves the thermal conductivity of PS-PEG polymers. The findings suggest that thermal conductivity increases as the PbO ratios of 0–60%, 90–70%, and 90% increase. NPs, including BN and PbO, were incorporated into the PS-PEG polymer to obtain improved thermal conductivity values for nanocomposites. **Figure 4** shows Nps and the nanocomposites [69].

Thermal conductivity values of the cement-HB(PAE-bPCL)-PU Plaster-PbO or arsenic oxide PCMs are between 3.22 and 3.90 $W\ (m\ K)^{-1}$. Primarily, the composite materials were formed into a disk. The disks of our samples were fabricated at a room temperature of 22°C utilizing a hydraulic press at a pressure of 10 MPa for 15 minutes. The disks possess a diameter of 13 mm, and their thickness was assessed using a BTS 12051 μm (0–25 MMX 0.01 mm). As the amount of HB(PAE-b-PCL)-PU plaster increased, the thermal conductivity values of composites decreased. The cement-HB(PAE-b-PCL)-PU plaster composites with insulation properties were obtained with the reduced thermal conductivity values. The polymer, arsenic oxide, and PbO in our samples undoubtedly contribute to thermal conductivity. When Yilmaz et al. examine all the cement-HB(PAE-b-PCL)-PU plaster PCMs, they can conclude that the highest RPE values and best absorption of gamma photons are due to the lead content

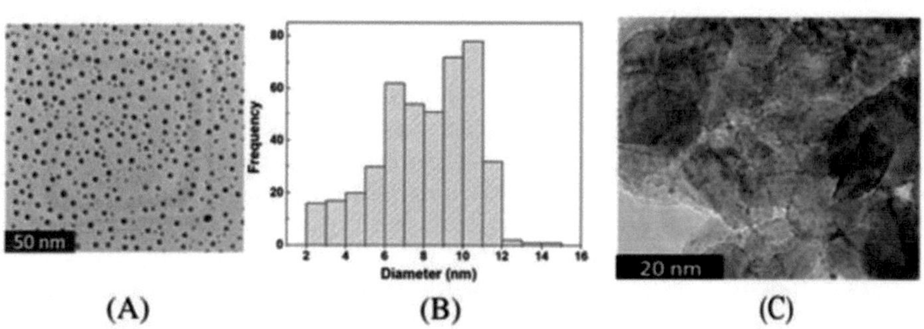

Figure 4.
(A) TEM images of PbO NPs and (B) the calculated size distribution of PbO NPs from the TEM images. (C) NCPSPbBN21 [69].

in the samples. Our research substantiates that the cement HB(PAE-b-PCL)-PU plaster-PbO and arsenic oxide-doped PCMs are good choices for radiation protective purposes for gamma irradiations that are especially utilized as a shielding material for the transport of radioactive sources and an isolation material for the radioactive waste management facilities or the construction industry [68]. Yilmaz et al. studied how crosslinked polystyrene-b-polyethyleneglycol block copolymers (PS-b-PEG) and nanostructured selenium dioxide (SeO_2) and boron nitride (BN) particles protect against gamma rays. Adding nanostructured SeO_2 and BN particles to PS-b-PEG copolymer-based composite materials makes them better at blocking radiation and protecting against harmful gamma rays. The SeO_2 ratio's ability to block radiation was improved to 53.8 wt%, which was the best value in the dataset [81].

Mutlu et al. [70] indicated that the phase change properties, thermal stability, and thermal conductivity of various nanocomposites can be enhanced through the combination of PS-PEG copolymers with SeO_2 and BN NPs. The powdered PCM nanocomposite was introduced into a compression mold. Subsequent to the fixation of the powder within the compression mold using the compression tube, it was processed under hydraulic pressure at a tension of 10 MPa for 20 minutes at 22°C. The tablet's thickness is 13 mm, contingent upon the inner diameter of the compression mold chamber utilized. The thickness of the tablets we manufactured was measured using a BTS instrument, yielding a value of 12,051 μm. The tablets, measuring 13 mm in diameter, display varying thicknesses ranging from 0.8 to 5 mm, contingent upon the qualities of the materials employed.

Metal oxide nanomaterials are highly promising candidates for the fabrication of polymer nanocomposites characterized by enhanced thermal and chemical stability due to their exceptional properties. Preparation of PS-PEG PCM nanocomposite tablets incorporates BN and SeO_2 nanoparticles, along with images of the tablets and nanocomposites utilized in this investigation. Ortaç et al. [69] similarly synthesized PCM nanocomposites using PbO nanoparticles, BN nanoparticles, and PS-PEG copolymers. The thermal conductivity of PS-PEG polymers is enhanced by doping with PbO nanoparticles. The results indicate that thermal conductivity rises with increasing PbO ratios of 0–60, 90–70, and 90%. Enhanced thermal conductivity values were achieved in nanocomposites with the incorporation of nanoparticles such as BN and PbO into the PS-PEG polymer.

High-performance nanocomposites of Cu-MOF with conducting polymers such as polyaniline (PANI) and polypyrrole (PPy) have been made using an *in situ* hydrothermal method. PANI and PPy boost the conductivity of the nanocomposite while establishing supplementary MOF-PANI-MOF transport channels, thereby facilitating efficient electrolyte ion transport and improving overall electrochemical performance. The specific capacitances of the Cu-MOF/PANI and Cu-MOF/PPy nanocomposites are higher than those of pure Cu-MOF (104.8 F/g) [82]. They are 160.5 and 132.5 F/g, respectively, at a current density of 0.5 A/g. Sheberla et al. [83] produced the conductive metal-organic framework Ni_3(2,3,6,7,10,11-hexaiminotriphenylene)$_2$ ($Ni_3(HITP)_2$), achieving an real capacitance of 18 $\mu F/cm^2$. The constructed symmetrical supercapacitor exhibited a capacitance of 107 F/g. Zhou et al. [84] synthesized conductive Ni-MOF on cellulose nanofibers for supercapacitor applications. The sample exhibited a capacitance of 125 F/g at a discharge rate of 0.33 A/g. A separate article details the synthesis of Cu-CAT conducting MOF, exhibiting a specific capacitance of 202 F/g for supercapacitor applications. The symmetrical supercapacitor demonstrated an energy density of 2.6 Wh/kg at a power density of 0.2 kW/kg when built.

The researchers led by Sheta created a label-free biosensor using a Ni-MOF composite made of PANI and nickel to measure hepatitis C virus (HCV) nucleic acid. MOFs are great for making nucleic acid biosensors because they are cheap, have a lot of surface area, and can be made to have different levels of porosity. Moreover, MOFs offer substantial loading capacity for ssDNA probes and inhibit degradation. The primary drawback of MOFs is their limited electrical conductivity. This study involved the synthesis of Ni-MOF using a simple technique, which was then combined with a conjugated PANI polymer to create a Ni-MOF@PANI nanocomposite [85].

A Zr MOF, NU-1003, in conjunction with TEMPO (2,2,6,6-tetramethylpiperidine-N-oxyl), is developed as a cooperative photocatalytic system for blue light-driven aerobic sulfoxidation. Adding 4 mol% of TEMPO improves hole transport and, as a result, aerobic sulfoxidation through NU-1003 photocatalysis. This shows how important a redox mediator is in changing a known MOF [86].

2.3 Radiation shielding

2.3.1 Preparation of radiation shielding composites

In this section, preparation methods of composites used in radiation shielding applications are presented. UIO-66-OH/epoxy composites were fabricated using the resin transfer molding technique [87]. Nanosized UiO-66 and UiO-66-NH_2 were effectively incorporated into epoxy resin (EP) by a solution casting method, resulting in MOF/EP composites. The glasses, including the formula $xTiO_2$-$xCeO_2$-(30-x) PbO-(70-x)B_2O_3, where $2 \leq x < 10$ mol%, are synthesized using the melt quenching process. The tablets of PbO nanoparticle-doped crosslinked PS-b-PEG block copolymer composites were fabricated at room temperature (22°C) using hydraulic compression at 10 MPa for 20 minutes. The tablets possess a diameter of 12 mm, and their thickness was assessed utilizing a BTS 12051 μm (0–25 MMX 0.01 mm) [88].

2.3.2 Applications

In their study, Shuai Wang et al. [87] looked at how radiation affected epoxy that had UIO-66-OH metal-organic framework added to it. UiO-66-OH nanoparticles were synthesized from Zr-n-propanol (71 μ) in 7 mL DMF and 4 mL acetic acid at 130°C for 2 hours. Subsequently, 2-hydroxyterephthalic acid (75 mg) was included and heated at 130°C for 18 hours. The mixture underwent centrifugation at 10,000 rpm, followed by solvent exchange with acetone. The samples were subsequently dried in a vacuum oven at 80°C for 12 hours.

By employing UIO-66-OH NPs as radical scavengers to inhibit epoxy breakdown, they achieved exceptional radiation resistance. The samples were exposed to a 60Co radiation source at a dosage rate of 25 kGy/h for 40 hours at ambient temperature. The cumulative irradiation dose was 1 MGy. The irradiation medium of the sample is air. UiO-66-OH decreases the concentration of free radicals produced post-irradiation by 88.17%. UiO-66-OH enhances the mechanical and thermal characteristics of epoxy both pre- and post-irradiation. The scavenging radicals method of hydrogen atom transfer in UIO-66-OH is more prevalent than that in GO and carbon nanotubes. The UiO-66-OH NPs' resistance to radiation is linked to their porous structure and hydroxyl functionalization. UiO-66-OH NPs may offer considerable benefits as anti-radiation fillers.

Chen et al. [89] picked epoxy resin (EP) and two distinct MOFs, UiO-66 and UiO-66-NH_2, to create UiO-66/EP and UiO-66-NH_2/EP composites. UiO-66 and UiO-66-NH_2 nanocrystals were produced using the hydrothermal technique. UiO-66 crystals with an average size of approximately 220 nm and UiO-66-NH_2 at about 230 nm were successfully synthesized. They studied how these composites got tougher and gave information on how UiO-66 NPs can be used to make materials resistant to irradiation. UiO-66-NH_2/EP is thought to provide superior radiation resistance compared to pure EP.

Lee et al. [90] studied how well three MOFs (MIL-100(Fe), MIL-101(Cr), and UiO-66(Zr)) could separate xenon and krypton. The study revealed that UiO-66(Zr) is the most promising adsorbent among the three options; however, the assessment of its radiation stability under a low radiation dose of only 2 kGy does not apply to the practical separation of Xe/Kr in nuclear reprocessing facilities. The research revealed that UiO-66(Zr) is the most beneficial adsorbent among the three options; however, the assessment of its radiation stability at a low radiation dose of merely 2 kGy does not apply to the practical separation of Xe/Kr in nuclear reprocessing facilities. As part of their ongoing research to find materials with better Kr adsorption and selectivity, Sameh K. Elsaidi et al. [31] made, tested, and studied the SIFSIX-3-M series (M = Zn, Cu, Ni, Co, Fe) of MOFs (MOFs). SIFSIX^{-3}-M is a type of isoreticular hybrid ultra-microporous materials (HUMs) made up of SiF^{-2} pillars and saturated metal centers (SMCs). SIFSIX-3-M can be optimized for krypton separation by substituting various metal centers such as zinc, cobalt, copper, nickel, or iron. The MOF materials were subjected to gamma radiation (^{60}Co source) ranging from 0 to 200 kGy, and their stability was assessed by powder X-ray diffraction (PXRD) to verify the preservation of crystallinity following γ and β irradiations. The comparable radiation damage tolerance dose between ^{85}Kr β and ^{60}Co γ-rays has been well established; for instance, "no significant difference in the breakdown yield was observed" in fluorocarbons subjected to beta and gamma radiation of the same dose [91].

The advancement of innovative MOFs for radiation detection and luminescence-based dosimetry is a current focal point of interest. Shailesh Joshi et al. [92] have synthesized and studied a uranyl MOF that exhibits an optically stimulated luminescence (OSL) phenomenon triggered by ionizing radiation. Dosimetric tests utilizing OSL were conducted on Uranyl MOF, demonstrating a linear dose-response up to 100 Gy, with a minimum detection dose of 1.1 Gy. Xie and colleagues have created a photoluminescent uranium-based metal-organic framework for the accurate detection of X-rays and gamma rays as low as 0.1 mGy [93]. The same group also created a photo-responsive uranyl-based coordination polymer with a detection limit one order of magnitude lower (0.01 mGy) than the prior version [94]. A Pb-based MOF was described by Liu and colleagues as a radiophotoluminescent (RPL) dosimeter that could measure high doses (10 kGy), showing better dosimetric performance than previous high-dose Mg^{2+}-doped LiF RPL dosimeters [95].

Suryaji N. Mane and colleagues synthesized Zn-BDC MOF using a modified solvothermal method, combining 3.1 g of zinc nitrate hexahydrate ($Zn(NO_3)_2{\cdot}6(H_2O)$) and 0.83 g of 1,4-benzenedicarboxylic acid ($C_8H_6O_4$) in a mixture of 50 ml dimethylformamide and 50 ml distilled water within a 100 ml autoclave, followed by heating for 24 hours at 170°C in a furnace. The white crystalline powder of Zn-BDC is employed for subsequent investigations [96]. The investigation focused on the exposure of Zn-BDC MOF to gamma (γ) radiation doses ranging from 0 to 50 kGy. A Co-60 facility (Russian Type: 87 ROBO) was utilized for γ irradiation, delivering a dosage rate of 15 Gy/min. The complete irradiated sample was divided into three components: 15,

35, and 50 kGy, for structural characterization and adsorption measurements. It is noted that zinc metal likely absorbs γ radiation, and the adsorption of γ does not alter the structure of the MOF, indicating that the MOF structure restores its structural stability at elevated γ radiation doses. For the first time, they have reported that the stability of Zn-BDC MOF reaches up to 35 kGy, and they have observed that irradiation influences the increased porosity of the MOF. The current results significantly indicate the radiation-sensitive characteristics of Zn-BDC MOF.

The radiation stability of various MOFs containing different metals (Al, Zr, Cu, and Zn) has been examined at varying gamma doses, revealing that MIL-100 (Al) has the highest radiation endurance, specifically 2 MGy [97]. A metal-organic framework (MOF) made of thorium was made by Gilson et al. [98] and can withstand γ radiation at a dose of 4 MGy and α particle exposure at a dose of 25.5 MGy. Nenoff et al. [99] looked at how gamma dose rates (0.78 Gy/min for 3 days and 423.3 Gy/min for 23 minutes) affected the stability of NU-1000 and UiO-66. They found that NU-1000 is more stable because it has more linker connectivity and fewer nodes.

The nickel-doped bismuth borate glass system was synthesized using a traditional melt-quenching method. Cobalt-60 served as an irradiation source to activate the thermoluminescence in the glasses. The kinetic and trapping properties of the glasses were analyzed, and their radiation shielding efficacy was evaluated by simulations using the user-friendly Phy-X/PSD program. The shielding parameters, including the Mass Attenuation Coefficient (MAC), Linear Attenuation Coefficient (LAC), Half Value Layer (HVL), and Tenth Value Layer (TVL), were computed for the analyzed glasses. The sample with a 2% NiO content exhibited the maximum radiation shielding efficiency. The glass shielding properties were compared with commercially available concretes, including ordinary, serpentine, barite, and chromite concretes, revealing significant promise for the glasses in radiation shielding applications. At IUAC, New Delhi, they irradiated the powdered glass samples at ambient temperature using a Co-60-irradiator (Gamma Chamber-1200) at a dosage rate of 7.5 kGy/h. The samples were subjected to varying dosages between 10 kGy and 200 kGy. Optical absorption data for gamma-irradiated glass samples were acquired using a Shimadzu-1601 double-beam UV-Visible spectrophotometer. Additionally, they obtained thermoluminescent (TL) glow curves by heating 5 mg of each sample at a rate of 5°C/s, using a Harshaw TLD reader, Model 3500, equipped with a 931B photomultiplier tube (PMT). The deconvolution of the data received from irradiated glass samples for thermoluminescence experiments was conducted using the "peak analyzer" module of Origin 8.5 software. I.G. Geidam et al. and Gurinder et al. [100, 101] utilized the user-friendly online Phy-X/PSD program to determine the radiation shielding properties of the currently examined NiO-doped glass samples.

Researchers led by Yilmaz looked into how well gamma irradiation worked and how well PbO-doped crosslinked polystyrene-b-polyethyleneglycol (PS-b-PEG) blocked copolymers and PS-b-PEG-BN nanocomposite materials acted as shields. The composite tablets were fabricated at a room temperature of 22°C and subjected to hydraulic compression at a load of 10 MPa for 20 minutes. The gamma irradiation attenuation factors of the studied composites were obtained over a broad range of energy spectra radiated from the $E\ 152u$ source using a High Purity Germanium (HPGe) detection system. Observations indicate that variations in the measured and calculated values correspond with different polymer types utilized to enhance gamma radiation shielding materials. Composites were prepared in varying proportions through the mixing of polymers with PbO. Upon examination of these composites, the optimal HVL value was determined to be 0.490 cm at 121.782 keV radiation in the PS-PEG (1000)-S3 sample, which contains 90% lead, as anticipated. The PS-PEG (1500)-S3 sample, which contains 90% PbO, yielded an optimal HVL value of 5.856 cm at 1408.006 keV. The calculated TVL value yielded a maximum of 121.782 keV radiation from the PS-PEG (1000)-S3 sample at a thickness of 1.629 cm. The PS-PEG (1500)-S3 sample achieved an optimal value of 1408.006 keV, measuring 19.451 cm. Furthermore, it was noted that the radiation protection capabilities of the samples enhanced with variations in the percentages of PbO or BN in the prepared materials. Upon examining the HVL, TVL, MFP, and RPE results of all samples from our research, it is evident that these samples provide significant and reliable insights for radiation shielding studies [88, 102].

Atom Transfer Reaction and Condensation Polymerization methods were employed to synthesize a novel Hyperbranched Poly Amino Ester-block-Poly Caprolactone-Polyurathane copolymer. The composite PCMs, which consisted of 89, 87, 69, and 67% Portland cement, 1 and 3% PU-Plaster, 10 and 30% weight percent PbO, and arsenic oxide, were exposed to gamma ray photons in the following energy regions: 121.78, 344.28, 778.90, 964.08, 1085.87, 1112.07, and 1408.01 keV for duration of 3600 seconds. Subsequently, the specimens of concrete—the Hyperbranched Poly Amino-Ester-block-Poly Caprolactone-Polyurathane copolymers—were experimentally examined for linear attenuation coefficients, mass attenuation coefficients, half-value layer, tenth value layer, mean free path, radiation protection efficiency, and gamma-ray absorption [68].

Researchers investigated graphene and CNT-based materials for their intriguing structural properties to create radiation shielding materials against ionizing radiation, specifically X-rays and gamma rays, by several methods [103]. Hashemi et al. [104] investigated PbO-decorated GO/epoxy composites for X-ray radiation shielding. A PbO-doped GO/epoxy composite for X-ray radiation shielding is documented, employing a modified Hummers method through a multi-step synthesis process. GO was first synthesized using the modified Hummers process and subsequently functionalized with Pb_3O_4 at varying weight ratios to produce a GO-Pb_3O_4 composite. The GO-Pb_3O_4 composite was subsequently synthesized with different filler loadings (5 and 10 wt%) by altering the thicknesses (4 and 6 mm) using the vacuum shock process. The particle size of Pb_3O_4 ranges from 1 to 4 μm, while the particle size of GOPb_3O_4 varies from 380 nm to 6 μm. The X-ray attenuation investigation verified that the reinforcement of epoxy-based composites with GOPb_3O_4 resulted in a significant increase in the overall attenuation rate of the X-ray beam. The 10 wt% GO-Pb_3O_4 mixture, with a thickness of 6 mm, demonstrated an equivalent aluminum thickness of 4.06 mm at 40 kVp, 4.83 mm at 60 kVp, and 3.91 mm at 80 kVp. This resulted in an attenuation rate improvement of 124.3, 124.6, and 103.6% compared to pure epoxy resin. The simulation results indicated that elevating the X-ray beam voltage diminishes the X-ray shielding efficacy of the GO-Pb_3O_4 composite. The composites exhibited enhanced performance when Pb_3O_4 particles were affixed to the surface of GO particles. The dispersion enhanced in epoxy resins with an improved composite interface, rendering it an effective shielding material. Another research team, Viegas et al. [105] enhanced X-ray attenuation efficacy of graphene-based nanocomposite They synthesized a nanocomposite utilizing PVDF included with 1.88 wt % of diverse functionalized materials, including multiwalled carbon nanotubes (MWCNT), pyrolytic graphite (P.G), GOs, and amorphous carbon (soot). Subsequently, they analyzed their mass attenuation coefficient. Upon irradiation of the nanocomposite with 6.9 keV photons, the X-ray attenuation of the PVDF/GO nanocomposite was fourfold greater (84.3 cm^2/g) than that of PVDF/MWCNT, PVDF/PG, PVDF/Soot, and PVDF. An analytical investigation was conducted to determine the distribution of various graphitic materials and their impact on the crystalline structure of PVDF. It clarifies the potential for varied interactions between PVDF chains and functionalized GO nanosheets, inhibiting the formation of large crystalline formations. It was demonstrated that thin films of ferroelectric PVDF/1.88 wt % GO nanocomposite, with a thickness of 0.1 mm, attenuate 82.9 and 48.5% of X-ray beams at energies of 6.9 and 8.1 keV, respectively. Consequently, lead-free, lightweight PVDF/GO radiopaque films may serve as a superior option for effective protection against X-ray radiation in radiological applications. Silva et al. [106] investigated the enhancement of X-ray shielding characteristics in PVDF/$BaSO_4$ nanocomposites augmented with GO. The use of a small amount of GO in the polymer-based nanocomposite demonstrated enhanced X-ray attenuation efficiency. A composite mixture was created by initially combining barium sulfate nanoparticles with PVDF homopolymer. Upon analyzing the attenuation property, it was determined that the PVDF/BaSO4 nanocomposite exhibited 9.14% attenuation when subjected to a 20 kV X-ray beam. An addition of 4 wt % GO nanosheets to the nanocomposite enhanced the X-ray attenuation efficiency to around 24.56%. The X-ray beams with elevated energy (100 kV) progressively reduce X-ray attenuation to 6.71 and 17.62%, respectively. Additionally, Ghaseminejad and his colleagues [107] documented the X-ray attenuation characteristics of modified PbO incorporated with graphene in the epoxy polymer. Initially, GO was synthesized using Hummers' method and then modified with PbO. Subsequently, epoxy-based composites were created utilizing fillers such as GO and PbO-GO by adjusting the weight ratios. Composite samples were constructed with thicknesses of 2, 4, and 6 mm. The X-ray and mass attenuation coefficients were determined by positioning the composite 60 cm from the diagnostic X-rays, measured at 20, 40, 60, and 80 kVp potentials. The findings indicated that 10 wt % of ER-PbOGO exhibits the most favorable X-ray shielding properties compared to the other composites.

2.4 Magnetic molecules and MOFs quantum information in qubits quantum computing

The Bloch sphere provides a graphical representation of a single qubit. The Bloch sphere is a graphical representation of a qubit that exhibits geometric properties similar to the unit circle in trigonometry. Each point on the Bloch sphere represents a distinct potential superposition of a single qubit. The qubit's two observable states, $|0\rangle$ and $|1\rangle$, are represented by the upper and lower sections of the sphere. An arrow on the Bloch sphere, orientable toward various locations on its surface, represents the current state of a qubit. **Figure 5** indicates how quantum computers work on a chemical system.

A proposal for a magnetic quantum processor, including individual molecular spins linked to superconducting coplanar resonators and transmission lines, is meticulously analyzed. They formulate a straightforward Hamiltonian for magnetic quantum electrodynamics to elucidate the fundamental physics. These hybrid devices are capable of executing arbitrary operations on individual spin qubits and facilitating adjustable interactions between any two of them. The integration of these two procedures guarantees that the processor can execute universal quantum calculations. Realistic calculations grounded in the parameters

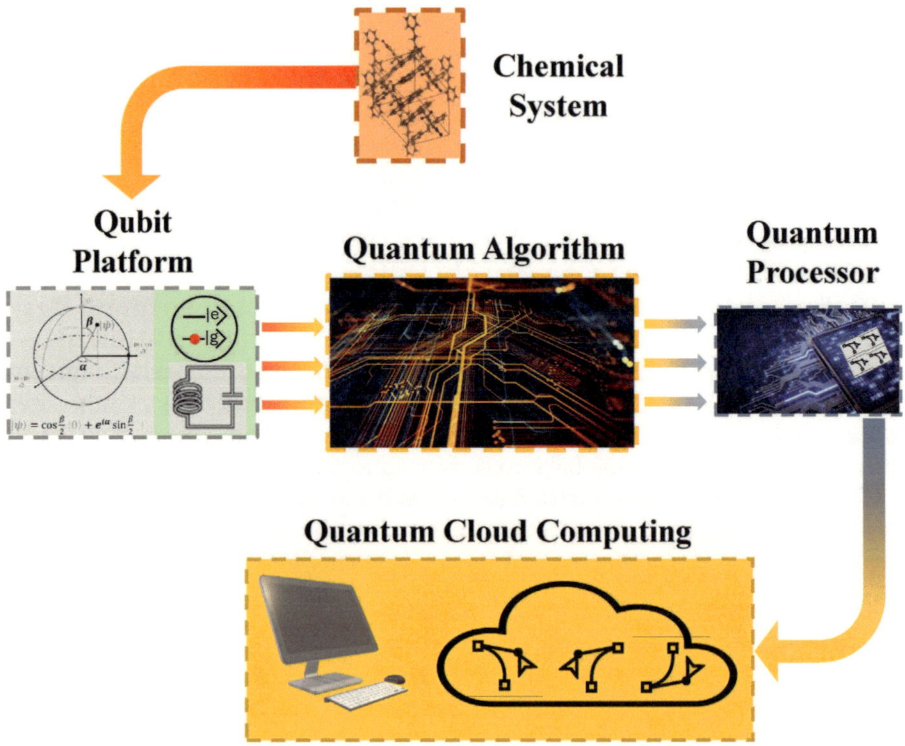

Figure 5.
How work quantum computers on a chemical system?

of current devices and molecular qubits rigorously examine the viability of this approach. The results show that the method could work, but only if molecules with sufficiently long coherence durations are developed and precisely integrated into the right parts of the device [108].

MOFs are regarded as molecular crystals from a physical perspective, as they possess hierarchical structures that are essential for the development of future applications in the fields of energy transfer, data processing and storage, high-energy physics, and light manipulation. Yuri A. Mezenov and his colleagues [109] showed that they explore the perspectives of MOFs as a novel family of functional materials in contemporary physics. This includes optical superstructures, materials for particle physics, and molecular-scale mechanical metamaterials, as well as porous metals and superconductors, topological insulators, and classical and quantum memory elements. A description of how such artificial materials have extended their impact on applied physics to become the mainstream in material science is provided, based on the complementary properties of crystallinity, elasticity, organic-inorganic nature, and complex hierarchy.

The utilization of the crystalline materials for next-generation memory and trigger devices is facilitated by the soft [110], porous, and organic–inorganic character of MOFs. Because of their softness, MOFs exhibit dynamic, reversible, and nonreversible structural changes in response to external stimuli such as pressure, electric and magnetic fields, and light [111, 112], among others. They can be deployed as active binary systems for the recording of information. The organic nature of MOFs confers specific dielectric properties (low-k or high-k) that facilitate their use as passive components in microcircuits and transistors. In addition, their complex organic makeup (bonded ligands and solvents) makes it easier for certain structural changes that are needed to make ultrafast optical data processing and storage devices in the near future. In contrast, the inorganic nature of MOFs offers distinctive states, including quantum qubits, that can be employed to record information with an unprecedented density. In this section, they examine the perspectives of classical and quantum memory and trigger devices that are based on MOFs.

Quantum computers use a qubit, the smallest physical element, to store information. A spin qubit records information by observing the orientation of spins in coherent superposition. Recent research has examined MOFs to implement this idea. Within the framework of an organometallic structure, an array of clocklike qubits was generated (**Figure 6a**).

Studies have verified a clocklike transition for cobalt (II) spins in a two-dimensional environment. $[(TCPP)Co0.07Zn0.93]_3[Zr_6O_4(OH)_4(H_2O)_6]_2$(T CPP = 5,10,15,20-tetrakis(carboxyphenyl)porphyrin) framework. Despite the local copious nuclear rotation of the spins, the clocklike spin qubits had a lifetime of 14 μs [114]. It was also demonstrated that $[[CuTCPP]\ Zn_2(H_2O)_2]$, an isolated molecule, retains quantum coherence and is a potential spin qubit as a node in the 2D MOF. Phase coherence in this material is temperature-independent up to 30 K. The qubit lifetime is 2.24 μs at 6 K and decreases to 0.85 μs at 60 K, which may be attributed to thermal processes associated with the softening of the frozen solution and the relaxation of the spin lattice [115]. In general, spin qubits are suitable for quantum processing or as data storage elements in the future due to their adherence to quantum physics laws and their operation at the minimal scale of single atoms.

In the future, they anticipate that additional research in the field of physics of metal-organic frameworks will concentrate on the following areas: classical and quantum data storage with increased rates and record densities [116, 117], lasing within the visible [118] and THz ranges, spintronics [119], excitonics [120, 121], valleytronics, and the absorption of light, microwaves, and shock waves [122].

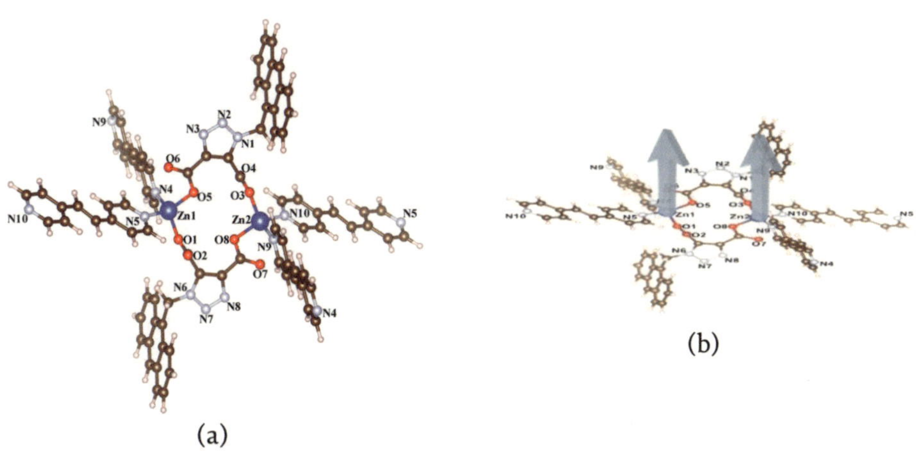

Figure 6.
(a) Illustration crystal structure of MOF [113]. (b) MOF structure with a central paramagnetic ion as a clock qubit.

Molecular nanomagnets comprise a magnetic core that contains one or more magnetic ions, encased and stabilized by organic ligands. Approximately 10 years ago, the demonstration that various molecule spin orientations and their quantum superpositions can represent qubit states led to their inclusion in the quantum hardware candidate list [123, 124]. A single chemical reaction can synthesize large quantities of identical molecules, and chemical design can tailor their magnetic characteristics, which determines the qubit frequency and states [125]. Chemistry also facilitates the formation of stiff molecular structures with a low density of nuclear spins. This method has resulted in remarkable advancements, increasing spin coherence times to near-millisecond maxima [126, 127]. Additionally, isolated molecular qubits can be affixed to surfaces [128] and other nanostructures, such as carbon nanotubes [129], and can also be incorporated into nanoelectronic devices, including nanocontacts created through electromigration [130]. This capability has enabled the detection of the reversal of an individual molecular spin, as well as the readout and coherent manipulation of its nuclear spin state, utilizing either magnetic or electric RF fields [131]. All molecules must be identical and properly separated in order to move them into small, predetermined groups. The periodicity of 2D networks, where the spin qubit acts as a node, can accomplish this. The underlying covalent or metal-organic framework (designated as COF and MOF, respectively) will ensure the precise identity and uniform orientation of all molecules/nodes, while an appropriate adjustment of the dilution with a non-magnetic analogous node can facilitate the requisite control over the density of qubits per unit area. The surface-confined assembly of two-dimensional architectures is currently a focus of rigorous research. Researchers have successfully synthesized on-surface MOFs [132, 133], demonstrating a high degree of structural order that extends to the micrometer scale. The previous materials offer enhanced thermal and chemical durability; nevertheless, they often do not ensure error correction during assembly due to the irreversible establishment of covalent bonds [134].

In the multidisciplinary field of molecular magnetism, many research groups have been working on making MOFs that become magnetized on their own below a critical temperature (T_c) [135]. Among the functional guest molecules suitable for incorporation into MOFs, the so-called single-ion magnets (SIMs) represent highly promising options. SIMs have become very important. Lanthanides [136], actinides [137], and highly anisotropic transition-metal ions [138] are just a few examples. These work as molecular nanomagnets. Molecular spintronics is a new field that is growing quickly. One of the biggest problems it faces right now is how to manipulate and combine individual molecules on different substrates so that they can be used in high-density magnetic memories and quantum computers. Their superiority in high-density information storage relative to conventional ferro- or ferrimagnetic materials is evident due to their molecular size. Drawing from Fujita's [139] foundational research on the encapsulation of guest molecules in porous coordination networks *via* targeted host-guest interactions, they propose a synthetic approach that effectively integrates well-ordered SIMs into the pores of a magnetic MOF through a single-crystal-to-single-crystal (SC to SC) post-synthetic method. This method combines the long-range magnetic ordering of the magnetic MOF with the gradual magnetic relaxation of the integrated SIM into a single solid. This could cause both magnetic phenomena to interact.

When MOFs contain paramagnetic three-dimensional transition metal nodes with the requisite diamagnetic organic linkers, they acquire the ability to exhibit magnetic properties [140]. To a large extent, the development of porous molecular magnets that are capable of modifying their spin quantum number and magnetic anisotropy has been considerably affected by first-row MOFs that are composed of transition metals such as V, Cr, Mn, Fe, Co, Ni, and Cu. In a domain that is not constrained by any boundaries, the quantity of variables creates significant hurdles. In this particular application, the weak magnetic interaction of proximity shell ligands, which comprise polycarboxylic acid linkers as well as oxo-, cyano-, and azido-bridging ligands, is what makes it conceivable. Another factor that determines the magnetic activity of MOFs is their structural framework, which is characterized by layered geometry and reduced conjugated distances between metal clusters at the same time. The synthesis of MMOFs has been accomplished by the utilization of organic linkers. These MMOFs are distinguished by the magnetic properties of the radicals that are present in the linkers. Yu and Wang reported [141] that the manganese-based metal-organic framework, $Mn_3(THQ)_2$ (THQ = tetrahydroxyquinone), which was synthesized utilizing an electrostatic gating approach, demonstrated a high Curie temperature of 250 K in the neutral monolayer and 300 K in the hole-doped monolayer. This was stated in their paper. The presence of a ligand of a minimal size and the coordination of MnO in a square-planar configuration were shown to be responsible for this high temperature. A ferromagnetic activity was seen in the Ni-Glutarate-based MOF $[Ni_{20}(H_2O)_8(C_5H_6O_4)_{20}{\cdot}40H_2O]$ when it was subjected to a Curie temperature of 4 K. The weak ferromagnetic interactions that occurred along the Ni-O-Ni angle were thought to be responsible for this activity.

Lanthanides are able to build magnetic MOFs in addition to the three-dimensional transition metals that they usually produce. $Er_4(bpa)_4(H_2O)_6$ and $Dy_2(bpa)_2(H_2O)_3$ are two examples: (H_2O) are the formulas for two different lanthanide organic frameworks (LnOF) that function magnetically. They are made up of three-dimensional structures that are made up of bpa_3-ligands and one-dimensional rod-shaped metal carboxylate secondary building units (SBUs). These structures display ferromagnetic interactions between LnII and LnIII. The copper(II)/zinc(II)-MOF-74 composite is an example of a magnetic MOF family that is widely used. In order to manufacture it, a mechano-synthetic approach was utilized, and the linker 2,5-dioxido-1,4-benzenedicarboxylic acid (H_4 dobdc) was utilized as a bridging agent. According to the findings of the magnetic performance analysis, copper(II)/zinc(II)-MOF-74 is affected by subtle ferromagnetic interchain interactions as well as antiferromagnetic intrachain interactions. When compared to the interactions of Cu-MOF-74, these interactions are analyzed and investigated. For the purpose of magnetic solid phase extraction (MSPE) of pesticides from water, a metal-organic framework (MOF) based on iron has been shown to be an extremely effective platform. The Fe_3O_4@NH_2-MIL-101 MOF, which was produced through a solvothermal reaction that involved Fe_3O_4@NH_2, H_2BTC (where BTC denotes benzenetricarboxylic acid), and MIL-101 (Cr), demonstrated an incredibly high surface area of 1012.7 $m^2\ g^{-1}$, a substantial pore volume of 0.64 cm^3g^{-1}, and a notable magnetic responsiveness of 30.0 $emug^{-1}$. As part of their investigation on UV sensitivity, researchers were able to create a linear range that extended up to 1000 μgL^{-1}. The linear range was made possible by the good repeatability (1.8–3.1%) and low detection limits (S/N = 3) that were found between 0.13 and 0.86 μgL^{-1}. An effective analysis of several pesticides in an aqueous media employing Fe_3O_4@NH_2-MIL-101 resulted in a recovery range of 70.5–119.8%, with relative standard deviations at or below 8.1% and satisfactory recoveries [142].

In their study, Yamauchi and his colleagues [143] are the first to show that they can use hybridized triplet qubits in flexible MOFs to sense different neutral chemicals. They used the photoexcited triplet, a qubit with a long coherence time (T2) and a room temperature setup. The MOF, which can change its structure when different analyte molecules bind to it, was used to make the qubits more responsive. Given the various instances of MOFs preferentially adsorbing particular guests or exhibiting sudden adsorption at specified pressures, they might develop a variety of quantum sensors capable of detecting the kind and concentration of molecules. Adding different triplet qubits to rigid and flexible MOFs as guest molecules or ligands could also help make a "quantum nose" that can tell the difference between different types of molecules. The regulation of MOF orientation is anticipated to facilitate enhanced quantum sensing through intricate qubit manipulation. Optical methods such as phosphorescence and delayed fluorescence can identify triplets, leading to highly sensitive detection and single-molecule quantum sensing [144, 145]. The triplet-MOF complex will offer an exceptionally sensitive and highly selective quantum sensing platform, influencing various domains such as analytical chemistry, life sciences, and medicine.

The propagation of intense laser light in nonlinear optical materials can produce output photon sources that possess quantum entanglement across several degrees of freedom, rendering this process a crucial instrument in optical quantum technology. The restricted diversity of efficient optical crystals for entangled light generation significantly constrains the potential for advancing next-generation protocols in quantum sensing, communication, and computation. To address this, Raj et al. [146] devised and executed a multi-scale first-principles modeling approach for the computational identification of innovative nonlinear optical devices utilizing metal-organic framework (MOF) materials capable of effectively producing entangled light through spontaneous parametric down-conversion (SPDC). Raj et al. conducted a computational screening of a database comprising synthesized MOF materials, utilizing collinear degenerate type-I SPDC as a case study, to identify correlations between the structural and chemical composition of MOFs and the brightness and coherence characteristics of entangled photon pairs. A subset of 49 non-centrosymmetric mono-ligand MOF crystals exhibiting strong chemical and optical stability is identified, which generates entangled photon pairs with intrinsic G(2) correlation durations τc of about 10–30 fs and pair generation rates between 10^4 and 10^{10} s^{-1} mW^{-1} mm^{-1} at 1064 nm. Optimal type-I phase matching conditions are specified for each MOF, and the correlations between pair brightness, crystal band gap, and optical birefringence are examined. Correlations between the optical characteristics of crystals and their constituent molecular ligands are also provided. The research facilitates the computational design of MOF-based devices for optical quantum technologies. First-principles screening of metal-organic frameworks is used for entangled photon pair generation.

3. Future directions

3.1 Energy management

Energy research uses MOFs that have to be able to maintain structural integrity and functionality over a wide temperature range. Using metal centers and ligands with high thermal stability in MOFs helps one to achieve durability in demanding temperatures. MOFs should be engineered to possess high heat capacity and rapid heat dissipation properties for application as heat sinks or cooling layers in electronic devices. MOFs in reactors can regulate exothermic and endothermic reactions, thereby mitigating overheating and enhancing energy efficiency. Multidisciplinary collaboration is essential to connect theoretical MOF design with effective thermal management solutions. The resolution of scalability and performance challenges in MOF-based thermal management necessitates the use of computational modeling, innovative synthesis methods, and experimental validation. These projects will facilitate the actual application of MOFs across several areas, including energy storage and electronic cooling, among others.

3.2 Phase change materials

Future research must solve numerous important integration-related issues if MOF-based phase PCMs for useful applications are to be developed. Particularly interesting for thermal energy storage, smart building materials, and temperature control technologies are MOF-based PCMs with their adjustable porosity and thermal characteristics. High thermal conductivity design of MOFs with improved thermal characteristics could help to improve heat transfer and storage capacity during phase shifts. Pore diameters and geometry have to be maximized if we are to capture MOF-based PCMs and stop leakage during migration. Developed as a supporting matrix, composite materials should help to stabilize MOF-PCMs and lower phase separation or deterioration. It should enhance the interaction with embedded PCMs to raise thermal stability and latent heat storage by including functional groups into MOF cages. By raising the latent heat capacity, one can improve the possibility of MOF-based PCMs to compete with conventional phase change materials. Improved thermal cycling stability of MOFs will help to sustain consistent performance throughout several phase transitions. Advanced encapsulation methods have to be developed to totally encapsulate PCMs utilizing MOFs and eradicate leakage during phase shift transitions. The phase change behavior should be enhanced and progressively changed by means of study and optimization of the interactions between MOFs and embedded PCMs. It has to be made to modify the melting and freezing points of MOF-PCMs, so customizing them for particular temperature ranges in uses such as textiles, electronics, or buildings. The development of scalable and reasonably priced techniques is necessary to synthesize consistent quality MOF PCM compounds. For sustainable MOF-PCM uses, biodegradable or recyclable techniques should be investigated together with ecologically friendly MOF synthesis approaches. The obtained MOF-PCM composites should not compromise structural integrity or thermal performance. Measurement and optimization of the thermal conductivity of MOF-PCM systems must find better techniques. Designed for energy storage in solar power plants, heating, ventilation, and air conditioning (HAVC) systems, MOF-based PCMs should be MOF-PCMs that have to be incorporated into windows, walls, or roofs for passive temperature control in buildings. The development of lightweight MOF-PCMs for clothes or wearable devices will help to control body temperature in very hostile environments. Thermal management in batteries or electrical devices should be handled by MOF-based PCMs to extend lifespan and performance. Modeling the heat transfer and energy storage dynamics in MOF-PCMs helps one to maximize design criteria. The development of hybrid MOF-PCMs combining thermal storage with further purposes including gas adsorption, photoresponsibility, or chemical sensing is made possible. Investigated for usage in advanced devices are MOF-PCMs displaying changes in electrical or magnetic characteristics during phase change transitions. Increasing the long-term durability and recyclability of MOF-based PCMs will help to lower environmental effects. Environmentally safe systems can be biobased or non-toxic phase change materials MOFs. The development of MOF-PCM composites as latent heat cells for off-grid energy storage and night power supply is in progress. Food, pharmaceutical, or vaccine transportation can all benefit from MOF-PCMs produced for temperature-regulated packaging. Practical uses of temperature control systems made possible by MOF-based PCM technology. These characteristics and studies will open the path for effective and environmentally friendly uses in several sectors.

3.3 Radiation shielding

Future investigations will concentrate on creating MOF-based and metal-metal oxide nanoparticle radiation shielding materials that efficiently attenuate or absorb detrimental radiation while also tackling issues related to scalability, stability, and multifunctionality. The incorporation of heavy metals and metal oxides into MOF cages enhances the shielding efficacy against high-energy radiation, including gamma and X-rays. Flexible and lightweight MOFs with superior protective capabilities can be engineered to minimize weight while preserving structural integrity, particularly for aircraft applications. The interaction of radiation with the MOF structure must be examined to comprehend the mechanisms of energy absorption, scattering, and attenuation at the atomic level. By synthesizing MOF composites by the integration of MOFs with polymers, nanometals, and nanometal oxides or ceramics, their protective powers and mechanical characteristics require enhancement. Layered MOF structures can maximize the attenuation of various radiation types. Enhancing thermal and radiation stability is essential for preserving the performance of MOFs in high-radiation conditions. Formulate economical and scalable techniques for the mass production of MOF-based radiation shielding materials. Utilize 3D printing and advanced manufacturing techniques to create intricate MOF shielding geometries customized for particular applications. Integrating MOFs with PCMs can yield a radiation barrier that simultaneously regulates temperature, particularly suited for space and high-energy conditions. By investigating the application of phase transitions in MOFs, self-healing capabilities can be engaged for prolonged application in high-radiation settings. Monte Carlo machine learning and theoretical optimization methods should be employed to model the interactions of MOFs with various forms of radiation, study, and design MOFs with optimal shielding properties, and estimate the attenuation coefficients and shielding efficacy of MOF materials. MOFs ought to be engineered with recyclable components to mitigate environmental impact post-materials' end-of-life. MOFs ought to be engineered for radiation shielding in spacecraft, satellites, and spacesuits to ensure radiation protection in nuclear reactors, waste containment, and decontamination systems. MOFs should be incorporated into protective materials for radiography and nuclear medicine to safeguard patients and healthcare professionals.

MOFs must be engineered for nuclear waste management to concurrently collect radiation and absorb radioactive gasses, such as radon, krypton, or iodine. MOFs can be integrated with luminescent or conductive materials to provide systems that both shield against radiation and facilitate real-time detection and monitoring of radiation levels. To enhance safety in nuclear or high-radiation facilities, MOF-based radiation shielding must be incorporated into construction materials.

3.4 Quantum computation

Future research should address the development of MOF-based systems that can function as quantum information storage devices, logic gate systems, or catalysts examined using quantum computing. MOFs must be engineered with precisely defined cavities or lattices capable of stabilizing qubits, such as magnetic ions or organic radicals, exhibiting high spin coherence times. In the synthesis of MOFs, it is essential to investigate modular and scalable designs that facilitate the systematic organization of qubit active centers. Engineered MOFs with adjustable magnetic or electronic states, activated by external stimuli (e.g., electric fields, light, or pressure), are essential for dependable qubit initialization, manipulation, and readout. Research should focus on hybrid systems that enable qubit coupling through the integration of MOFs with superconducting circuits or photonic structures. Understanding spin-spin interactions, coherence, and relaxation mechanisms will be crucial for utilizing quantum computers to accurately replicate the electrical and magnetic properties of MOFs. The emphasis should be on engineering MOFs with minimal spin mismatch ratios to reduce nuclear spin noise and enhance lattice symmetry. Quantum algorithms must be formulated to enhance reaction pathways that facilitate the creation of highly efficient systems for the energy applications of MOFs or their synthesis. It must tackle the difficulty of synthesizing MOFs to ensure their stability under quantum computing operational circumstances. Collaboration with experimentalists is essential to test theoretical predictions and create prototypes of MOF-based quantum systems.

4. Conclusion

It is anticipated that the development of metal-organic framework-water working pairs will be a significant component of the upcoming significant advancement in modern sorption-based thermal energy storage and conversion systems. At present, there is a substantial increase in global energy consumption as a result of the growing apprehension regarding the depletion of petroleum reserves and the degradation of the environment. Thermal energy storage and conversion (TESC) systems have attracted substantial attention due to their capacity to reduce CO_2 emissions by increasing the utilization of renewable and residual thermal energy and reducing reliance on fossil fuels. Three-dimensional porous materials known as MOFs have recently garnered significant attention due to their potential for thermal energy storage applications. These materials are composed of metal centers that are connected by organic connectors. In comparison with conventional porous adsorbents, MOFs offer a variety of benefits, such as the ease of chemical modifications, remarkably large surface areas, precisely defined porous structures, and superb crystallinity.

MOFs, which are more commonly known as metal-organic frameworks, offer a unique platform for the development of radiation shielding materials of the future due to the malleability of their design. In specific applications, MOF-based materials have the potential to surpass or build upon conventional shielding materials. This is due to the meticulous selection of metal centers, the incorporation of neutron-absorbing elements, the modification of porosity and density, and the utilization of their composite-forming capabilities. These are merely a few of the methods by which MOF-based materials have the potential to enhance or surpass traditional shielding materials.

A significant amount of research has been conducted on MOFs as adsorbents for the purpose of capturing radionuclides and cleaning up nuclear waste. This is due to their high level of chemical stability, tunable functional groups that enhance selectivity, and a large specific surface area.

Transition metal ions or lanthanides with single electrons are examples of metal centers that are paramagnetic. In certain MOFs, these can function as localized spin carriers. Positioning these spins in a well-defined lattice within the ordered, crystalline framework of a MOF has the potential to accomplish the "bottom-up" architecture for qubit arrays. Magnetic ions that are embedded in a crystalline lattice can be used to create spin qubits, which rely on spin states as their computational foundation. The rigid and tunable structure of MOFs enables the meticulous design of spin environments, thereby enabling the precise tuning of qubit-qubit interactions and the possibility of extended coherence times [147].

Acknowledgements

The authors thank to Ibrahim ERDEMOGLU for their support. In addition, the authors also would like to express gratitude to Karadeniz Technical University and Bilkent University UNAM for their kind hospitality.

Author details

Bülend Ortaç[1], Saliha Mutlu[1,2], Ahmet Hakan Yilmaz[3], Nergis Arsu[4] and Sevil Savaskan Yilmaz[1,2]*

1 Bilkent University, UNAM-National Nanotechnology Research Center and Institute of Materials Science and Nanotechnology, Ankara, Turkey

2 Faculty of Sciences, Department of Chemistry, Karadeniz Technical University, University Avenue, Trabzon, Turkey

3 Karadeniz Technical University, Faculty of Sciences, Department of Physics, University Avenue, Trabzon, Turkey

4 Department of Chemistry, Davutpasa Campus, Yildiz Technical University, Istanbul, Turkey

*Address all correspondence to: sevily@ktu.edu.tr

IntechOpen

© 2025 The Author(s). Licensee IntechOpen. This chapter is distributed under the terms of the Creative Commons Attribution License (http://creativecommons.org/licenses/by/4.0), which permits unrestricted use, distribution, and reproduction in any medium, provided the original work is properly cited.

References

[1] Evans JD, Sumby CJ, Doonan CJ. Post-synthetic metalation of metal–organic frameworks. Chemical Society Reviews. 2014;**43**:5933-5951. DOI: 10.1039/C4CS00076E

[2] Rösler C, Fischer RA. Metal–organic frameworks as hosts for nanoparticles. CrystEngComm. 2015;**17**:199-217. DOI: 10.1039/C4CE01251H

[3] Saikia M, Bhuyan D, Saikia L. Facile synthesis of Fe_3O_4 nanoparticles on metal organic framework MIL-101(Cr): Characterization and catalytic activity. New Journal of Chemistry. 2015;**39**:64-67. DOI: 10.1039/C4NJ01312C

[4] He L, Dumée LF, Liu D, Velleman L, She F, Banos C, et al. Silver nanoparticles prepared by gamma irradiation across metal–organic framework templates. RSC Advances. 2015;**5**:10707-10715. DOI: 10.1039/C4RA10260F

[5] Doherty CM, Buso D, Hill AJ, Furukawa S, Kitagawa S, Falcaro P. Using functional nano- and microparticles for the preparation of metal–organic framework composites with novel properties. Accounts of Chemical Research. 2014;**47**:396-405. DOI: 10.1021/ar400130a

[6] Zondag H, Kikkert B, Smeding S, de Boer R, Bakker M. Prototype thermochemical heat storage with open reactor system. Applied Energy. 2013;**109**:360-365. DOI: 10.1016/j.apenergy.2013.01.082

[7] An G, Xia X, Wu S, Liu Z, Wang L, Li S. Metal–organic frameworks for ammonia-based thermal energy storage. Small. 2021;**17**:2102689. DOI: 10.1002/smll.202102689

[8] Palash ML, Jahan I, Rupam TH, Harish S, Saha BB. Novel technique for improving the water adsorption isotherms of metal-organic frameworks for performance enhancement of adsorption driven chillers. Inorganica Chimica Acta. 2020;**501**:119313. DOI: 10.1016/j.ica.2019.119313

[9] Zhou Y, Wang X, Liu X, Sheng D, Ji F, Dong L, et al. Polyurethane-based solid-solid phase change materials with halloysite nanotubes-hybrid graphene aerogels for efficient light- and electro-thermal conversion and storage. Carbon. 2019;**142**:558-566. DOI: 10.1016/j.carbon.2018.10.083

[10] Chen X, Gao H, Yang M, Xing L, Dong W, Li A, et al. Smart integration of carbon quantum dots in metal-organic frameworks for fluorescence-functionalized phase change materials. Energy Storage Materials. 2019;**18**:349-355. DOI: 10.1016/j.ensm.2018.08.015

[11] Zheng Q, Hao M, Miao R, Schaadt J, Dames C. Advances in thermal conductivity for energy applications: A review. Progress in Energy. 2021;**3**:012002. DOI: 10.1088/2516-1083/abd082

[12] Islamov M, Babaei H, Wilmer CE. Influence of missing linker defects on the thermal conductivity of metal–organic framework HKUST-1. ACS Applied Materials & Interfaces. 2020;**12**:56172-56177. DOI: 10.1021/acsami.0c16127

[13] Babaei H, DeCoster ME, Jeong M, Hassan ZM, Islamoglu T, Baumgart H, et al. Observation of reduced thermal conductivity in a metal-organic framework due to the presence of adsorbates. Nature Communications. 2020;**11**:4010. DOI: 10.1038/s41467-020-17822-0

[14] DeSantis D, Mason JA, James BD, Houchins C, Long JR, Veenstra M. Techno-economic analysis of metal–organic frameworks for hydrogen and natural gas storage. Energy & Fuels. 2017;**31**:2024-2032. DOI: 10.1021/acs.energyfuels.6b02510

[15] Mason JA, Oktawiec J, Taylor MK, Hudson MR, Rodriguez J, Bachman JE, et al. Methane storage in flexible metal–organic frameworks with intrinsic thermal management. Nature. 2015;**527**:357-361. DOI: 10.1038/nature15732

[16] Babaei H, McGaughey AJH, Wilmer CE. Effect of pore size and shape on the thermal conductivity of metal-organic frameworks. Chemical Science. 2017;**8**:583-589. DOI: 10.1039/C6SC03704F

[17] Wang C, Hua L, Yan H, Li B, Tu Y, Wang R. A thermal management strategy for electronic devices based on moisture sorption-desorption processes. Joule. 2020;**4**:435-447. DOI: 10.1016/j.joule.2019.12.005

[18] Tang Z, Gao Y, Cheng P, Jiang Y, Xu J, Chen X, et al. Metal-organic framework derived magnetic phase change nanocage for fast-charging solar-thermal energy conversion. Nano Energy. 2022;**99**:107383. DOI: 10.1016/j.nanoen.2022.107383

[19] Chen X, Gao H, Tang Z, Wang G. Metal-organic framework-based phase change materials for thermal energy storage. Cell Reports Physical Science. 2020;**1**:100218. DOI: 10.1016/j.xcrp.2020.100218

[20] Wu Z, Ye Z, Shi S, Fang Y, Li L, Yuan W. TiN doped Ni–Co bimetal organic framework based-composite phase change material with enhanced photothermal conversion performance. Solar Energy Materials and Solar Cells. 2022;**248**:112019. DOI: 10.1016/j.solmat.2022.112019

[21] Yan D, Li M. TaON doped metal-organic frameworks-based composite phase change materials with excellent photo-thermal conversion performance. Solar Energy Materials and Solar Cells. 2022;**243**:111790. DOI: 10.1016/j.solmat.2022.111790

[22] Chen X, Yu H, Gao Y, Wang L, Wang G. The marriage of two-dimensional materials and phase change materials for energy storage, conversion and applications. EnergyChem. 2022;**4**:100071. DOI: 10.1016/j.enchem.2022.100071

[23] Lin R-B, Xiang S, Zhou W, Chen B. Microporous metal-organic framework materials for gas separation. Chem. 2020;**6**:337-363. DOI: 10.1016/j.chempr.2019.10.012

[24] Knebel A, Caro J. Metal–organic frameworks and covalent organic frameworks as disruptive membrane materials for energy-efficient gas separation. Nature Nanotechnology. 2022;**17**:911-923. DOI: 10.1038/s41565-022-01168-3

[25] Xu J, Li T, Chao J, Wu S, Yan T, Li W, et al. Efficient solar-driven water harvesting from arid air with metal–organic frameworks modified by hygroscopic salt. Angewandte Chemie International Edition. 2020;**59**:5202-5210. DOI: 10.1002/anie.201915170

[26] Chen X, Tang Z, Chang Y, Gao H, Cheng P, Tao Z, et al. Toward tailoring chemistry of silica-based phase change materials for thermal energy storage. iScience. 2020;**23**:101606. DOI: 10.1016/j.isci.2020.101606

[27] Nozari V, Azar ANV, Sajzew R, Castillo-Blas C, Kono A, Oschatz M, et al.

Observation of a reversible order-order transition in a metal-organic framework – Ionic liquid nanocomposite phase-change material. Small. 2024;**20**:2303315. DOI: 10.1002/smll.202303315

[28] Seidi F, Jouyandeh M, Taghizadeh M, Taghizadeh A, Vahabi H, Habibzadeh S, et al. Metal-organic framework (MOF)/epoxy coatings: A review. Materials. 2020;**13**:2881. DOI: 10.3390/ma13122881

[29] Ma C, Wolterbeek HT, Denkova AG, Serra Crespo P. Porphyrinic metal–organic frameworks as molybdenum adsorbents for the 99 Mo/99m Tc generator. Inorganic Chemistry Frontiers. 2023;**10**:2239-2249. DOI: 10.1039/D2QI01919A

[30] Ding M, Cai X, Jiang H-L. Improving MOF stability: Approaches and applications. Chemical Science. 2019;**10**:10209-10230. DOI: 10.1039/C9SC03916C

[31] Elsaidi SK, Mohamed MH, Helal AS, Galanek M, Pham T, Suepaul S, et al. Radiation-resistant metal-organic framework enables efficient separation of krypton fission gas from spent nuclear fuel. Nature Communications. 2020;**11**:3103. DOI: 10.1038/s41467-020-16647-1

[32] Hastings AM, Fairley M, Wasson MC, Campisi D, Sarkar A, Emory ZC, et al. Role of metal selection in the radiation stability of isostructural M-UiO-66 metal–organic frameworks. Chemistry of Materials. 2022;**34**:8403-8417. DOI: 10.1021/acs.chemmater.2c02170

[33] Gilson SE, Fairley M, Hanna SL, Szymanowski JES, Julien P, Chen Z, et al. Unusual metal–organic framework topology and radiation resistance through Neptunyl coordination chemistry. Journal of the American Chemical Society. 2021;**143**:17354-17359. DOI: 10.1021/jacs.1c08854

[34] Doty FP, Bauer CA, Skulan AJ, Grant PG, Allendorf MD. Scintillating metal-organic frameworks: A new class of radiation detection materials. Advanced Materials. 2009;**21**:95-101. DOI: 10.1002/adma.200801753

[35] Ma C, Sánchez-García I, Wang R, Galán H, Denkova AG, Serra Crespo P. Radiation resistant metal–organic frameworks for the production of high specific activity 51Cr by the Szilard-Chalmers effect. Separation and Purification Technology. 2025;**352**:128212. DOI: 10.1016/j.seppur.2024.128212

[36] Li X, Wang J, Liu X, Liu L, Cha D, Zheng X, et al. Direct imaging of Tunable crystal surface structures of MOF MIL-101 using high-resolution electron microscopy. Journal of the American Chemical Society. 2019;**141**:12021-12028. DOI: 10.1021/jacs.9b04896

[37] Qin M, Zhang L, Wu H. Dielectric Loss mechanism in electromagnetic wave absorbing materials. Advanced Science. 2022;**9**:2105553. DOI: 10.1002/advs.202105553

[38] Xie LS, Skorupskii G, Dincă M. Electrically conductive metal–organic frameworks. Chemical Reviews. 2020;**120**:8536-8580. DOI: 10.1021/acs.chemrev.9b00766

[39] Heidari P, Masoudpanah SM, Ong CK. Metal-organic framework (MOF) derived porous $NiCo_2O_4$ and $ZnCo_2O_4$ spinels with high microwave absorption performance for environmentally friendly EMI shielding applications. Ceramics International. 2023;**49**:6678-6687. DOI: 10.1016/j.ceramint.2022.10.291

[40] Yan J, Huang Y, Yan Y, Ding L, Liu P. High-performance electromagnetic wave absorbers based on two kinds of nickel-based MOF-derived Ni@C microspheres. ACS Applied Materials & Interfaces. 2019;**11**:40781-40792. DOI: 10.1021/acsami.9b12850

[41] Qiang R, Du Y, Zhao H, Wang Y, Tian C, Li Z, et al. Metal organic framework-derived Fe/C nanocubes toward efficient microwave absorption. Journal of Materials Chemistry A. 2015;**3**:13426-13434. DOI: 10.1039/C5TA01457C

[42] Huang L, Liu X, Yu R. An efficient Co/C microwave absorber with Tunable Co nanoparticles derived from a ZnCo bimetallic zeolitic imidazolate framework. Particle & Particle Systems Characterization. 2018;**35**:1800107. DOI: 10.1002/ppsc.201800107

[43] Lv S, Luo H, Wang Z, Yu J, Cheng Y, Chen F, et al. Size regulated N-doped carbon encapsulated NiFe alloys/Ni phosphide composites derived from bimetallic Prussian blue analogues for effective microwave absorption. Carbon. 2024;**218**:118668. DOI: 10.1016/j.carbon.2023.118668

[44] Xiong J, Xiang Z, Zhao J, Yu L, Cui E, Deng B, et al. Layered NiCo alloy nanoparticles/nanoporous carbon composites derived from bimetallic MOFs with enhanced electromagnetic wave absorption performance. Carbon. 2019;**154**:391-401. DOI: 10.1016/j.carbon.2019.07.096

[45] Liang X, Quan B, Chen J, Tang D, Zhang B, Ji G. Strong electric wave response derived from the hybrid of lotus roots-like composites with tunable permittivity. Scientific Reports. 2017;**7**:9462. DOI: 10.1038/s41598-017-09985-6

[46] Zhang X, Tian X, Wu N, Zhao S, Qin Y, Pan F, et al. Metal-organic frameworks with fine-tuned interlayer spacing for microwave absorption. Science Advances. 2024;**10**:1-10. DOI: 10.1126/sciadv.adl6498

[47] Chen X, Jia Z, Feng A, Wang B, Tong X, Zhang C, et al. Hierarchical Fe_3O_4@carbon@MnO_2 hybrid for electromagnetic wave absorber. Journal of Colloid and Interface Science. 2019;**553**:465-474. DOI: 10.1016/j.jcis.2019.06.058

[48] Zhang M, Ma W, Cui J, Wu S, Han J, Zou Y, et al. Hydrothermal synthesized UV-resistance and transparent coating composited superoloephilic electrospun membrane for high efficiency oily wastewater treatment. Journal of Hazardous Materials. 2020;**383**:121152. DOI: 10.1016/j.jhazmat.2019.121152

[49] Li S, Toprak MS, Jo YS, Dobson J, Kim DK, Muhammed M. Bulk synthesis of transparent and homogeneous polymeric hybrid materials with ZnO quantum dots and PMMA. Advanced Materials. 2007;**19**:4347-4352. DOI: 10.1002/adma.200700736

[50] Xiong H-M, Wang Z-D, Xia Y-Y. Polymerization initiated by inherent free radicals on nanoparticle surfaces: A simple method of obtaining ultrastable (ZnO) polymer core–shell nanoparticles with strong blue fluorescence. Advanced Materials. 2006;**18**:748-751. DOI: 10.1002/adma.200501899

[51] Tarasi S, Morsali A. Fabrication of transparent ultraviolet blocking films using nanocomposites derived from metal-organic frameworks. Journal of Alloys and Compounds. 2021;**868**:158996. DOI: 10.1016/j.jallcom.2021.158996

[52] Li J, Miao P, Chen K-J, Cao J, Liang J, Tang Y, et al. Highly effective

electromagnetic wave absorbing prismatic Co/C nanocomposites derived from cubic metal-organic framework. Composites Part B: Engineering. 2020;**182**:107613. DOI: 10.1016/j.compositesb.2019.107613

[53] Hussain MZ, Schneemann A, Fischer RA, Zhu Y, Xia Y. MOF derived porous ZnO/C nanocomposites for efficient dye photodegradation. ACS Applied Energy Materials. 2018;**1**:4695-4707. DOI: 10.1021/acsaem.8b00822

[54] Rojas S, Horcajada P. Metal–organic frameworks for the removal of emerging organic contaminants in water. Chemical Reviews. 2020;**120**:8378-8415. DOI: 10.1021/acs.chemrev.9b00797

[55] Yuan S, Feng L, Wang K, Pang J, Bosch M, Lollar C, et al. Stable metal–organic frameworks: Stable metal–organic frameworks: Design, synthesis, and applications. Advanced Materials. 2018;**30**:1704303. DOI: 10.1002/adma.201870277

[56] Ma C, Vasileiadis A, Wolterbeek HT, Denkova AG, Serra Crespo P. Adsorption of molybdenum on Zr-based MOFs for potential application in the 99Mo/99mTc generator. Applied Surface Science. 2022;**572**:151340. DOI: 10.1016/j.apsusc.2021.151340

[57] Li X, Warden D, Bayazitoglu Y. Analysis to evaluate multilayer shielding of galactic cosmic rays. Journal of Thermophysics and Heat Transfer. 2018;**32**:525-531. DOI: 10.2514/1.T5292

[58] Nambiar S, Yeow JTW. Polymer-composite materials for radiation protection. ACS Applied Materials & Interfaces. 2012;**4**:5717-5726. DOI: 10.1021/am300783d

[59] Wagh A, Manjunath K, Hegde V, Kamath SD. Gamma irradiation on bismuth borate glasses doped by Eu^{3+} ions: Structural, optical and mechanical investigations. Optik. 2018;**160**:298-306. DOI: 10.1016/j.ijleo.2018.01.115

[60] Wei K, Shi Y, Tan X, Shalash M, Ren J, Faheim AA, et al. Recent development of metal-organic frameworks and their composites in electromagnetic wave absorption and shielding applications. Advances in Colloid and Interface Science. 2024;**332**:103271. DOI: 10.1016/j.cis.2024.103271

[61] Miao P, Zhang T, Wang T, Chen J, Gao T, Wang Y, et al. A two-dimensional semiconductive metal-organic framework for highly efficient microwave absorption. Chinese Journal of Chemistry. 2022;**40**:467-474. DOI: 10.1002/cjoc.202100660

[62] Liu W, Tan S, Yang Z, Ji G. Enhanced low-frequency electromagnetic properties of MOF-derived cobalt through Interface design. ACS Applied Materials & Interfaces. 2018;**10**:31610-31622. DOI: 10.1021/acsami.8b10685

[63] Liu Z, Chen J, Que M, Zheng H, Yang L, Yuan H, et al. 2D Ti3C2T MXene/MOFs composites derived CoNi bimetallic nanoparticles for enhanced microwave absorption. Chemical Engineering Journal. 2022;**450**:138442. DOI: 10.1016/j.cej.2022.138442

[64] Balamurugan KS, Sivakami A, Mathankumar M, Prasad YJDS, Ahmad I. Quantum computing basics, applications and future perspectives. Journal of Molecular Structure. 2024;**1308**:137917. DOI: 10.1016/j.molstruc.2024.137917

[65] Rupam TH, Steenhaut T, Palash ML, Filinchuk Y, Hermans S, Saha BB. Thermochemical energy applications of green transition metal doped MIL–100(Fe). Chemical Engineering

Journal. 2022;**448**:137590. DOI: 10.1016/j.cej.2022.137590

[66] Matter ME, Tagnon C, Stache EE. Recent applications of photothermal conversion in organic synthesis. ACS Central Science. 2024;**10**:1460-1472. DOI: 10.1021/acscentsci.4c00545

[67] Zhao X, Zou D, Wang S. Flexible phase change materials: Preparation, properties and application. Chemical Engineering Journal. 2022;**431**:134231. DOI: 10.1016/j.cej.2021.134231

[68] Cinan ZM, Baskan T, Erol B, Mutlu S, Misirlioglu Y, Yilmaz SS, et al. Gamma irradiation, thermal conductivity, and phase change tests of the cement-hyperbranched poly amino-ester-block-poly cabrolactone-polyurathane plaster-lead oxide and arsenic oxide composite for development of radiation shielding m. International Journal of Energy Research. 2021;**45**:20729-20762. DOI: 10.1002/er.7136

[69] Ortaç B, Mutlu S, Baskan T, Savaskan Yilmaz S, Yilmaz AH, Erol B. Thermal conductivity and phase-change properties of boron nitride–Lead oxide nanoparticle-doped polymer nanocomposites. Polymers. 2023;**15**:2326. DOI: 10.3390/polym15102326

[70] Mutlu S, Ortaç B, Baskan T, Yilmaz SS, Yilmaz AH. Thermally conductive phase change nanocomposites presenting PS-PEG/SeO_2 and PS-PEG/BN/SeO_2. Journal of Alloys and Compounds. 2024;**1007**:176291. DOI: 10.1016/j.jallcom.2024.176291

[71] Tang J, Chen X, Zhang L, Yang M, Wang P, Dong W, et al. Alkylated meso-macroporous metal–organic framework hollow tubes as nanocontainers of octadecane for energy storage and thermal regulation. Small. 2018;**14**:1801970. DOI: 10.1002/smll.201801970

[72] Liu S, Cao C, Liu D, Pan S, Ma A, Yuan H, et al. Novel phase-change material based on stearic-acid@MOFs@SiO_2 with perfect impermeability promoted by coordination with unsaturated metal sites. Journal of Energy Storage. 2024;**95**:112486. DOI: 10.1016/j.est.2024.112486

[73] Li M, Wang C. Preparation and characterization of GO/PEG photo-thermal conversion form-stable composite phase change materials. Renewable Energy. 2019;**141**:1005-1012. DOI: 10.1016/j.renene.2019.03.141

[74] Zhang Q, Liu J. Sebacic acid/CNT sponge phase change material with excellent thermal conductivity and photo-thermal performance. Solar Energy Materials and Solar Cells. 2018;**179**:217-222. DOI: 10.1016/j.solmat.2017.11.019

[75] Zhou Y, Wang X, Liu X, Sheng D, Ji F, Dong L, et al. Multifunctional ZnO/polyurethane-based solid-solid phase change materials with graphene aerogel. Solar Energy Materials and Solar Cells. 2019;**193**:13-21. DOI: 10.1016/j.solmat.2018.12.041

[76] Yang J, Tang L-S, Bao R-Y, Bai L, Liu Z-Y, Yang W, et al. An ice-templated assembly strategy to construct graphene oxide/boron nitride hybrid porous scaffolds in phase change materials with enhanced thermal conductivity and shape stability for light–thermal–electric energy conversion. Journal of Materials Chemistry A. 2016;**4**:18841-18851. DOI: 10.1039/C6TA08454K

[77] Qian T, Zhu S, Wang H, Li A, Fan B. Comparative study of single-walled carbon nanotubes and graphene nanoplatelets for improving the

thermal conductivity and solar-to-light conversion of PEG-infiltrated phase-change material composites. ACS Sustainable Chemistry & Engineering. 2019;**7**:2446-2458. DOI: 10.1021/acssuschemeng.8b05335

[78] Bao Z, Bing N, Yao H-R, Zhang Y, Xie H, Yu W. Three-dimensional interpenetrating network phase-change composites with high Photothermal conversion and rapid heat storage and release. ACS Applied Energy Materials. 2021;**4**:7710-7720. DOI: 10.1021/acsaem.1c01061

[79] Xue S, Lei C, Liu D, Wang K, Wu K, Fu Q. Thermo-conductive phase change materials with binary fillers of core-shell-like distribution. Composites Part A: Applied Science and Manufacturing. 2021;**144**:106326. DOI: 10.1016/j.compositesa.2021.106326

[80] Liu P, Huang M, Chen X, Gao Y, Li Y, Dong C, et al. Polypyrrole-boosted photothermal energy storage in MOF-based phase change materials. Interdisciplinary Materials. 2023;**2**:423-433. DOI: 10.1002/idm2.12092

[81] Cinan ZM, Erol B, Baskan T, Mutlu S, Ortaç B, Savaskan Yilmaz S, et al. Radiation shielding tests of crosslinked polystyrene-b-polyethyleneglycol block copolymers blended with nanostructured selenium dioxide and boron nitride particles. Nanomaterials. 2022;**12**:297. DOI: 10.3390/nano12030297

[82] Dubey P, Shrivastav V, Sundriyal S, Maheshwari PH. Sustainable Nanoporous metal–organic framework/conducting polymer composites for supercapacitor applications. ACS Applied Nano Materials. 2024;**7**:18554-18565. DOI: 10.1021/acsanm.4c01697

[83] Sheberla D, Bachman JC, Elias JS, Sun C-J, Shao-Horn Y, Dincă M. Conductive MOF electrodes for stable supercapacitors with high areal capacitance. Nature Materials. 2017;**16**:220-224. DOI: 10.1038/nmat4766

[84] Zhou S, Kong X, Zheng B, Huo F, Strømme M, Xu C. Cellulose nanofiber @ conductive metal–organic frameworks for high-performance flexible supercapacitors. ACS Nano. 2019;**13**:9578-9586. DOI: 10.1021/acsnano.9b04670

[85] Sheta SM, El-Sheikh SM, Osman DI, Salem AM, Ali OI, Harraz FA, et al. A novel HCV electrochemical biosensor based on a polyaniline@Ni-MOF nanocomposite. Dalton Transactions. 2020;**49**:8918-8926. DOI: 10.1039/D0DT01408G

[86] Zeng B, Wang Y, Xiong K, Zhang K, Zhang S, Lang X. Zirconium metal–organic framework photocatalysis with TEMPO for blue light-powered aerobic sulfoxidation. Chemical Communications. 2024;**60**:14061-14064. DOI: 10.1039/D4CC05173D

[87] Wang S, Song L, Liu S, Pei X, Zhao Y, Min C, et al. Metal-organic framework nanoparticles as a free radical scavenger improving the stability of epoxy under high dose gamma irradiation. Polymer Degradation and Stability. 2023;**207**:110231. DOI: 10.1016/j.polymdegradstab. 2022.110231

[88] Cinan ZM, Erol B, Baskan T, Mutlu S, Savaskan Yilmaz S, Yilmaz AH. Gamma irradiation and the radiation shielding characteristics: For the Lead oxide doped the crosslinked polystyrene-b-polyethyleneglycol block copolymers and the polystyrene-b-polyethyleneglycol-boron nitride nanocomposites. Polymers. 2021;**13**:3246. DOI: 10.3390/polym13193246

[89] Hu C, Xiao J-D, Mao X-D, Song L-L, Yang X-Y, Liu S-J. Toughening

mechanisms of epoxy resin using aminated metal-organic framework as additive. Materials Letters. 2019;**240**:113-116. DOI: 10.1016/j.matlet.2018.12.123

[90] Lee S-J, Yoon T-U, Kim A-R, Kim S-Y, Cho K-H, Hwang YK, et al. Adsorptive separation of xenon/krypton mixtures using a zirconium-based metal-organic framework with high hydrothermal and radioactive stabilities. Journal of Hazardous Materials. 2016;**320**:513-520. DOI: 10.1016/j.jhazmat.2016.08.057

[91] Yamamoto T, Ootsuka N. Radiation damage of fluorocarbon by Krypton-8 5 beta-rays, (IV) irradiation experiments using Krypton-85 Beta-rays on trichlorofluoromethane in batch and flow system. Journal of Nuclear Science and Technology. 1982;**19**:903-917. DOI: 10.1080/18811248.1982.9734236

[92] Joshi S, Panda M, Annalakshmi O, Srinivas CV, Venkatraman B. Metal–organic frameworks as optically stimulated luminescence dosimeter. Journal of Radioanalytical and Nuclear Chemistry. 2024;**333**:5949-5956. DOI: 10.1007/s10967-024-09761-7

[93] Xie J, Wang Y, Liu W, Yin X, Chen L, Zou Y, et al. Highly sensitive detection of ionizing radiations by a photoluminescent uranyl organic framework. Angewandte Chemie International Edition. 2017;**56**:7500-7504. DOI: 10.1002/anie.201700919

[94] Xie J, Wang Y, Liu W, Liang C, Zhang Y, Chen L, et al. A uranyl based coordination polymer showing response to low-dosage ionizing radiations down to 10−5 Gy. Science China Chemistry. 2020;**63**:1608-1612. DOI: 10.1007/s11426-020-9817-9

[95] Liu H, Qin H, Shen N, Yan S, Wang Y, Yin X, et al. Emergence of a radical-stabilizing metal–organic framework as a radio-photoluminescence dosimeter. Angewandte Chemie International Edition. 2020;**59**:15209-15214. DOI: 10.1002/anie.202006380

[96] Mane SN, Kolekar SK, Sapnar KB, Dhole SD. Gamma radiation induced modification in metal–organic framework: A case study. Journal of Radioanalytical and Nuclear Chemistry. 2024;**333**:3115-3122. DOI: 10.1007/s10967-024-09430-9

[97] Gendy EA, Oyekunle DT, Ali J, Ifthikar J, El-Motaleb Mosad Ramadan A, Chen Z. High-performance removal of radionuclides by porous organic frameworks from the aquatic environment: A review. Journal of Environmental Radioactivity. 2021;**238-239**:106710. DOI: 10.1016/j.jenvrad.2021.106710

[98] Gilson SE, Fairley M, Julien P, Oliver AG, Hanna SL, Arntz G, et al. Unprecedented radiation resistant thorium–binaphthol metal–organic framework. Journal of the American Chemical Society. 2020;**142**:13299-13304. DOI: 10.1021/jacs.0c05272

[99] Hanna SL, Rademacher DX, Hanson DJ, Islamoglu T, Olszewski AK, Nenoff TM, et al. Structural features of zirconium-based metal–organic frameworks affecting radiolytic stability. Industrial & Engineering Chemistry Research. 2020;**59**:7520-7526. DOI: 10.1021/acs.iecr.9b06820

[100] Geidam IG, Matori KA, Halimah MK, Chan KT, Muhammad FD, Ishak M, et al. Oxide ion polarizabilities and gamma radiation shielding features of TeO_2–B_2O_3–SiO_2 glasses containing Bi_2O_3 using Phy-X/PSD software. Materials Today Communications. 2022;**31**:103472. DOI: 10.1016/j.mtcomm.2022.103472

[101] Singh GP, Singh J, Kaur P, Singh T, Kaur R, Kaur S, et al. Impact of TiO_2 on radiation shielding competencies and structural, physical and optical properties of CeO_2–PbO–B_2O3 glasses. Journal of Alloys and Compounds. 2021;**885**:160939. DOI: 10.1016/j.jallcom.2021.160939

[102] Hakan Yilmaz A, Ortaç B, Savaskan YS. Boron and boron compounds in radiation shielding materials. In: Boron, Boron Compounds and Boron-Based Materials and Structures. London: IntechOpen; 2024. Available from: https://www.intechopen.com/chapters/87315

[103] Verma S, Sarma B, Chaturvedi K, Malvi D, Srivastava AK. Emerging graphene and carbon nanotube-based carbon composites as radiations shielding materials for X-rays and gamma rays: A review. Composite Interfaces. 2023;**30**:223-251. DOI: 10.1080/09276440.2022.2094571

[104] Hashemi SA, Mousavi SM, Faghihi R, Arjmand M, Sina S, Amani AM. Lead oxide-decorated graphene oxide/epoxy composite towards X-ray radiation shielding. Radiation Physics and Chemistry. 2018;**146**:77-85. DOI: 10.1016/j.radphyschem.2018.01.008

[105] Viegas J, Silva LA, Batista AMS, Furtado CA, Nascimento JP, Faria LO. Increased X-ray attenuation efficiency of graphene-based nanocomposite. Industrial & Engineering Chemistry Research. 2017;**56**:11782-11790. DOI: 10.1021/acs.iecr.7b02711

[106] Silva LA, Batista AMS, Serodre T, Neto ATB, Furtado CA, Faria LO. Enhancement of X-ray shielding properties of PVDF/BaSO4 nanocomposites filled with graphene oxide. MRS Advances. 2019;**4**:169-175. DOI: 10.1557/adv.2018.651

[107] Ghaseminejad M, Gholamzadeh L, Ostovari F. Investigation of x-ray attenuation property of modification PbO with graphene in epoxy polymer. Materials Research Express. 2021;**8**:035008. DOI: 10.1088/2053-1591/abecea

[108] Jenkins MD, Zueco D, Roubeau O, Aromí G, Majer J, Luis F. A scalable architecture for quantum computation with molecular nanomagnets. Dalton Transactions. 2016;**45**:16682-16693. DOI: 10.1039/C6DT02664H

[109] Mezenov YA, Krasilin AA, Dzyuba VP, Nominé A, Milichko VA. Metal–organic frameworks in modern physics: Highlights and perspectives. Advanced Science. 2019;**6**:1900506. DOI: 10.1002/advs.201900506

[110] Horike S, Shimomura S, Kitagawa S. Soft porous crystals. Nature Chemistry. 2009;**1**:695-704. DOI: 10.1038/nchem.444

[111] Sato O. Dynamic molecular crystals with switchable physical properties. Nature Chemistry. 2016;**8**:644-656. DOI: 10.1038/nchem.2547

[112] Deng H, Olson MA, Stoddart JF, Yaghi OM. Robust dynamics. Nature Chemistry. 2010;**2**:439-443. DOI: 10.1038/nchem.654

[113] Mutlu S, Ortaç B, Ozbey DH, Durgun E, Savaskan Yılmaz S, Arsu N. Laser-driven rapid synthesis of metal-organic frameworks and investigation of UV-NIR optical absorption, luminescence, photocatalytic degradation, and gas and ion adsorption properties. Polymers. 2024;**16**:217. DOI: 10.3390/polym16020217

[114] Lyndon R, Konstas K, Ladewig BP, Southon PD, Kepert PCJ, Hill MR. Dynamic photo-switching in

metal-organic frameworks as a route to low-energy carbon dioxide capture and release. Angewandte Chemie - International Edition. 2013;**52**:3695-3698. DOI: 10.1002/anie.201206359

[115] Urtizberea A, Natividad E, Alonso PJ, Andrés MA, Gascón I, Goldmann M, et al. A porphyrin spin qubit and its 2D framework Nanosheets. Advanced Functional Materials. 2018;**28**:1-15. DOI: 10.1002/adfm.201801695

[116] Ding G, Wang Y, Zhang G, Zhou K, Zeng K, Li Z, et al. 2D metal–organic framework Nanosheets with time-dependent and multilevel memristive switching. Advanced Functional Materials. 2019;**29**:1-11. DOI: 10.1002/adfm.201806637

[117] Pan L, Ji Z, Yi X, Zhu X, Chen X, Shang J, et al. Metal-organic framework nanofilm for mechanically flexible information storage applications. Advanced Functional Materials. 2015;**25**:2677-2685. DOI: 10.1002/adfm.201500449

[118] Medishetty R, Nalla V, Nemec L, Henke S, Mayer D, Sun H, et al. A new class of lasing materials: Intrinsic stimulated emission from nonlinear optically active metal-organic frameworks. Advanced Materials. 2017;**29**:1-7. DOI: 10.1002/adma.201605637

[119] Yamada MG, Fujita H, Oshikawa M. Designing Kitaev spin liquids in metal-organic frameworks. Physical Review Letters. 2017;**119**:1-6. DOI: 10.1103/PhysRevLett.119.057202

[120] Son HJ, Jin S, Patwardhan S, Wezenberg SJ, Jeong NC, So M, et al. Light-harvesting and ultrafast energy migration in porphyrin-based metal-organic frameworks. Journal of the American Chemical Society. 2013;**135**:862-869. DOI: 10.1021/ja310596a

[121] Milichko VA, Makarov SV, Yulin AV, Vinogradov AV, Krasilin AA, Ushakova E, et al. Van der Waals metal-organic framework as an excitonic material for advanced photonics. Advanced Materials. 2017;**29**:1606034. DOI: 10.1002/adma.201606034

[122] Zhou X, Miao Y, Suslick KS, Dlott DD. Mechanochemistry of metal–organic frameworks under pressure and shock. Accounts of Chemical Research. 2020;**53**:2806-2815. DOI: 10.1021/acs.accounts.0c00396

[123] Leuenberger MN, Loss D. Quantum computing in molecular magnets. Nature. 2001;**410**:789-793. DOI: 10.1038/35071024

[124] Ardavan A, Rival O, Morton JJL, Blundell SJ, Tyryshkin AM, Timco GA, et al. Will spin-relaxation times in molecular magnets permit quantum information processing? Physical Review Letters. 2007;**98**:057201. DOI: 10.1103/PhysRevLett.98.057201

[125] Shiddiq M, Komijani D, Duan Y, Gaita-Ariño A, Coronado E, Hill S. Enhancing coherence in molecular spin qubits via atomic clock transitions. Nature. 2016;**531**:348-351. DOI: 10.1038/nature16984

[126] Bader K, Dengler D, Lenz S, Endeward B, Jiang S-D, Neugebauer P, et al. Room temperature quantum coherence in a potential molecular qubit. Nature Communications. 2014;**5**:5304. DOI: 10.1038/ncomms6304

[127] Atzori M, Morra E, Tesi L, Albino A, Chiesa M, Sorace L, et al. Quantum coherence times enhancement in vanadium(IV)-based potential

molecular qubits: The key role of the Vanadyl moiety. Journal of the American Chemical Society. 2016;**138**:11234-11244. DOI: 10.1021/jacs.6b05574

[128] Mannini M, Pineider F, Danieli C, Totti F, Sorace L, Sainctavit P, et al. Quantum tunnelling of the magnetization in a monolayer of oriented single-molecule magnets. Nature. 2010;**468**:417-421. DOI: 10.1038/nature09478

[129] Ganzhorn M, Klyatskaya S, Ruben M, Wernsdorfer W. Strong spin–phonon coupling between a single-molecule magnet and a carbon nanotube nanoelectromechanical system. Nature Nanotechnology. 2013;**8**:165-169. DOI: 10.1038/nnano.2012.258

[130] Perrin ML, Burzurí E, van der Zant HSJ. Single-molecule transistors. Chemical Society Reviews. 2015;**44**:902-919. DOI: 10.1039/C4CS00231H

[131] Thiele S, Balestro F, Ballou R, Klyatskaya S, Ruben M, Wernsdorfer W. Electrically driven nuclear spin resonance in single-molecule magnets. Science. 2014;**344**:1135-1138. DOI: 10.1126/science.1249802

[132] El Garah M, Ciesielski A, Marets N, Bulach V, Hosseini MW, Samorì P. Molecular tectonics based nanopatterning of interfaces with 2D metal–organic frameworks (MOFs). Chemical Communications. 2014;**50**:12250-12253. DOI: 10.1039/C4CC03622K

[133] Bartels L. Tailoring molecular layers at metal surfaces. Nature Chemistry. 2010;**2**:87-95. DOI: 10.1038/nchem.517

[134] Bartolomé J, Fernando Luis JFF, editors. Physics and applications. In: Hoogdalem Van K, Stepanenko D, Loss D, editors. Molecular Magnets for Quantum Information. Molecular Magnets. Heidelberg: Springer; 2014

[135] Zeng M-H, Yin Z, Tan Y-X, Zhang W-X, He Y-P, Kurmoo M. Nanoporous cobalt(II) MOF exhibiting four magnetic ground states and changes in gas sorption upon post-synthetic modification. Journal of the American Chemical Society. 2014;**136**:4680-4688. DOI: 10.1021/ja500191r

[136] Yamashita A, Watanabe A, Akine S, Nabeshima T, Nakano M, Yamamura T, et al. Wheel-shaped Er III Zn II 3 single-molecule magnet: A macrocyclic approach to designing magnetic anisotropy. Angewandte Chemie International Edition. 2011;**50**:4016-4019. DOI: 10.1002/anie.201008180

[137] Mougel V, Chatelain L, Pécaut J, Caciuffo R, Colineau E, Griveau J-C, et al. Uranium and manganese assembled in a wheel-shaped nanoscale single-molecule magnet with high spin-reversal barrier. Nature Chemistry. 2012;**4**:1011-1017. DOI: 10.1038/nchem.1494

[138] Martínez-Lillo J, Mastropietro TF, Lhotel E, Paulsen C, Cano J, De Munno G, et al. Highly anisotropic rhenium(IV) complexes: New examples of mononuclear single-molecule magnets. Journal of the American Chemical Society. 2013;**135**:13737-13748. DOI: 10.1021/ja403154z

[139] Inokuma Y, Yoshioka S, Ariyoshi J, Arai T, Hitora Y, Takada K, et al. X-ray analysis on the nanogram to microgram scale using porous complexes. Nature. 2013;**495**:461-466. DOI: 10.1038/nature11990

[140] Kurmoo M. Magnetic metal–organic frameworks. Chemical Society Reviews. 2009;**38**:1353. DOI: 10.1039/b804757j

[141] Yu Q, Wang D. Room-temperature magnetism in two-dimensional metal–organic frameworks enabled by electrostatic gating. Journal of Materials Chemistry A. 2023;**11**:5548-5558. DOI: 10.1039/D2TA08540B

[142] Li D, Yadav A, Zhou H, Roy K, Thanasekaran P, Lee C. Advances and applications of metal-organic frameworks (MOFs) in emerging technologies: A comprehensive review. Global Challenges. 2024;**8**:2300244. DOI: 10.1002/gch2.202300244

[143] Yamauchi A, Fujiwara S, Kimizuka N, Asada M, Fujiwara M, Nakamura T. et al. Quantum chemical sensing using molecular triplet qubits in a flexible metal–organic framework. 2022. Available from: https://chemrxiv.org/engage/chemrxiv/article-details/63049dabd858fb6380592476

[144] Christensen JA, Zhou J, Tcyrulnikov NA, Krzyaniak MD, Wasielewski MR. Spin-polarized molecular triplet states as qubits: Phosphorus hyperfine coupling in the triplet state of Benzoisophosphinoline. The Journal of Physical Chemistry Letters. 2020;**11**:7569-7574. DOI: 10.1021/acs.jpclett.0c01912

[145] Wrachtrup J, von Borczyskowski C, Bernard J, Orrit M, Brown R. Optical detection of magnetic resonance in a single molecule. Nature. 1993;**363**:244-245. DOI: 10.1038/363244a0

[146] Raj S, Paiva S, Fritz RA, Herrera F, Colón YJ. First-principles screening of metal–organic frameworks for entangled photon pair generation. Materials for Quantum Technology. 2024;**4**:015404. DOI: 10.1088/2633-4356/ad2c96

[147] Wang Z, Cohen SM. Postsynthetic modification of metal–organic frameworks. Chemical Society Reviews. 2009;**38**:1315. DOI: 10.1039/b802258p